Quality Management

Creating and Sustaining Organizational Effectiveness

Second Edition

Donna C. S. Summers
University of Dayton

PEARSON

Prentice Hall

Upper Saddle River, New Jersey
Columbus, Ohio

Library of Congress Cataloging-in-Publication Data
Summers, Donna C. S.
    Quality management : creating and sustaining organizational effectiveness / Donna C. S. Summers.—
2nd ed.
        p. cm.
    Includes bibliographical references and index.
    ISBN-13: 978-0-13-500510-1
    ISBN-10: 0-13-500510-8
    1. Total quality management.   2. Quality control.   3. Organizational effectiveness.   I. Title.
    HD62.15.S85 2009
    658.4′013—dc22

                                                                                2008001026

Vice President and Executive Publisher: Vernon R. Anthony
Acquisitions Editor: Eric Krassow
Editorial Assistant: Sonya Kottcamp
Project Manager: Maren L. Miller
Production Coordination: Rebecca K. Giusti, GGS Book Services
Design Coordinator: Diane Y. Ernsberger
Cover Designer: Jason Moore
Operations Specialist: Laura Weaver
Director of Marketing: David Gesell
Marketing Manager: Leigh Ann Simms
Marketing Coordinator: Alicia Dysert

This book was set in Goudy by GGS Book Services. It was printed and bound by Courier Companies, Inc.
The cover was printed by Courier Companies, Inc.

Case Studies in Chapters 3, 4, 5, 6, 7, 9, 11, and 15 are used with the permission of Remodeling Designs, Inc.

Pearson Education Ltd.   Pearson Education Australia Pty. Limited
Pearson Education Singapore Pte. Ltd.   Pearson Education North Asia Ltd.
Pearson Education Canada, Ltd.   Pearson Educación de Mexico, S.A. de C.V.
Pearson Education—Japan   Pearson Education Malaysia Pte. Ltd.

9 10 11 12 V092 16 15 14
ISBN-13: 978-0-13-500510-1
ISBN-10:   0-13-500510-8

To my beloved Karl,
who is always seeking a higher level of entropy

Preface

In a kaleidoscope, bits of glass or other objects are shown in ever-changing symmetrical patterns caused by their reflection from two or more mirrors. Saving the patterns is impossible, as any movement disrupts the pattern and forms a new one. Running a business is a bit like a kaleidoscope. Decisions are made, strategies are defined and deployed, patterns are set, and then something moves the entire system, and the process begins all over again. Quality management—its philosophies, tools, and techniques—helps organizations manage their business effectively. Knowledge of quality management principles is what enables leaders of effective organizations to remain flexible and adaptable in an ever-changing market. This text, based on the Malcolm Baldrige National Quality Award criteria, seeks to provide the foundation for running an effective business, one that gets it right the first time. It shows readers how to pay attention to detail and lets them know that the real measure of performance is not how you do on your best day but how you do when you are at your worst. Running an effective organization is complex. This text takes a broad view and, besides quality management philosophies, includes business management tools; human resource practices; and lean, industrial engineering, quality, and Six Sigma tools. The success of any business depends on how leaders recognize the need for improvement, commit to improving, create a customer focus, create and deploy strategic plans, manage their supply chain, measure their successes, manage their processes, and focus on business results.

TEXT GOALS

The goal of this text is to enable the reader to recognize the cornerstones of creating and sustaining an effective organization. The information is presented in a format that enables the reader to understand the general principles behind creating an organizational atmosphere that focuses on the customer. The selection of topics is based on the Malcolm Baldrige National Quality Award, the teachings of W. Edwards Deming, the Six Sigma methodology, and lean manufacturing concepts. By the end of the text, the reader should be able to answer the following questions:

- What does an organization need to do to remain competitive in today's global economy?
- How can an organization improve its processes and integrate its functions so that it provides customer satisfaction through the best use of available resources?
- How does an organization create an internal environment that enables everyone who works for it to perform to the best of his or her ability?
- How does an organization know what the customer wants?
- How does an organization know that it is doing the right things in the right way?

The second edition of this text brings with it many changes. Chapters on supply chain management and lean principles have been added. Benchmarking has become a separate chapter. The reader will note significant additions to the material in order to reflect an ever-increasing focus on customers, value chains, process improvement, ethics in business, and lean and supply chain management. The text contains more examples too. One special feature is "How Do We Know It's Working?" Found in each chapter, this feature provides insight into one real organization's pursuit of excellence. The unifying features found in each chapter of this text are:

> Chapter Objectives
> Achieving Organizational Effectiveness Chart
> Malcolm Baldrige National Quality Award Criteria Applications
> JQOS "How Do We Know It's Working?"
> Are You a Quality Management Person?
> Lessons Learned

TEXT FORMAT

As the contents reveal, each chapter has been laid out in a question-and-answer format. Each chapter begins with objectives and a quote that identify the material to be covered. Examples taken from industry are used throughout the text to support chapter material. At the end of each chapter, questions are presented to encourage readers to consider how they would know that an organization is operating effectively in a particular area. These questions are based on the Malcolm Baldrige National Quality Award criteria. Following "Lessons Learned," readers are encouraged to determine whether or not they are a quality management person by responding to a series of chapter-related questions. At the end of Chapters 3 through 7, 9, 11, and 15, a detailed case describes an organization that has been reviewed using the Malcolm Baldrige National Quality Award criteria. This case allows readers to clearly understand what is required for an effective organization to be effective. For a term assignment, instructors may have their students review a nearby organization using the Malcolm Baldrige National Quality Award criteria. If so, Appendix 3, as well as the questions at the end of the chapter, are useful when designing assignments. Other cases and readings follow many of the other chapters.

KEY TEXT FEATURES

This text serves as an introduction to the activities and philosophies related to creating and managing an effective organization. The key features of this text help readers develop a greater understanding of the eight critical areas: customers; leadership; strategic planning; human resource development and management; measurement, analysis, and knowledge management; process management; project management; and business results. Emphasis is on the practical application of quality management principles, tools, and techniques. Industry and business examples and a case study provide insight into organizational effectiveness. The text focuses on the following:

- Key quality initiatives, including:
  Six Sigma
  Malcolm Baldrige National Quality Award
  ISO 9000
  Lean manufacturing

- Key topics, including:
    - Customer focus creation
    - Value creation
    - Leadership
    - Process improvement and management
    - Strategic planning
    - Measures of performance
    - Supply chain management
    - Human resource management
    - Measurement, analysis, and knowledge management
    - Project management
    - Lean principles
    - Problem solving
    - Business results
- Overview of quality philosophies, including those of:
    - Dr. Armand Feigenbaum
    - Dr. Walter Shewhart
    - Dr. W. Edwards Deming
    - Dr. Joseph Juran
    - Philip Crosby
    - Genichi Taguchi
- Overview of process improvement techniques, including:
    - $\overline{X}$ and R charts
    - Check sheets
    - Cause-and-effect diagrams
    - Pareto diagrams
    - Histograms
    - WHY-WHY diagrams

ONLINE INSTRUCTOR'S RESOURCES

To access supplementary materials online, instructors need to request an instructor access code. Go to **www.pearsonhighered.com/irc,** where you can register for an instructor access code. Within 48 hours after registering you will receive a confirming e-mail, including an instructor access code. Once you have recieved your code, go to the site and log on for full instructions on downloading the materials you wish to use.

A SPECIAL THANKS

I would like to express my deep appreciation to the individuals who helped create this text—my colleagues and my students. I would like to extend a special thank you to the following people for their contributions and advice: Karl Summers, David Sweeney, Erich Eggers, and the employees of Remodeling Designs, Inc. Your input was invaluable. I would also like to thank my editors at Prentice Hall: Maren Miller, Eric Krassow, and Vernon Anthony.

Contents

Case Studies

Quality
Management

1 Organizational Effectiveness

What is an effective organization?
How is an effective organization created?
What benefits can be gained from creating an effective organization?
Lessons Learned

Learning Opportunities

1. To become familiar with the reasons why organizations pursue excellence

2. To introduce the structure of the text

3. To become familiar with the benefits of creating an effective organization
Consumer demands

Confronted by enormous competition from overseas, leaders of Homer Laughlin China Co., a fifth-generation, privately owned manufacturer of dinnerware, faced a difficult decision: close down operations or find ways to significantly improve their organizational effectiveness. Company leaders chose to listen to their customers and develop a strategic plan based on what they had heard. The plan involved investing $500,000 in new equipment and reconfiguring their manufacturing processes. The elimination of dozens of existing manufacturing steps resulted in a reduction of production time from five days to one. The faster production time enabled the company to cut inventories by 75% and reduce costs by 15%. Within a year they were able to offer their customers prices similar to those offered by importers while generating a profit for themselves. Knowing that their higher wages won't enable them to be the lowest-cost operator, company leaders now focus their efforts on their customers' demands for customized and unique items. The company's quick turnaround time on orders and small-batch capabilities—benefits of their reconfigured processes—are part of the unique services they offer their customers. Because of their significantly improved organizational effectiveness, this company is making plans to take itself into the next generation. In 2007, they are still going strong, having launched new product lines for the food service industry. They also have been awarded "Best Buy" certificate from Consumer Digest Magazine.

Paraphrased from:
"Keep It Trendy," Forbes, July 18, 1994

WHAT IS AN EFFECTIVE ORGANIZATION?

Competition. Outsourcing. Technological developments. Global threats. Headlines containing these words fill the media. It seems new and threatening for U.S. industries. Is it? Subtract one hundred years and consider that in 1900 fewer than 8,000 automobiles were on U.S. roads. By 1910 that number had increased to nearly 500,000. The Wright Brothers flew the first heavier-than-air craft for 12 seconds in 1903. By 1909, Frenchman Louis Bleriot flew the 26 miles across the English Channel. The first decade of the previous century ushered in such marvels as the phonograph, wireless telegraphy, turbine-powered steamships, electric lighting, the Kodak Brownie camera, motion pictures, and X-rays. Steam power replaced sails on ocean travel, shortening the trip across the North Atlantic from more than a month to less than a week. Trains traveled at speeds of up to 70 miles per hour. Scientists gained an understanding of the genes, hormones, and vitamins. Einstein presented $E = mc^2$. Marie Curie unlocked the secrets of radium.

Just like today, companies in the first decade of the 1900s had to deal with industrial, economic, and social changes never witnessed before. Companies that faced up to these challenges survived, and those that did not went the way of the buggy whip. In this new century, industries across the world face technological challenges and global competition. Survival will depend on how well they face these challenges. Now, just as then, survival skills must be assimilated into how an organization conducts its business. Regardless of the times, effective organizations compete through value creation, delivery, and market performance. By introducing basic foundation information, this text seeks to provide a basis for exploring what it takes to create and optimize a successful organization. Its focus is on learning how to optimize performance through quality management and continuous improvement.

In today's highly competitive, global marketplace, a company that excels is one that continually strives to identify and focus on factors critical to its customers and improve its processes in order to provide the highest-quality product or service possible. Their task is far from simple. Effective organizations must make excellence a part of their organization's DNA. Work isn't a chore; it's a passion key to their organization's effectiveness. To achieve organizational excellence, they recognize that it takes continual daily efforts to improve their products, services, and the business processes that provide them. They put in place methods and standards that enable them to achieve excellence every time.

Companies have a variety of strategies that can be used to enhance their position in the marketplace—for example, teams, quality assurance, just-in-time, quality management, Six Sigma, lean manufacturing, and others. In some cases, like pieces of a puzzle that won't fall into place, many of these efforts have not created an overall alignment of an entire organizational system that focuses on the factors critical to the organization's success. Instead, efforts are disjointed programs that fail to reach their full potential, generating success in some areas and disillusionment in others. In truth, what is needed is an organization-wide approach that improves and enhances the whole process of providing a product or service, enabling the organization to exceed customer expectations, each time and every time. Effective organizations recognize the elements of success (Figure 1.1). They have a passion to seek and implement better ways to do things. Combined with this

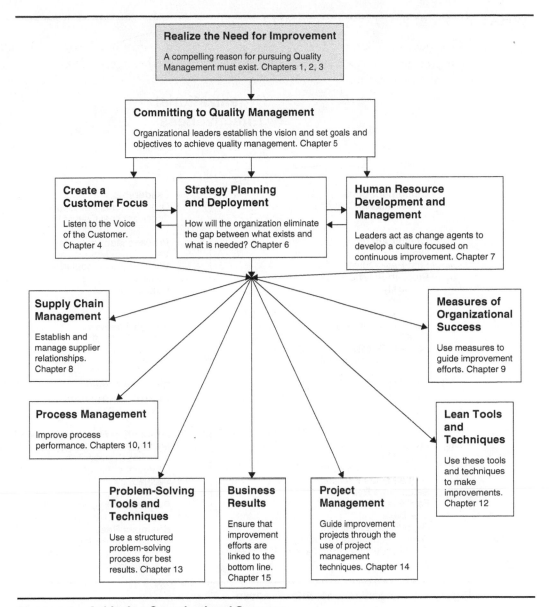

Figure 1.1   Achieving Organizational Success

urge to improve is an attentiveness to details and a willingness to come up with creative solutions by thinking outside the box. Every aspect of their business becomes part of their competitive strategy.

Effective organizations seek to improve their company's value. As you will see in this text, enhancing value has its roots in product and service design improvements, process

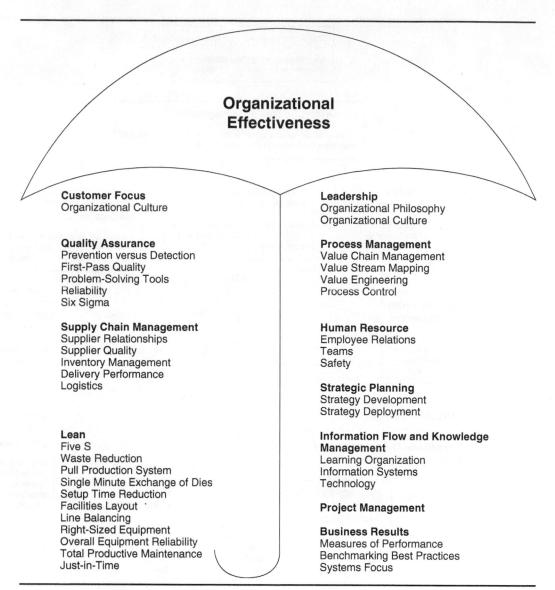

Organizational Effectiveness

Customer Focus
Organizational Culture

Quality Assurance
Prevention versus Detection
First-Pass Quality
Problem-Solving Tools
Reliability
Six Sigma

Supply Chain Management
Supplier Relationships
Supplier Quality
Inventory Management
Delivery Performance
Logistics

Lean
Five S
Waste Reduction
Pull Production System
Single Minute Exchange of Dies
Setup Time Reduction
Facilities Layout
Line Balancing
Right-Sized Equipment
Overall Equipment Reliability
Total Productive Maintenance
Just-in-Time

Leadership
Organizational Philosophy
Organizational Culture

Process Management
Value Chain Management
Value Stream Mapping
Value Engineering
Process Control

Human Resource
Employee Relations
Teams
Safety

Strategic Planning
Strategy Development
Strategy Deployment

Information Flow and Knowledge Management
Learning Organization
Information Systems
Technology

Project Management

Business Results
Measures of Performance
Benchmarking Best Practices
Systems Focus

Figure 1.2   Organizational Effectiveness

improvements, supplier and investor relationships, and the recruiting and training of their workforce. Creating an effective organization requires an understanding of what makes an organization tick. Although no step-by-step process or instant-pudding approach for creating and sustaining organizational effectiveness exists, certain key characteristics have been identified (Figure 1.2). Throughout this text, these characteristics will be introduced and discussed:

- Defining organizational effectiveness—Chapter 1
- Developing an organizational philosophy—Chapter 2
- Utilizing a quality management system—Chapter 3
- Creating a customer focus and customer satisfaction—Chapter 4
- Implementing organizational leadership—Chapter 5
- Creating strategic plans—Chapter 6
- Developing and managing the organization's human element—Chapter 7
- Managing the Supply Chain—Chapter 8
- Measuring organizational success—Chapter 9
- Benchmarking—Chapter 10
- Improving processes—Chapter 11
- Lean Tools and Techniques—Chapter 12
- Utilizing quality tools—Chapter 13
- Managing projects—Chapter 14
- Tracking key business results—Chapter 15

In an effective organization the focus is on the key processes that provide the organization's customers with products or services. An organization is a compilation of a wide variety of activities. Any one or group of those activities can be world-class operationally, but if the other activities within the firm are performing at a suboptimal level, then the organization as a whole is not effective. For instance, if an organization manufactures the best-performing microwave ovens on the market but has difficulties marketing them to the customer, the end result for the overall organization is less-than-optimal performance. Some companies focus very well on the microactivities of their organization, but they forget about the macroaspects—that is, how all the functions need to work together to support each other. An organization needs to be looked at from all perspectives, from manufacturing, marketing, and information technology to research and development. All of these functions must work well together in an effective organization, and their focus must be on the elements of the organization that make the greatest contribution to the organization's success, however that is measured. Effective organizations adhere to a customer-centered philosophy that includes paying attention to organizational, strategic, environmental, and people factors (Figure 1.3). As this text unfolds, the tools for organizational effectiveness and success will be discussed. These fundamental tools can be found in many philosophies and methodologies, from quality management and continuous improvement to Kaizen, lean manufacturing, and Six Sigma.

Every company has its own organizational structure consisting of policies, procedures, and processes. Integration of these components results in business strategies, measures of organizational performance, and problem-solving methods particular to the firm. When developing strategic plans for the organization's future, company leaders will consider strategic factors such as the marketplace, their present and future product lines, existing and new technology, current and potential customers, and supplier relationships. For instance, Chapter 4 discusses the importance of establishing a customer focus. Strategic planning for effective organizations will be covered in greater detail in Chapter 6. Topics related to process and knowledge management are covered in Chapters 9 and 11.

Strategic Factors
Market
Product
Technology
Customers
Suppliers
Vision
Value Creation Strategy

Organizational Factors
Strategy Deployment
Effective Leadership
Policies
Procedures
Processes
Continuous Improvement Culture
Problem-solving
Commitment
Culture
Measures of Performance
Trust, Honesty and Ethical Behavior

People Factors
Employee Involvement
Education
Training
Internal Supplier-customer Relations
Motivation
Teamwork
Communication
Safety

Environmental Factors
Social
Economic
Competitive
Technology

Figure 1.3   Factors Affecting Organizational Success

Supporting the organizational structure are the culture and the commitment of the employees. Organizations are only as strong as the people who work in them. How well motivated these individuals are will depend on internal relationships between people and their leaders. Open lines of communication, as well as the type of education and training provided to employees, will have an effect on organizational effectiveness. Leadership skills are discussed in Chapter 5. Human resource development and management, a key factor in organizational effectiveness, will be covered in Chapter 7. Environmental factors that an organization must be aware of include existing social issues in the countries and locations where the company has facilities and customers, economic conditions in these areas, existing as well as future competition for their products and services, and technological advancements related to their products and services.

How Do We Know It's Working?

One of Dr. Deming's favorite questions was "How do we know?" He would ask that question whenever he was trying to determine whether or not some change or action had worked in a way that benefited the organization, its employees, or its customers. The *How Do We Know It's Working?* series of examples will provide insight into how one company focuses on applying quality management tools and techniques to answer the question "How do we know?"

Chuck began his career as a mechanical engineering graduate. He gained experience in the chemical processing industry in his hometown before moving on to consumer products manufacturing at a major multinational firm. It was there that his desire to move up the corporate ladder first appeared. Following graduate school in manufacturing management, he joined a small and growing metal stamping firm. Beginning with production control, he worked his way up and through a variety of positions including information technology, quality assurance, engineering design, manufacturing, and sales to become the plant manager and then sales manager. From there, he became vice president of sales, then vice president of manufacturing, and finally president. Beginning with his first exposure to quality assurance and continuous improvement ideas in graduate school, Chuck carefully studied and implemented improvement ideas appropriate for his company. During his tenure as vice president and president, he took the firm from $15 million in sales to over $70 million in sales. Under him, the workforce grew from 40 to 250 and the number of plants from one to three nationwide. Then he felt it was time to do it all over again at another company.

Following a careful search, he located a precision shaft manufacturer that was for sale. Together with another investor and the helpful financing of a bank, he purchased the company. During his first week at his new company, he performed two critical tasks. First, he established and quickly enforced a safety policy modeled after the Occupational Safety and Health Administration's requirements for metal finishing manufacturers. By the end of the week, all employees were wearing their eye, ear, hand, and feet protection where required. All glaring safety hazards had been eliminated or minimized. Over time, more improvements to safety were made. His second task during this time was to create a rather daunting "to-do" list of the changes needed in order to make this manufacturing organization a viable competitor in a tough market.

Steeped in quality management knowledge of the need to establish a focus on customers and to continually improve, Chuck began the long journey to organizational effectiveness. Throughout this text, his story will support chapter material by providing examples of how his organization strives to achieve organizational excellence and effectiveness. It's a true story that began in September 2005 and continues today.

HOW IS AN EFFECTIVE ORGANIZATION CREATED?

You are part of the equation for an organization's success. Imagine that you worked for the Homer Laughlin China Co., discussed in the introduction. What would you do if your boss walked into your office or team meeting today and said, "Improve things"? Would you know where to start? Would you continue with business as usual? What would you change? Requests such as this one can make many people feel uncomfortable. What such people may be lacking is a sense of how to pull a variety of knowledge and skills together to make improvements in a complex work environment. The larger the scope of the task, the more difficult it becomes to decide the appropriate plan of action. How did Homer Laughlin identify areas for improvement? How did they make appropriate changes? How did they know which changes would enable them to excel? How did they become an effective organization?

When you are lost, part of knowing where to go involves knowing where you are. Asking someone for directions to a particular location isn't of much use unless you are able to provide that person with information about where you are. For the same reason, to become more effective, an organization must first know where it stands in the present. Knowing the organization's current levels of performance provides a foundation on which to stand when developing strategic plans for the future. An effective organization develops a customer-oriented approach, studying how its product or service is used from the moment a customer first becomes aware of the product or service until the moment the product is disposed of or the service is complete. Organizations seeking to optimize business processes take a systems approach, emphasizing the improvement of systems and processes that enable a company to provide products or services for its customers. Systemic problems, inherent to the way the organization does business, hinder organizational effectiveness. People don't come to work to do a bad job—their activities are hampered by poor internal communication, faulty processes, and a lack of coordination. System failures, such as missing a customer's order due date or dispensing a medication incorrectly, are caused not by people trying to make mistakes but rather by problems that are the result of poor systems within an organization. To combat these problems, organizations must adopt new methods of managing. These new methods establish a customer focus; encourage management by facts by utilizing measures of performance and key critical success factors; cultivate people-based management through teamwork, education, and training; and utilize continuous improvement through the prevention of defects and improvement of processes.

Effective organizations learn to recognize and remove sources of waste. The Toyota Production System emphasizes seven forms of waste: overproduction, waiting, transportation, processing, inventory, motion, and defects. Ineffective organizations often recognize only the waste found in defects. Missing the other six forms of waste means missed opportunities to become more effective and more competitive. Overproduction occurs when organizations make more than what they can sell. Waiting refers to people or products that sit idle because of a lack of demand. Transportation waste occurs when people or products are moved unnecessarily. Processing waste occurs when processes are incorrectly constructed leading to errors, missed opportunities, or inefficient activities. Motion waste can be found in human activities, such as people taking unnecessary steps or movements. Combined, all these forms of waste mean that products or services do not reach the customer without delay. Later chapters will describe how these forms of waste can be attacked and removed from processes and activities.

Effective organizations are always asking themselves questions about who they are and what business they are really in. They are constantly asking customers about what they are doing right and what needs to be improved. When these questions are asked consistently and when the answers are acted on, the result is a focused organization. Throughout this text we will be studying how answering questions such as the ones below can help create and maintain an effective organization:

1. What is the mission of the organization?
2. What business is the organization really in?
3. What are the primary products or services the organization provides to their customers?
4. Who are their customers?
5. What do their customers expect and need?
6. How does the organization know what the customers' needs and expectations are?
7. How well does the organization meet the needs and expectations of their customers?
8. How does the organization know how well it is doing? What is the proof? What are the indicators?
9. Do management's strategies and actions support the business and support the organization in meeting the customers' needs and expectations?
10. Do the employees know how the work they do specifically benefits the ultimate, external customer?
11. What improvements have been made based on the answers to these questions?
12. What is management doing to support improvement efforts?

Organizational goals, internal processes, and individual efforts are intrinsically related to organizational effectiveness. Effective organizations create a total system that allows and

Recognize	Define	Organize	Improve	Control	Sustain
Leaders recognize the need for change and focus on achieving organizational effectiveness.	Leaders define the strategy, goals, and objectives necessary to pursue organizational effectiveness.	Steps are taken to measure and analyze the organization's current position.	Deploy strategy by selecting key problem areas, training people, establishing measures, developing solutions, and implementing improvements.	Determine if the improvements are working; if so, establish controls to maintain this new level of performance.	Integrate improvements throughout the organization and standardize best practices. Select new areas for improvement.

Figure 1.4   The Journey to Organizational Effectiveness

encourages these three to work together to create the product or service provided to the customer (Figure 1.4). Alignment between the customers' expectations and an organization's strategic plan, processes, and activities ultimately results in the right things being done right.

EXAMPLE 1.1 One Company's Journey to Continuous Improvement

When PLC Inc. began operations three decades ago, their job shop specialized in machining large forgings into finished products. At that time, they utilized three separate inspections as their primary method of ensuring the quality of their products. The first inspection occurred after the initial machining operations (grinding, milling, and boring) were completed and before the part was sent to a subcontractor for heat treatment. After the part returned from heat treatment, key dimensions were checked at the second inspection. A final inspection was conducted before the finished part left the plant. With these three inspections, discrepancies between actual part dimensions and the specifications were found only after the part had completed several machining operations. Although this method resulted in significant scrap and rework costs, PLC continued to use it with only one minor modification: They determined that occasional forgings were not up to standard, so they added an incoming materials inspection to ensure the quality of the blank forgings.

Even with these four inspections, scrap and rework costs were still very high. If the forging passed incoming inspection and began to progress through the machining operations, for a typical part, four to six operations had been completed before any errors were caught during the in-process inspection that occurred before the part was shipped out to be heat-treated. The work done after the operation where the error occurred was wasted because each subsequent machining operation was performed on a faulty part. Beyond a few minor measurements taken once operators had completed their work on the part, the operators were not responsible for checking actual part dimensions against specifications. This type of inspection scheme was very costly to PLC Inc. because it involved not only the defective aspect of these large parts but also the labor cost of performing work on a defective part done by later workstations.

In an attempt to correct this situation, PLC Inc. established a quality control department. The members of this department developed a documented quality control program. The program was designed to ensure conformance to established standards for each product. The department also initiated a corrective action plan that required a root-cause analysis and corrective action for each nonconformance to standards. Following this plan enabled them to implement corrective action plans that prevented future similar errors.

By the mid-1980s, the companies with whom PLC did business began requiring statistical process control information. Statistical process control was a new concept for PLC. Fortunately they realized that the prevention of defects could make a significant impact on their profit performance. With this in mind, they set up control charts to monitor key characteristics over the long term and within a particular run.

Key characteristics such as critical safety dimensions, working diameters, ID/OD for mating parts, radius, and any tight tolerances set by the designer were charted. Besides having each operator inspect his own work, for each part production run the chief inspector checked the first piece for all critical dimensions on a coordinate-measuring machine. Once the first part was approved, the operator had permission to run the rest of the parts in the lot. The "first part" designation applied to any part following a change to the process, such as a new operator, a new setup, a setup after a broken tool, and so on. By tracking the critical part dimensions, those monitoring the processes were able to identify changes that affected the quality of the product and were able to adjust the process accordingly.

PLC sought additional ways to reduce the variation present in the process that prevented them from producing parts as close to the nominal dimension as possible. Thus, emphasis shifted away from inspecting quality into the parts toward making process improvements by designing and machining quality into the product.

Realizing that in a job shop, the small lot sizes made significant use of statistical process control techniques difficult, PLC took a good hard look at the way they did business. In order to stay in this highly competitive business, PLC needed to determine what market need they were going to meet and how they were going to fill that need in an error-free, customer service–oriented manner. With this in mind, they began studying the ideas and concepts surrounding total quality management and continuous improvement. This led them to consider all of their business operations from a process point of view instead of a part-by-part focus. Over the next few years, they applied continuous improvement concepts to how they managed their business.

Their continuous improvement efforts led to an increase in their competitive position. In the 1990s, PLC began specializing in machining large, complex parts for the aviation industry. Soon the shop was full of axles, pistons, steering collars, braces, and other parts for landing gear. The forgings brought in were made from many different kinds of materials, including steel, a variety of alloys, and aluminum. Since the parts could cost anywhere from $6,500 to $65,000 each, not including material and heat-treat or coating costs, PLC had to develop effective methods to run such a wide variety of material and part types on their machines. They realized that they needed to shift their focus away from part inspection to controlling the processes they used to make the parts.

When PLC asked themselves what they really needed to improve in order to please their customers, they realized that reducing the number of tags on parts for out-of-specification conditions was critical. In order to do this effectively over the long term, PLC focused on processes, specifically designing processes, equipment, fixtures, and tooling to meet the needs of the product. This approach resulted in fewer setups and reduced the number of times a part needed to be handled, which in turn reduced the number of times a part could be damaged. Fewer setups also reduced the number of opportunities for mistakes in setups. Better fixtures and tooling enabled the machining process to hold part dimensions throughout the part. Changes such as these resulted in improved quality and throughput.

Their efforts paid off. As PLC improved their part uniformity, rework and defect rates fell and machine uptime increased, as did labor productivity. These factors enabled them to increase output because less time was spent on fixing problems. Instead, the focus shifted to where it belonged: making parts right the first time. Increased output enabled PLC to meet ship dates predictably. They were even able to reduce their prices while maintaining their profitability. As word got out to their customers about pricing, quality, and delivery, PLC was able to attract more and more business. Two plant expansions occurred as their competitive position improved with increasing customer satisfaction. Management at PLC felt that their increased understanding of the processes utilized in making parts enabled them to make better decisions and enhanced their focus on their customers.

Their continuous improvement changes included the following.

Customer-Focused Changes

Equipment: PLC acquired new machines of better design for machining long, relatively thin parts. For instance, they replaced a milling machine that had a cantilevered work-holding system with a horizontal milling machine. Having the part hang as a cantilever allowed vibration and tool movement to play a role in the machining of the part, affecting the ability to hold tolerance. This new horizontal milling machine significantly lowers rework due to the reduction in the variation inherent in the older process.

Machining: PLC made significant investments in their other machining operations. Their equipment now includes numerically controlled turning and three-axis contour machining centers, and boring, honing, milling, grinding, and drilling machines. Most machines are able to perform multiple machining functions in one setup. The equipment is functionally grouped for efficient work flow and close tolerance control.

Job Tracking: PLC implemented a new system of job tracking. First, a manufacturing plan is created and reviewed with the customer. The manufacturing plan provides key information, including part dimensions and the sequence of operations that the part will complete. Once approved, this information is converted to dimensional part drawings for each applicable workstation. Also included is a "traveler," a bill of material that moves through the operations with each part. The traveler must be signed and dated by each operator as he completes his work. Inspections for the part are also noted on the traveler.

Internal Process Changes

Supplier Involvement: When quoting a job, sales engineers at PLC involve their tooling suppliers. The supplier is able to help select the best cutting tools for the type of job. Improved cutting tool technology enables the machining operations to run at higher speeds while holding part dimensions more accurately.

Gage Control System: A new gage verification system has resulted in fewer gage-related errors. Each gage is now calibrated regularly and participates in a preventive maintenance program. These changes have significantly reduced the possibility of measurement error.

Inventory Control Systems: A new tooling inventory control system was recently installed. Using this system has resulted in fewer tooling selection errors which in the past had caused production delays. Inventory control is easier, because taking inventory is now a visual task that has reduced the potential of not having the correct tool when it is needed. Cost savings are expected with this system because inventory can be monitored more easily, thus reducing the potential for lost or misplaced tools.

Smaller Lot Sizes: Recently, PLC has been moving toward single-piece production runs rather than multiple-part runs. This is possible with the new machining and cutting tool technology that allows the machine to run much faster. Now a single piece can be machined in less than one third the time required previously. Not only is this a time-saver, but if something goes wrong, only one part will be damaged. The single-piece lot size also enables PLC to be very reactive to small customer orders.

Human Resources Changes

Cross-Functional Involvement: PLC was quick to realize that an early understanding of what it would take to machine a part resulted in higher-quality parts produced more efficiently. When quoting jobs, sales engineers work with machine tool designers, as well as operators and tooling suppliers, to establish the best machining practices for holding and cutting each particular part based on its material type.

Communication: Prior to the continuous improvement efforts at PLC, machine operators were not considered a valuable source of information. Now, management at PLC is working to increase operator involvement and enhance communication so that an operator will tell management when opportunities for improvement arise. Operators work together with engineers to conduct root-cause analysis investigations and implement corrective actions. Operators are also involved in audits of proper use of procedures.

Many of the improvements listed above came about because PLC had an understanding of the material presented in this text, including creating a customer focus (Chapter 4) and process improvement (Chapters 8, 9, 10, 11, 12, and 13). As a result of their continuous improvement efforts, PLC has increased its efficiency and effectiveness. During the last three years, its growth rate has been 10% to 15% annually, without adding any additional employees. For PLC, continuous improvement begins with the design of processes, tooling, machining centers, and fixtures that support producing quality parts. Well-designed processes and procedures combine to make quality parts and successful customers.

WHAT BENEFITS CAN BE GAINED FROM CREATING AN EFFECTIVE ORGANIZATION?

An effective organization is able to produce more with its existing resources through an improved customer focus and streamlined work processes. With increased awareness of its internal and external customers, organizations have a greater focus on what really needs to be accomplished in order to meet their customers' needs and expectations. As the example of the Homer Laughlin China Co. at the beginning of this chapter showed, an organization that increases its effectiveness will see improvements in its profitability through increased customer retention. Being able to meet customer expectations the first time and every time will enable an organization to increase its market share as new customers seek out the company. Since an effective organization has focused and streamlined its work processes, it will benefit from lower costs because of reduced waste and less rework. One of the major savings that occurs is fewer customer complaints and fewer warranty claims. More successful customers result in greater market share.

Improvements will also exist internal to the organization. Because of improved communication and teamwork, effective organizations have good management-employee relations. As problems are solved and the organization begins to run more smoothly, employee involvement and satisfaction will increase, lowering turnover and absenteeism. Benefits of creating and maintaining an effective organization are summarized in Figure 1.5.

Improved profitability
Increased customer retention
Reduced customer complaints and warranty claims
Reduced costs through less waste, rework, and so on
Greater market share
Increased employee involvement and satisfaction, lower turnover
Increased ability to attract new customers
Improved competitiveness
Improved customer satisfaction
Improved management–employee relations
Improved focus on key goals
Improved communication
Improved teamwork
Improved employee morale
Improved company image
Improved revenue
Improved service to internal and external customers
Improved effectiveness
Greater adaptability
Improved planning
Improved work environment
Improved decision making

Figure 1.5   Benefits of Creating and Maintaining an Effective Organization

LESSONS LEARNED

Today's consumers have come to expect quality as an essential dimension of the product or service they are purchasing. Effective organizations recognize that it is the quality of managing that equates to business success. Effective organizations respond to their customers' expectations by focusing their company's value chain on providing quality products and services for their customers. As will become apparent throughout this text, effective organizations embrace such concepts as optimization of processes, elimination of waste, value chain management, and creation of a customer focus. In today's global marketplace, nearly everything provided by a company can be duplicated, such as products or services, guarantees, hours of operation or delivery schedules, or order-filling speed. Effective organizations work to provide greater value to their customers, finding new sources of customer delight more rapidly than their competitors. An effective organization concentrates on what is important: meeting both their internal and external customers' needs and reasonable expectations; encouraging teamwork and cooperation; tracking key indicators of performance; maintaining a long-term focus on continuous improvement; making decisions based on facts; and finding solutions, not fault. Such focus means that everyone is involved in the process of creating and maintaining an effective organization. As you read this text, you will learn that in order to delight and excite their customers, effective organizations listen carefully to their customers. They seek to constantly and forever enrich and improve their business processes in order to optimize the performance of their products and services.

1. Effective organizations compete on value creation and market performance.
2. Effective organizations make excellence part of their organization's DNA.
3. The organizational effectiveness umbrella (Figure 1.2) covers a wide variety of topics, tools, and techniques.
4. Effective organizations perform well in the market because their activities provide for a long list of benefits (Figure 1.5), including increased profitability and customer retention, less waste, improved overall customer and employee satisfaction, and improved communication.
5. Effective organizations create an overall alignment of the entire organizational system that focuses on the factors critical to the organization's success.

Are You a Quality Management Person?

It's the little day-to-day activities that make a quality management person. We'd all like to think that we are, but how do you know? Usually, QM people are the type of people who are always asking questions. These questions can be organizational or personal in nature. At the end of each chapter, check out the quality management person questions and find out whether or not you are a quality management person. How do you answer them on a scale of 1 to 10?

Do you keep your commitments?

| 1 | 2 | 3 | 4 | 5 | 6 | 7 | 8 | 9 | 10 |

Are you on time?

| 1 | 2 | 3 | 4 | 5 | 6 | 7 | 8 | 9 | 10 |

Are you more productive than you were 12 months ago?

| 1 | 2 | 3 | 4 | 5 | 6 | 7 | 8 | 9 | 10 |

Have you learned from your experiences? How?

| 1 | 2 | 3 | 4 | 5 | 6 | 7 | 8 | 9 | 10 |

Are you open to change?

| 1 | 2 | 3 | 4 | 5 | 6 | 7 | 8 | 9 | 10 |

Do you like to learn new things?

| 1 | 2 | 3 | 4 | 5 | 6 | 7 | 8 | 9 | 10 |

Do you like challenges?

| 1 | 2 | 3 | 4 | 5 | 6 | 7 | 8 | 9 | 10 |

Do you constantly strive to improve?

| 1 | 2 | 3 | 4 | 5 | 6 | 7 | 8 | 9 | 10 |

Chapter Questions

1. Why would an organization want to be effective?
2. How would you describe or define *organizational effectiveness?*
3. Who is in charge of creating organizational effectiveness?
4. If you initiated changes to improve organizational effectiveness, how would you verify the improvement?
5. How would you measure the relative overall organizational effectiveness of any organization?
6. What drove PLC Inc. (Example 1.1) to make changes?
7. Describe an organization you have worked for. How effective was the organization? How were you able to draw that conclusion?
8. What types of efforts have you seen organizations make in order to become more effective?

2 Organizational Philosophy

What are the key philosophical elements to delighting customers and generating organizational success?

Who are the individuals and what are their philosophies?

Dr. Armand Feigenbaum

Dr. Walter Shewhart

Dr. W. Edwards Deming

Dr. Joseph M. Juran

Philip Crosby

Dr. Kaoru Ishikawa

Dr. Genichi Taguchi

Lessons Learned

Learning Opportunities

1. To become familiar with the quality masters
2. To understand the philosophies of quality management and continuous improvement

Marshall Field, who in 1880 created one of the United States' first department stores, was an involved manager. Each day during his 40-year reign as owner, he would walk through his flagship store on State Street in Chicago, Illinois, and observe his customers and their interactions with his employees. The authors of Give the Lady What She Wants describe the situation that created the store's famous philosophy:

> One day while making his rounds, Marshall Field came upon an assistant retail manager involved in a heated argument with a female customer. He pulled the employee aside and asked him what he was doing. The employee responded that he was resolving a customer complaint concerning returning merchandise. Marshall Field retorted with, "No you're not! Give the lady what she wants!" Marshall Field had established a policy of accepting merchandise returns from customers, a procedure not followed by any other department store at the time.

Paraphrased from:
Give the Lady What She Wants, Lloyd Wendt and Herman Kogan, 1952

WHAT ARE THE KEY PHILOSOPHICAL ELEMENTS TO DELIGHTING CUSTOMERS AND GENERATING ORGANIZATIONAL SUCCESS?

Why is it that some leaders seem to know instinctively just what to do to make their customers happy? How do they enable their organization to delight their customers and cause their customers to have a love affair with their products or services? These same leaders are also ones who create a warm and productive working environment for their employees. What are the philosophies that guide these leaders in their efforts to create and sustain organizational effectiveness?

Marshall Field took the opinions of his internal and external customers so seriously that each working day throughout his life, he walked through his store and met daily with customers and employees. His philosophy, *Give the Lady What She Wants*, continues to be a guiding force of Marshall Field's department stores 100+ years later (although it is less gender specific today). Leaders of today's effective organizations have studied the theories of quality management gurus such as Armand Feigenbaum, W. Edwards Deming, Joseph Juran, and others to gain an understanding of the underlying philosophy necessary to provide customer satisfaction. Although each of these men's approaches to creating effective organizations differ, the key elements remain the same. In order to consistently provide for customer satisfaction, effective organizations must:

- Determine who their customers are
- Determine the critical key success factors for meeting their customers' needs, requirements, and expectations
- Establish effective processes that enable them to provide products and services that meet their customers' needs, requirements, and expectations
- Focus on process measurement and improvement
- Provide the management involvement and commitment required for organizational success (Figure 2.1)

Leaders of effective organizations study and apply the ideas of the individuals discussed below, as well as the ideas of leaders of effective organizations, such as Marshall Field, to improve their own processes and delight their customers. The key philosophical elements listed above will be covered throughout this text beginning with this chapter about several important individuals and their philosophies.

How Do We Know It's Working?

Two weeks after taking ownership of JR Precision Shafts, an old friend visited the facility. In just two short weeks, several changes, most related to safety, were in evidence. During her tour of the plant, the shop foreman, an employee from the previous management team, pulled her aside and said, "Yeah, Chuck has a lot of energy right now, but we'll get him calmed down and things will return to normal." Having seen Chuck in action before, she knew that this man was in for a big surprise. Under active and dedicated leadership, the future would be full of more and more improvements.

The philosophy espoused by any organization's leadership shapes and directs the future of that organization. Effective leaders absorb the philosophies of quality masters and make value creation and market performance a fundamental way of managing throughout their organizations. Quality is fundamental to their business strategy, guiding and directing their day-to-day operations and ongoing strategy. Chuck, like other effective leaders, practices what the quality advocates preach.

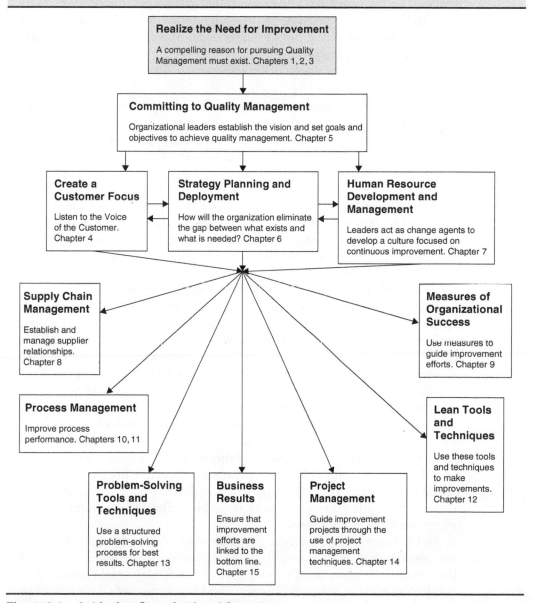

Figure 2.1   Achieving Organizational Success

WHO ARE THE INDIVIDUALS, AND WHAT ARE THEIR PHILOSOPHIES?

Dr. Armand Feigenbaum

Armand Feigenbaum (1920–) is considered to be the originator of the total quality movement. Dr. Feigenbaum defined quality based on a customer's actual experience with the product or service. His landmark text, *Total Quality Control*, first published in 1951 and updated regularly since then, has significantly influenced industrial practices. In his original text, he predicted that quality would become a significant customer-satisfaction issue, even to the point of surpassing price in importance in the decision-making process. As he predicted, consumers have come to expect quality as an essential dimension of the product or service they are purchasing.

In his text, Dr. Feigenbaum defines quality as follows:

> . . . a customer determination which is based on the customer's actual experience with the product or service, measured against his or her requirements—stated or unstated, conscious or merely sensed, technically operational or entirely subjective—always representing a moving target in a competitive market.

Note that Dr. Feigenbaum's definition of quality is broad reaching. His definition is as sound in today's highly technical world as it was in the 1950s when he wrote it. Its very flexibility is what makes it so useful to effective organizations. Regardless of their product or service, effective organizations recognize the key point of Feigenbaum's definition: quality is what the customer says, not what the company thinks. It stresses that quality is a customer determination; that is, only a customer can decide if and how well a product or service meets his or her needs, requirements, and expectations. These needs, requirements, and expectations may be stated or unstated, conscious or merely sensed, or technically operational or entirely subjective. Quality is also based on the customer's actual experience with the product or service throughout its life, from purchase to disposal. Dr. Feigenbaum's definition recognizes that quality, and therefore customer satisfaction, is a moving target in a competitive market. In today's competitive environment, merely providing a product or service that is defect free isn't enough. In a 2006 keynote address to the International Conference on Quality in Tokyo, Dr. Feigenbaum stressed that organizations that want to lead in the marketplace must continually increase the value of their products and services by finding ways to excite their customers. The complexity of Dr. Feigenbaum's definition is exactly what makes it an excellent definition of quality from the customer's point of view. In order to continually delight their customers, effective organizations must capture these stated or unstated, conscious or merely sensed, technically operational or entirely subjective needs, requirements, and expectations. How effective organizations do this is discussed in greater detail in Chapter 4.

To Dr. Feigenbaum, quality is more than a technical subject; it is an approach to doing business that makes an organization more effective. Throughout his life, Dr. Feigenbaum has consistently encouraged treating quality as a fundamental element of a business strategy. In his article "Changing Concepts and Management of Quality Worldwide," from the December 1997 issue of *Quality Progress* and again in the February 2007 issue of *Quality*

Progress, he asserts that quality is not a factor to be managed but a method of "managing, operating, and integrating the marketing, technology, production, information, and finance areas throughout a company's quality value chain with the subsequent favorable impact on manufacturing and service effectiveness." Effective leaders align all aspects of their organization, from design to service after the sale, according to what the customer values. Under their guidance, the processes within their organization must focus on maintaining and improving customer value. According to Dr. Feigenbaum, management is responsible for recognizing the evolution of the customer's definition of quality for an organization's products and services. Quality systems are a method of managing an organization to achieve higher customer satisfaction, lower overall costs, higher profits, and greater employee effectiveness and satisfaction. Leaders of effective organizations see quality as a fundamental business strategy that guides and directs their day-to-day operations and the long-term choices they make. Company leadership is responsible for creating an atmosphere that enables employees to provide the right product or service the first time, every time. Dr. Feigenbaum encourages companies to eliminate waste, which drains profitability, by determining the costs associated with failing to provide a quality product. Costs of quality are covered in greater detail in Chapter 9. Quality efforts should emphasize increasing the number of experiences that go well for a customer versus handling things when they go wrong. Statistical methods and problem-solving techniques should be utilized to effectively support business strategies aimed at achieving customer satisfaction. Note that Dr. Feigenbaum's definitions and philosophies cover all aspects of the business, from customers to employees and from products to processes. The newest edition of his text serves as a how-to guide for establishing a quality system.

Dr. Walter Shewhart

In his writings Dr. Walter Shewhart (1891–1967) points to two aspects of quality: the subjective aspect, what the customer wants; and the objective side, the physical properties of the goods or services, including value received for the price paid. During his lifetime Dr. Shewhart (Figure 2.2) worked to create statistical methods that control and improve the quality of the processes that provide goods and services. When an organization is translating customer requirements to actual products and services, statistical measures of key characteristics are important to ensure quality. While working at Bell Laboratories in the 1920s and 1930s, Dr. Shewhart was the first to encourage the use of statistics to identify, monitor, and eventually remove sources of variation found in repetitive processes.

Dr. Shewhart identified two sources of variation in a process. Controlled variation, also termed *common causes*, is variation present in a process due to the very nature of the process. This type of variation can be removed from the process only by changing the process. For example, consider a person who has driven the same route to work dozens of times and has determined that it takes about 20 minutes to get from home to work regardless of minor changes in weather or traffic conditions. If this is the case, then the only way the person can improve on this time is to change the process by finding a new route. Uncontrolled variation, also known as *special* or *assignable causes*, comes from sources external to the process. This type of variation is normally not part of the process. It can be

Figure 2.2   Dr. Walter Shewhart

identified and isolated as the cause of the change in the behavior of the process. For instance, the commuter described above would experience uncontrolled variation if a major traffic accident stopped traffic or a blizzard made traveling nearly impossible. Uncontrolled variation prevents the process from performing to the best of its ability.

It was Dr. Shewhart who put forth the fundamental principle of quality: that once a process is under control, exhibiting only controlled variation, future process performance can be predicted, within limits, on the basis of past performance. In his text *Economic Control of Quality of Manufactured Product* (Van Nostrand Reinhold, 1931, p. 6), he wrote:

A phenomenon will be said to be controlled when, through the use of past experience, we can predict, at least within limits, how the phenomenon may be expected to vary in the future. Here it is understood that prediction within limits means that we can state, at least approximately, the probability that the observed phenomenon will fall within the given limits.

Based on this principle, Dr. Shewhart developed the formulas and a table of constants used to create the most widely utilized statistical control charts in quality: the $\overline{X}$ and R charts (Chapter 13). These charts (Figure 2.3) first appeared in a May 16, 1924, memo written by Dr. Shewhart and later in his 1931 text, *Economic Control of Quality of Manufactured Product*, mentioned and quoted above. In this text Dr. Shewhart presented the foundation principles on which modern quality control is based. Control charts have three purposes: to define standards for the process, to aid in problem-solving efforts to attain the standards, and to serve to judge whether the standards have been met.

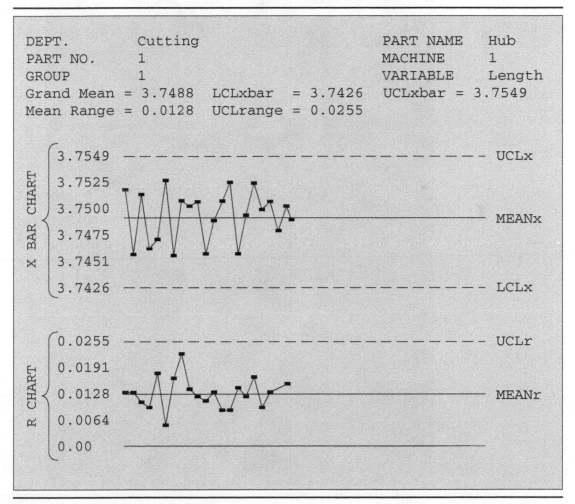

Figure 2.3   Typical $\overline{X}$ and R Charts

Although Dr. Shewhart concentrated his efforts on manufacturing processes, his ideas and charts are applicable to any process found in nonmanufacturing environments. Dr. Shewhart's tools and techniques required for improving processes and systems will be covered in greater detail in Chapter 13.

Dr. W. Edwards Deming

Dr. W. Edwards Deming (1900–1993) made it his mission to teach optimal management strategies and practices for effective organizations. Dr. Deming (Figure 2.4) encouraged top-level management to get involved in the process of creating an environment that supports continuous improvement. A statistician by training, Dr. Deming graduated from Yale

Figure 2.4   Dr. W. Edwards Deming

University in 1928. Following his work with the Bureau of the Census, he first began spreading his quality message shortly after World War II. In the face of American prosperity following the war, his message was not heeded in the United States. His work with the Bureau of the Census and other government agencies led to his eventual contacts with Japan as that nation began to rebuild. There he helped turn Japan into an industrial force to be reckoned with. His efforts resulted in his being awarded the Second Order of the Sacred Treasure from the Emperor of Japan. It was only after his early 1980s appearance on the TV program "If Japan Can, Why Can't We?" that Dr. Deming found an audience in the United States. Over time, he became one of the most influential experts on quality assurance.

Dr. Deming, who described his work as "management for quality," felt that the consumer is the most critical aspect in the production of a product or the provision of a service. Listening to the voice of the customer and then utilizing the information gathered to improve products and services is an integral part of his teachings. To Dr. Deming, quality must be defined in terms of customer satisfaction. Such a customer focus means that the quality of a product or service is multidimensional. It also means that there are different degrees of quality; a product that completely satisfies customer A may not satisfy customer B.

Dr. Deming considered quality and process improvement activities as the catalyst necessary to start an economic chain reaction. Improving quality leads to decreased costs, fewer mistakes, fewer delays, and better use of resources, which in turn leads to improved productivity, which enables a company to capture more of the market, which enables the company to stay in business, which results in providing more jobs (Figure 2.5). He felt that without quality improvement efforts to light the fuse, this process would not begin.

Dr. Deming's philosophies focus heavily on management involvement, continuous improvement, statistical analysis, goal setting, and communication. His message, in the form of 14 points, is aimed primarily at management (Figure 2.6). Dr. Deming's philosophy encourages company leaders to dedicate themselves and their companies to the long-term improvement of their products or services. Dr. Deming's first point is to *create a constancy of purpose toward improvement of product and service, with the aim to become competitive and to stay in business and to provide jobs*. This point encourages leadership to accept the obligation to constantly improve the product or service through innovation, research, education, and continual improvement in all facets of the organization. A company is like an Olympic athlete who must constantly train, practice, learn, and improve in order to attain a gold medal. Lack of constancy of purpose is one of the deadly diseases Dr. Deming warns about in his writings. Without dedication, the performance of any task cannot reach its best.

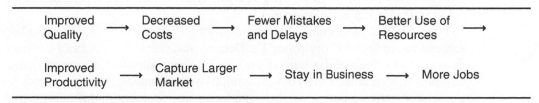

Figure 2.5 Deming's Economic Chain Reaction

1. Create a constancy of purpose toward improvement of product and service, with the aim to become competitive and to stay in business and to provide jobs.
2. Adopt the new philosophy.
3. Cease dependence on inspection to achieve quality.
4. End the practice of awarding business on the basis of price tag alone. Instead minimize total cost.
5. Constantly and forever improve the system of production and service.
6. Institute training on the job.
7. Institute leadership.
8. Drive out fear.
9. Break down barriers between departments.
10. Eliminate slogans, exhortations, and targets for the workforce.
11. Eliminate arbitrary work standards and numerical quotas. Substitute leadership.
12. Remove barriers that rob people of their right to pride of workmanship.
13. Institute a vigorous program of education and self-improvement.
14. Put everybody in the company to work to accomplish the transformation.

Figure 2.6   Deming's 14 Points
Source: Deming, W. Edwards, *Out of the Crisis*, pp. 23–24, © 2000 W. Edwards Deming Institute, by permission of The MIT Press.

Dr. Deming's second point—*adopt a new philosophy*—rejects "acceptable" quality levels and poor service as a way of life and supports continuous improvement in all that we do. The 12 other points ask management to rethink past practices, such as awarding business on the basis of price tag alone, using mass inspection, setting arbitrary numerical goals and quotas, enforcing arbitrary work time standards, allowing incomplete training or education, and using outdated methods of supervision. Mass inspection has limited value because quality cannot be inspected into a product. Quality can be designed into a product, and manufacturing processes can produce the product correctly. However, after the product has been made, quality cannot be inspected into it. Similarly, awarding business on the basis of price tag alone is shortsighted and fails to establish mutual confidence between the supplier and the purchaser. Low-cost choices may lead to losses in productivity elsewhere.

Leadership, along with the concepts of authority and responsibility, plays a significant role in all of Dr. Deming's points. Without leadership, an organization and the people working within it are rudderless. Without effective leadership, the organization and its people cannot reach their full potential. Throughout his life Dr. Deming encouraged leadership to create and manage systems that enable people to find joy in their work. Dr. Deming's point about driving out fear stresses the importance of communication between leadership and management.

Effective leaders welcome the opportunity to listen to their employees and act on valid suggestions and resolve key issues. Dr. Deming also points out the need to remove barriers that rob individuals of the right of pride in workmanship. Barriers are any aspect of a job that prevents employees from doing their jobs well. By removing barriers, leadership creates an environment supportive of their employees and the continuous improvement of their day-to-day activities. Improved management–employee interaction, as well as increased communication between departments, will lead to more effective solutions to the challenges of creating a product or providing a service.

Education and training also play an integral part in Dr. Deming's plan. Continual education creates an atmosphere that encourages the discovery of new ideas and methods. This translates to innovative solutions to problems. Training ensures that products and services are provided that meet standards established by customer requirements.

EXAMPLE 2.1 Dr. Deming's 14 Points

When products don't live up to a customer's expectations, customers shop elsewhere. At JQOS, costs associated with inspection and manufacturing replacement components mounted. The thought of losing a customer disturbed them. They recognized that Dr. Deming's philosophy of "creating a constancy of purpose toward improvement of product and service" would enable them to keep the component job, become more competitive, stay in business, and provide jobs. Since 100% inspection is expensive and ineffective, they wanted to enact process changes that would allow them to follow Dr. Deming's third point and "cease dependence on inspection to achieve quality."

Following Dr. Deming's fifth point, "constantly and forever improve the system of production and service," JQOS formed product improvement teams comprised of engineers and machine operators most closely associated with each component. Dr. Shewhart's $\overline{X}$ and R charts, recording process performance, formed the center of the improvement efforts.

The benefits of this approach to doing business were numerous. By "putting everyone in the company to work to accomplish the transformation," a change came over the production line. As operators learned to use and understand the $\overline{X}$ and R charts, they learned about their manufacturing processes. Process improvements, based on the knowledge gained from these charts, significantly improved component quality. Employee morale increased as they started to take an interest in their jobs. This was a big change for the union shop.

Changes happened on the individual level too. As management followed Dr. Deming's seventh, sixth, and eighth points "institute leadership" by "instituting training on the job," they were able to "drive out fear." One operator, who originally was very vocal about not wanting to be on the team, eventually ended up as a team leader.

Unbeknownst to his coworkers, this operator faced a significant barrier that "robbed him of his right to pride of workmanship" (Dr. Deming's 12th point). He had dropped out of school after sixth grade and had a very difficult time with math and reading. His understanding of math was so limited, he couldn't understand or calculate and average or a range. To hide his lack of math skills, he memorized which keys to use on the calculator. As he attended classes offered by the company, he learned how to plot and interpret data in order to make process adjustments based on trends and out of control points. The more involved he became in the improvement efforts, the more he realized how interesting his work had become. His involvement with the team inspired him to go back to school and get his High School Graduate Equivalency Degree at the age of 55. He followed Dr. Deming's 13th point, "institute a vigorous program of education and self-improvement."

Rather than rely on "slogans, exhortations, and targets for the workforce," JQOS manufacturing followed Dr. Deming's advice and eliminated "arbitrary work standards and numerical quotas." JQOS was able to "cease dependence on inspection to achieve quality." They "substituted leadership," earning awards for being the most improved supplier. With their ability to manufacture complex components to customer specifications with nearly zero scrap, the plant has become the most profitable of the entire corporation.

Dr. Deming defined quality as "non-faulty systems." At first glance this definition seems to be incomplete, especially when compared to that of Dr. Feigenbaum. Consider, however, what is meant by the term *systems*. Systems enable organizations to provide their customers with products and services. Faulty systems cannot help creating faulty products and services, resulting in unhappy customers. By focusing attention on the systems that create products and services, Dr. Deming is getting at the heart of the matter. Dr. Deming used the red bead experiment to help leaders understand how a process with problems can inhibit an individual's ability to perform at his or her best. He used this experiment to create an understanding of his point, "Remove barriers that rob people of their right to pride of workmanship." To conduct his experiment, Dr. Deming filled a box with 1,000 beads, 800 white and 200 red. Participants randomly scooped 100 beads from the box. The participants had no control over which beads the scoop picked up or what percentage of red beads were in the box. Given these constraints, 20% of the beads selected were red. Since only white beads were acceptable, Dr. Deming chastised those who scooped red beads from the box even though they had no control over their performance. Similarly, employees in an organization may often be blamed for faulty performance when in actuality it is the system that is faulty. The red beads represent problems in the system or process that can be changed only through leadership involvement. For Dr. Deming, it is the job of leaders to create non-faulty systems by removing the "red beads."

The importance of reducing the variation present in a system or process is one of the most critical messages Dr. Deming sent to leadership. To do this, he emphasized the use of the statistics and quality techniques espoused by Dr. Shewhart and covered in Chapter 13 in this text. According to Dr. Deming, process improvement is best carried out in three stages:

Stage 1: Get the process under control by identifying and eliminating the sources of uncontrolled variation. Remove the special causes responsible for the variation.

Stage 2: Once the special causes have been removed and the process is stable, improve the process. Investigate whether waste exists in the process. Tackle the common causes responsible for the controlled variation present in the process. Determine whether process changes can remove them from the process.

Stage 3: Monitor the improved process to determine whether the changes made are working.

Dr. Deming used a second experiment—the funnel experiment—to demonstrate how tampering with a process can actually make the performance of that process worse. For this experiment, beads were dropped one by one through a funnel over a target, while the target was held stationary and then the funnel was moved in different ways as shown in Figure 2.7.

Rule 1

No Compensation: Do not adjust the funnel position. Center the funnel over the target and leave it there for the duration of the experiment.

Rationale: Intuitively, we know that this is probably not the way to get the best results. However, this strategy will give us some baseline data. We can compare the results using one of the other rules with this baseline to measure our improvement. We could also be lucky enough to hit the target once in a while.

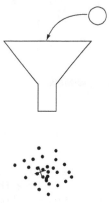

Rule 2

Exact Compensation: Measure the distance from the last drop to the target. Compensate for the error by moving the funnel the same distance, but in the opposite direction from its last position.

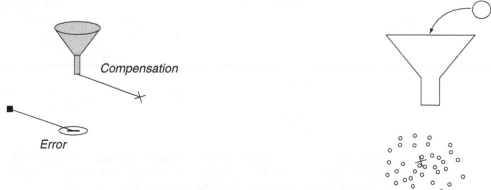

Rationale: This rule attempts to compensate for the inaccuracy of the funnel. If the funnel drops the bead off the target by a certain amount, it is reasonable to suppose that moving the funnel in the opposite direction by the same amount will improve the results. This rule requires us to remember the position of the funnel at the last drop.

Figure 2.7   Deming's Funnel Experiment

Rule 3

Overcompensation: Measure the distance from the last drop to the target. Center the funnel on the target; then move it the same distance from the target as the last drop, but in the opposite direction.

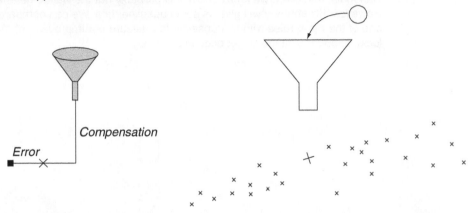

Rationale: In this case we use the target as a basis for our adjustment, rather than the last position of the funnel, as in Rule 2. This is probably our only recourse if we know only the position of the target and the last drop and not the position of the funnel.

Rule 4

Consistency: Center the funnel over the last drop.

Rationale: The objective of Rule 4 is to maintain consistent results. Even if we miss the target, the results should be consistent since we always aim for the position of the last drop. If we are off target, we can always take care of it later.

Figure 2.7   (*continued*)

The purpose of moving the funnel was to try to get the beads to cluster around the target, thus exhibiting little variation in where they landed. First the funnel was held stationary above the target as the beads were dropped, resulting in the bead pattern shown in Rule 1 in Figure 2.7. Next, as shown in Rule 2, the distance from the last dropped bead to the target was measured, and the funnel was moved the same distance from the target, but in the opposite direction from the last bead. Note the pattern that resulted. Then, as shown in Rule 3, the distance from the last dropped bead to the target was measured. The funnel was centered on the dropped bead and was moved the same distance from it as the last dropped bead, but again in the opposite direction, creating the pattern shown. Finally, as shown in Rule 4, the funnel was centered over the last dropped bead and the next bead was dropped, producing the pattern shown. Note that the smallest pattern (the one with the least variation around the target) is the first (Rule 1), where the funnel is not moved. Using this experiment, Dr. Deming showed that tampering with a process (that is, moving the funnel) can actually increase the variation and result in poorer performance.

Tampering can be avoided by isolating and removing the root causes of process variation through the use of the Plan-Do-Study-Act problem-solving cycle. When tackling process improvement, it is important to find the root cause of the variation. When seeking the causes of variation in the process, Dr. Deming encouraged the use of the Plan-Do-Study-Act (PDSA) cycle rather than a Band-Aid sort of fix (Figure 2.8). Originally developed by Dr. Walter Shewhart, the PDSA cycle is a systematic approach to problem solving. During the Plan phase, users of the cycle study a problem and plan a solution. This portion of the cycle should be the one that receives the most attention because good plans lead to well-thought-out solutions. The solution is implemented during the Do phase of the cycle. During the Study phase, the results of the change to the process are studied. Finally, during the Act phase, when the results of the Study phase reveal that the root cause of the problem has been isolated and removed from the process permanently, the changes are made permanent. If the problem has not been resolved, a return trip to the Plan portion of the cycle for further investigation is undertaken. The Plan-Do-Study-Act cycle of problem solving will be covered in detail in Chapter 13.

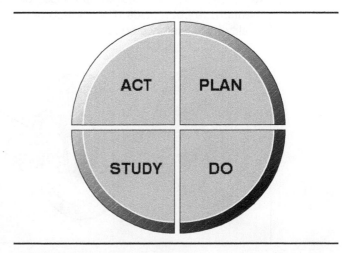

Figure 2.8   The Deming Cycle
Source: Deming, W. Edwards, *Out of the Crisis*, p. 88, © 2000 W. Edwards Deming Institute, by permission of The MIT Press.

EXAMPLE 2.2   Tampering with the Process

The Whisk Wheel Company has been notified by its largest customer, Rosewood Bicycle, Inc., that Whisk Wheel will need to dramatically improve the quality level associated with the hub operation. Currently the operation is unable to meet the specification limits set by the customer. Rosewood has been sorting the parts on the production line before assembly, but it wants to end this practice. Figure 2.9 shows the product in question, a wheel hub. The hub shaft is made of chrome-molybdenum steel. The dimension in question is the shaft length. The specification for the length is 3.750 inches, ±0.005 inch. The process involves taking 12-foot-long chrome-moly steel shafts purchased from a supplier, straightening them, and cutting them to 3.750-inch lengths.

In order to determine the root causes of variation in hub length, the engineers are studying the cutting operation and the operator. The operator performs the process in the following manner. Every 18 minutes he measures the length of six hubs. The length values for the six consecutively produced hubs are averaged, and the average is plotted on $\overline{X}$ and R charts. Periodically, the operator reviews the evolving data and makes a decision as to whether the process mean (the hub length) needs to be adjusted. These adjustments can be accomplished by stopping the machine, loosening some clamps, and jogging the cutting device back or forth, depending on the adjustment the operator feels is necessary. This process takes about five minutes and appears to occur fairly often.

Based on the engineers' knowledge of Dr. Deming's funnel experiment, they are quick to realize that the operator is adding variation to the process. He appears to be overcontrolling (overadjusting) the process because he cannot distinguish between common-cause variation and special-cause variation. The operator has been reacting to patterns in the data that may be inherent (common) to the process. The consequences of this mistake are devastating to a process. Each time an adjustment is made when it is not necessary, variation is introduced to the process that would not be there otherwise. Not only is quality essentially decreased (made more variable) with each adjustment, but production time is unnecessarily lost.

Use Figure 2.10 to compare the differences in the charts when no adjustment is made to the process. Note that the process has stabilized because no unnecessary adjustments have been made. The method of overcontrol has proved costly in terms of both quality (inconsistent product) and productivity (machine downtime, higher scrap).

Figure 2.9   Hub Assembly

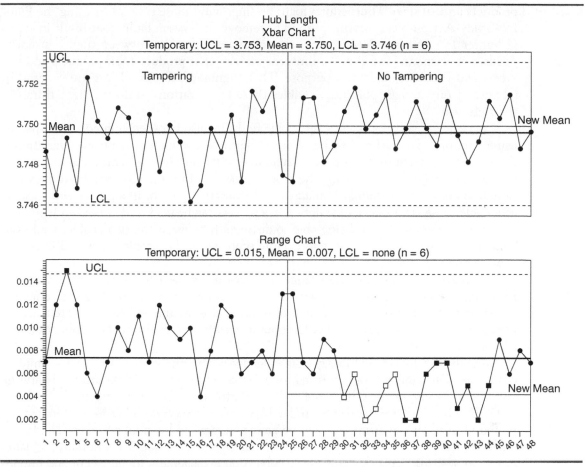

Figure 2.10 The Effects of Tampering on a System

In his final book, *The New Economics*, Dr. Deming tied much of his life's work together when he introduced the concept of profound knowledge. A system of profound knowledge has four interrelated parts:

- An appreciation for a system
- Knowledge about variation
- Theory of knowledge
- Psychology

Knowledge of all of these areas enables companies to expand beyond small process-improvement efforts and to optimize their systems in their entirety. Effective leaders have an appreciation for the systems that work together to create their organization's products and services. Their efforts focus on improving these systems by using the Plan-Do-Study-Act problem-solving method to remove the system faults that result in errors (Chapter 13). Effective leaders also seek to create alignment between their customers' needs, requirements, and expectations; the systems that produce their products and services; and their organization's purpose. This alignment, discussed in greater detail in Chapter 6 on strategic planning, enables these organizations to do the right things in the right way.

Knowledge of variation means being able to distinguish between common- and special-cause variation. First defined by Dr. Walter Shewhart, common or controlled variation is the variation present in a process or system due to its very nature. This natural variation can be removed only by changing the process or system in some way. Special-cause variation, also known as *uncontrolled variation*, is the variation present in a process due to some assignable cause. This source of variation in a process can be readily identified and removed from the system or process. Being able to distinguish between the two enables leaders to guide their system improvement efforts more proficiently. This topic is covered in greater detail later.

The theory of profound knowledge involves using data to understand situations. Dr. Deming encouraged the use of fact-based information when making decisions. Effective leaders gather and analyze information for trends, patterns, and anomalies before reaching conclusions. Chapters 11, 12, and 13 provide a variety of tools for data collection and analysis.

An understanding of psychology enables leaders to interact with customers and employees better. Creating and maintaining an effective organization is about understanding customers and employees. Chapters 4 and 7 cover these topics in greater detail.

Dr. Deming's influence continues today. Many of the concepts and ideas he espoused can be found in today's continuous improvement programs and international standards. For example, the 2000 revision of the International Organization for Standardization Quality Standard, ISO 9000, places significant emphasis on management involvement and responsibility, including communicating customer requirements, developing an integrated overall plan to support meeting customer requirements, measuring key product and service characteristics, providing ongoing training, and demonstrating leadership.

Living the continuous improvement philosophy is not easy. The level of dedication required to become the best is phenomenal. Dr. Deming warned against the "hope for instant pudding." Improvement takes time and effort and does not happen instantly. The hope for instant pudding is one that afflicts us all. After all, how many of us wouldn't like all of our problems to be taken care of by just wishing them away? Dr. Deming's philosophies cover all aspects of the business, from customers to leadership to employees and from products and services to processes. As evidenced by his fourteenth point (Figure 2.6)—"Put everyone in the company to work to accomplish the transformation"—Dr. Deming's quality system is actually an ongoing process of improvement. To him, quality must be an integral part of how a company does business. Organizations must continuously strive to improve; after all, the competition isn't going to wait for them to catch up! The philosophies of Dr. Deming will be used throughout this text to support discussions related to the characteristics of an effective organization.

Dr. Joseph M. Juran

Born December 24, 1904, Dr. Joseph M. Juran (1904–) immigrated from Romania to Minneapolis, Minnesota in 1912. In 1920, he enrolled in Electrical Engineering at the University of Minnesota. After earning his degree, he went to Western Electric as an engineer at the Hawthorne Manufacturing plant in Cicero, Illinois. There he served in one of the first inspection statistical departments in industry. During the Depression, he earned a law degree, just in case he needed an employment alternative. During World War II (WWII), he served in the Statistics, Requisitions, Accounts, and Control Section of the Lend-Lease administration. He was responsible for the procurement and leasing of arms, equipment, and supplies to WWII allies. Like Dr. Deming, Dr. Juran (Figure 2.11) played

Figure 2.11 Dr. Joseph Juran Wearing the Second Order of the Sacred Treasure

a significant role in the rebuilding of Japan following WWII. Based on their work, both he and Deming were awarded the Second Order of the Sacred Treasure from the Emperor of Japan.

Dr. Juran's approach involves creating awareness of the need to improve, making quality improvement an integral part of each job, providing training in quality methods, establishing team problem solving, and recognizing results. Dr. Juran emphasizes the need to improve the entire system. To improve quality, individuals in a company need to develop techniques and skills and understand how to apply them. Dr. Juran's definition of quality goes beyond the immediate product or moment of service. To Dr. Juran, quality is a concept that needs to be found in all aspects of business. As shown in Figure 2.12. Dr. Juran contrasts big Q and little q to show the broad applicability of quality concepts.

During his career, Dr. Juran significantly influenced the movement of quality from a narrow satistical field to quality as a management focus. He attributes his change in emphasis to having read Margaret Mead's book *Cultural Patterns and Technical Change* (first edition, UNESCO, 1955). The book describes how a clash of cultures leads to resistance to change, as demonstrated by resistance in developing nations to the United Nations efforts to improve conditions. Dr. Juran felt this resistance to change could also be seen in clashes between management and employees. His book, *Managerial Breakthrough* (McGraw-Hill, 1964), discusses cultural resistance and how to deal with it. He felt that managing for quality is an offshoot of general management but is a science in its own right. He followed this book with a trilogy that outlines three key components of managing for quality.

The Juran trilogy makes use of three managerial processes: Quality Planning, Quality Control, and Quality Improvement (Figure 2.13 and Table 2.1). By following Dr. Juran's approach, companies can reduce the costs associated with poor quality and remove chronic waste from their organizations. *Quality Planning* encourages the development of methods to stay in tune with customers' needs and expectations. *Quality Control* involves comparing products produced with goals and specifications. *Quality Improvement* involves the ongoing process of improvement necessary for the company's continued success.

	Content of Little q	Content of Big Q
Products and services	Manufactured goods Point of service	All products and services, whether for sale or not
Processes	Processes directly related to the manufacture of goods	All processes; manufacturing, support, business, etc.
Customer	Clients who buy the products	All who are affected, external and internal
Industries	Manufacturing	All industries; service, government, etc., whether for profit or not
Cost of poor quality	Costs associated with deficient manufactured goods	All costs that would disappear if everything were perfect

Figure 2.12   Big Q versus Little q

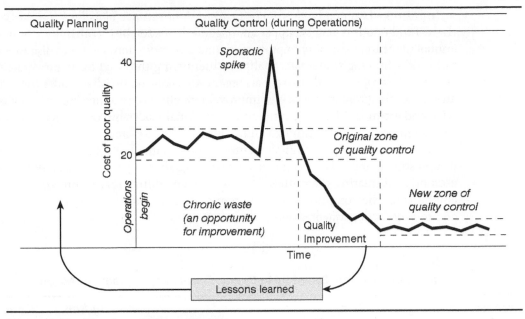

Figure 2.13   The Juran Trilogy Diagram
Source: Reprinted with the permission of the Free Press, a division of Simon & Schuster Adult Publishing Group, from *Juran on Leadership, for Quality: An Executive Handbook* by J. M. Juran. Copyright © 1989 by Juran Institute Inc.

Table 2.1   The Three Universal Processes of Managing for Quality

Quality Planning	*Quality Control*	*Quality Improvement*
Determine Who the Customers Are	Evaluate Actual Product Performance	Establish the Infrastructure
Determine the Needs of the Customers	Compare Actual Performance to Product Goals	Identify the Improvement Projects
Develop Product Features that Respond to Customers' Needs	Act on the Difference	Establish Project Teams
Develop Processes Able to Produce the Product Features		Provide the Teams with Resources, Training and Motivation to:
Transfer the Plans to the Operating Forces		Diagnose the Causes Stimulate Remedies Establish Controls to Hold the Gains

Source: Reprinted with the permission of the Free Press, a division of Simon & Schuster Adult Publishing Group, from *Juran on Leadership for Quality: An Executive Handbook* by J. M. Juran. Copyright © 1989 by Juran Institute Inc.

In his text *Juran on Leadership for Quality: An Executive Handbook.* Dr. Juran puts forth three fundamental tenets: upper management leadership, continuous education, and annual planning for quality improvement and cost reduction. Dr. Juran discusses the importance of achieving world-class quality by identifying the need for improvement, selecting appropriate projects, and creating an organizational structure that guides the diagnosis and analysis of the projects. Successful improvement efforts encourage breakthroughs in knowledge and attitudes. The commitment and personal leadership of top management must be assured in order to break through cultural resistance to change.

In the project-by-project implementation procedure (Table 2.2), project teams are set up to investigate and solve specific problems. To guide the project teams, the Juran program establishes a steering committee. The steering committee serves three purposes: to ensure emphasis on the company's goals, to grant authority to diagnose and investigate problems, and to protect deparmental rights.

Table 2.2   Juran's Journey from Symptom to Cause: Quality Improvement in Action

Process	Activity	Steering Arm	Diagnostic Arm
	Assign Priority to Projects	X	
	Pareto Analysis of Symptoms		X
Journey from Symptom to Cause	Theorize on Causes of Symptoms	X	
	Test Theories: Collect, Analyze Data		X
	Narrow List of Theories		
	Design Experiment(s)	X	
	Approve Design; Provide Authority		X
	Conduct Experiment; Establish Proof of Cause	X	
			X
	Propose Remedies		
Journey from Cause to Remedy	Test Remedy	X	
	Actions to Institute Remedy; Control at New Level		X
		X	

Source: Adapted with the permission of the Free Press, a division of Simon & Schuster Adult Publishing Group, from *Juran on Leadership for Quality: An Executive Handbook* by J. M. Juran, Copyright © 1989 by Juran Institute Inc.

The project teams should be composed of individuals with diverse backgrounds. Diversity serves several purposes. It allows for a variety of viewpoints, thus avoiding preconceived answers to the problem. Having a diversified group also aids in implementing the solutions found. Group members are more willing to implement the solution because they have a stake in the project. The different backgrounds of the group members can also assist in breaking down the cultural resistance to change.

Juran's project teams are encouraged to use a systematic approach to problem solving. Group members use a variety of investigative tools to clarify the symptoms and locate the true cause(s) of the problem. When the cause is determined, finding a solution becomes a process of proposing remedies, testing them, and instituting the remedy that most effectively solves the problem. Controlling the process once changes have been made is important to ensure that the efforts have not been wasted. Improvements continue as the groups study and resolve other problems.

Still active at over 100 years of age, in an interview with *Quality Progress* magazine (May 2004), Dr. Juran had this advice for people: "become bilingual; learn to communicate with senior managers by converting quality data into the language of business and finance." He is referring to the need to state quality goals in financial terms so that they can enhance the organization's overall business plan.

Philip Crosby

Philip Crosby's (1926–2001) message to management emphasizes four absolutes (Figure 2.14). The four absolutes of quality management set expectations for a continuous improvement process to meet. The first absolute defines quality as conformance to requirements. Crosby felt that it is necessary to define quality in order to manage quality. Customer requirements must translate to measurable characteristics for the organization's products or services. Crosby emphasized that effective organizations understand the importance of determining customer requirements, defining those requirements as clearly as possible, and then producing products or providing services that conform to the requirements as established by the customer.

Prevention of defects, the second absolute, needs to be in place in order to ensure that the products or services provided by a company meet the requirements of the customer. Prevention of quality problems is much more cost effective in the long run. Determining the root cause of defects and preventing their recurrence are integral to effective systems.

According to Crosby, the performance standard against which any system must be judged is zero defects—the third absolute. *Zero defects* refers to making products correctly the first time, with no imperfections. Traditional quality control centered on final inspection and

Figure 2.14 Crosby's Absolutes of Quality Management	Quality Definition: Conformance to Requirements Quality System: Prevention of Defects Quality Performance Standard: Zero Defects Quality Measurement: Costs of Quality

"acceptable" defect levels. Effective organizations must establish or improve systems that allow the worker to do it right the first time.

Crosby's fourth absolute, costs of quality, refers to the costs associated with providing customers with a product or service that conforms to their expectations. Quality costs, to be discussed in more detail in Chapter 8, are found in the costs associated with dissatisfied customers, rework, scrap, downtime, and material costs and costs involved anytime a resource has been wasted in the production of a quality product or the provision of a service. Once determined, costs of quality are used by effective organizations to justify investments in equipment and processes that reduce the likelihood of defects.

In several of his books, Crosby discusses the concepts of a successful customer versus a satisfied customer. To Crosby, a successful customer is one who receives a product or service that meets his or her expectations the first time. When a customer is merely satisfied, steps may have been taken to rework or redo the product or service until the customer is satisfied; for instance, a diner receives an undercooked piece of meat and then insists that the meal be taken off his or her bill. In the action of satisfying a customer whose expectations were not met the first time, the company has incurred quality costs.

Dr. Feigenbaum's definition of quality mentions the word *intangible*. By discussing five erroneous assumptions about quality, Crosby attempts to make quality more understandable and, therefore, tangible. The first erroneous assumption—that quality means goodness, or luxury, or shininess, or weight—makes *quality* a relative term. Only when quality is defined in terms of customer requirements can quality be manageable. The second incorrect assumption about quality is that it is intangible and therefore not measurable. If judged in terms of "goodness," then quality is intangible; however, quality is measurable by the costs of doing things wrong. More precisely, quality costs involve the costs of failures, rework, scrap, inspection, prevention, and loss of customer goodwill.

Closely related to the first two assumptions, the third assumption states that there exists "an economics of quality." Here again, one errs in thinking that quality means building luxuries into a product or service; rather, quality means that it is more economical to do things right the first time. Often workers are blamed for being the cause of quality problems. This is the fourth erroneous assumption about quality. Without the proper tools, equipment, raw materials, and training, workers cannot produce quality products or services. Management must ensure that the necessary items are available to allow workers to perform their jobs well. The final erroneous assumption that Crosby discusses is that quality originates in the quality department. According to Crosby, the quality department's responsibilities revolve around educating and assisting other departments in monitoring and improving quality.

Crosby's quality management philosophy supports creating a greater understanding of the complexities of managing an organization. Much of his focus was on simplifying the concepts surrounding the definition of quality and the need to design systems that support the concept of producing products or supplying services containing zero defects.

Dr. Kaoru Ishikawa

One of the first individuals to encourage total quality control was Dr. Kaoru Ishikawa (1915–1989). Dr. Ishikawa, a contemporary of Dr. Deming and Dr. Juran, transformed their early teachings into the Japanese approach to quality. Because he developed and delivered the first basic quality control course for the Union of Japanese Scientists and Engineers (JUSE) in 1949 and initiated many of Japan's quality programs, he is considered the focus of the quality movement in Japan. Dr. Ishikawa is also credited with initiating quality circles in 1962. Like Dr. Deming and Dr. Juran, his devotion to the advancement of quality merited him the Second Order of the Sacred Treasure from the Emperor of Japan.

To Dr. Ishikawa, quality must be defined broadly. Attention must be focused on quality in every aspect of an organization, including the quality of information, process, service, price, systems, and people. He played a prominent role in refining the application of different statistical tools to quality problems. Dr. Ishikawa felt that all individuals employed by a company should become involved in quality problem solving. He advocated the use of seven quality tools: histograms, check sheets, scatter diagrams, flowcharts, control charts, Pareto charts, and cause-and-effect, or fish-bone, diagrams. These tools, shown in Figure 2.15 are covered in detail in Chapter 13. Dr. Ishikawa developed the **cause-and-effect diagram** in the early 1950s. This diagram, used to find the root cause of problems, is also called the *Ishikawa diagram* after its creator or the *fish-bone diagram* because of its shape.

Dr. Ishikawa promoted the use of **quality circles,** *teams that meet to solve quality problems related to their own work.* The quality circle concept has been adapted and modified over time to include problem-solving team activities. Membership in a quality circle is often voluntary. Participants receive training in the seven tools, determine appropriate problems to work on, develop solutions, and establish new procedures to lock in quality improvements.

In order to refine organizations' approach to quality, Dr. Ishikawa encouraged the use of a system of principles and major focus areas as a holistic way to achieve business performance improvement. Customers and the processes that fulfill their needs, wants, and expectations were critical to Dr. Ishikawa. He felt that a focus on customer-oriented quality would break down the functional barriers that prevent the creation of defect-free products. In order to do this, processes should be analyzed from the viewpoint of the customer. As quoted in *Quality Progress* (April 2004), like others in the field, he felt that "quality should not be interpreted in the narrow sense but interpreted broadly, including price, delivery and safety, to satisfy consumer's needs."

As presented in *Quality Progress* (April 2004), his system includes six fundamentals that form the Japanese quality paradigm:

1. All employees should clearly understand the objectives and business reasons behind the introduction and promotion of company-wide quality control.
2. The features of the quality system should be clarified at all levels of the organization and communicated in such a way that the people have confidence in these features.
3. The continuous improvement cycle should be continuously applied throughout the whole company for at least three to five years to develop standardized work. Both statistical quality control and process analysis should be used, and upstream control for suppliers should be developed and effectively applied.
4. The company should define a long-term quality plan and carry it out systematically.
5. The walls between departments or functions should be broken down, and cross-functional management should be applied.
6. Everyone should act with confidence, believing his or her work will bear fruit.

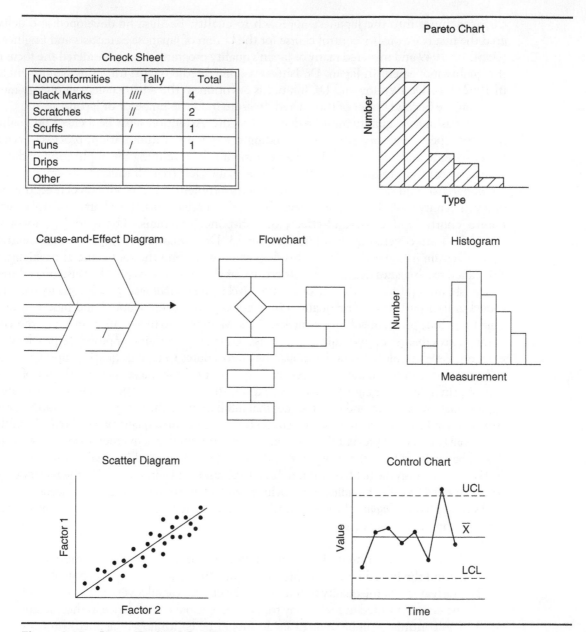

Figure 2.15   Seven Tools of Quality

The system also includes four major focus areas designed to influence quality through leadership:

1. ***Market-in quality:***   Leadership should encourage efforts that enable the organization to determine external customer needs, wants, requirements, and expectations.

By focusing on these elements and designing processes to deliver value to the market an organization can increase its business competitiveness.

2. ***Worker involvement:*** Quality improvement through the use of cross-functional teams enhances an organization's ability to capture improvements to the work processes. Appropriate training in problem-solving tools and techniques is a must.

3. ***Quality begins and ends with education:*** Education enhances an individual's ability to see the big picture. Education creates a deeper understanding of the activities that must take place in order for the organization to be successful.

4. ***Selfless personal commitment:*** Dr. Ishikawa lived his life as an example of selfless personal commitment. He encouraged others to do likewise, believing that improving the quality of the experience of working together helps improve the quality of life in the world.

Point 4 above summarizes Dr. Ishikawa's tireless, lifelong commitment to furthering the understanding and use of quality tools in order to better the processes that provide products and services for customers.

Dr. Genichi Taguchi

Dr. Genichi Taguchi (1924–) developed methods that seek to improve quality and consistency, reduce losses, and identify key product and process characteristics before production. Dr. Taguchi's methods emphasize consistency of performance and significantly reduced variation. Dr. Taguchi introduced the concept that the total loss to society generated by a product is an important dimension of the quality of a product. In his "loss function" concept, Dr. Taguchi expressed the costs of performance variation (Figure 2.16). Any deviation from target specifications causes loss, he said, even if the variation is within specifications. When the variation is within specifications, the loss may be in the form of poor fit, poor finish, undersize, oversize, or alignment problems. Scrap, rework, warranties, and loss of goodwill are all examples of losses when the variation extends beyond the specifications. Knowing the loss function helps designers to set product and manufacturing tolerances. Capital expenditure are

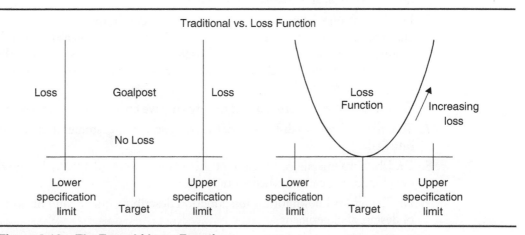

Figure 2.16   The Taguchi Loss Function

more easily justified by relating the cost of deviations from the target value to quality costs. Minimizing losses is done by improving the consistency of performance.

Dr. Taguchi is also known for his work in experiment design. Statistically planned experiments can identify the settings of product and process parameters that reduce performance variation. Dr. Taguchi's methods design the experiments to systematically weed out a product's or process's insignificant elements. The focus of experiment efforts is then placed on the significant elements. There are four basic steps:

1. Select the process/product to be studied.
2. Identify the important variables.
3. Reduce variation on the important variables through redesign, process improvement, and tolerancing.
4. Open up tolerances on unimportant variables.

The final quality and cost of a manufactured product are determined to a large extent by the engineering designs of the product and its manufacturing process.

LESSONS LEARNED

Most improvement strategies, methodologies, and standards have their foundation in the teachings of one or more of the men discussed in this chapter. Figure 2.17 provides a brief comparison of the philosophies of each man.

1. Dr. Shewhart developed statistical process control charts as well as the concepts of controlled and uncontrolled variation.
2. Dr. Deming is known for encouraging companies to manage for quality by defining quality in terms of customer satisfaction.
3. Dr. Deming created his 14 points as a guide to management.
4. Dr. Juran's process for managing quality includes three phases: quality planning, quality control, and quality improvement.
5. Dr. Feigenbaum defined quality as "a customer determination which is based on the customer's actual experience with the product or service, measured against his or her requirements—stated or unstated, conscious or merely sensed, technically operational or entirely subjective—always representing a moving target in a competitive market."
6. Crosby describes four absolutes of quality and five erroneous assumptions about quality.
7. To Crosby, there is a difference between a successful customer and one who is merely satisfied.
8. Dr. Ishikawa encouraged the use of the seven tools of quality, including the one he developed: the cause-and-effect diagram.
9. Dr. Taguchi is known for his loss function describing quality and his work in the area of design of experiments.

	Deming	Juran	Crosby	Feigenbaum	Taguchi	Baldrige	Six Sigma
Basic Orientation to Quality	Technical	Process	Motivational	Systematic	Technical	Motivational	Technical/Process
What Is Quality?	Non-faulty systems	Fitness for use	Conformance to requirements	Defined by customer	Customer performance requirements	Defined over seven categories	Defined by customer
Who Is Responsible for Quality?	Management	Management	Management	Everyone	Engineers	Management	Management
Goal of Quality	Meet/exceed customer needs; continuous improvement	Please customer; continuous improvement	Continuous improvement; zero defects	Meet/exceed customer needs; continuous improvement	Meet/exceed customer requirements; continuous improvement	Continuous improvement; customer satisfaction	Bottom-line results
Methodology	14-point program	Breakthrough projects; quality teams	Costs of quality	Statistics and engineering methods	Design of experiments	Award criteria	Defect reduction and waste elimination

Figure 2.17 Quality Advocates and Their Definitions of Quality

Are You a Quality Management Person?

What is your quality management philosophy? Check out the following questions and find out whether or not you are a quality management person. How do you answer them on a scale of 1 to 10?

Have you improved in the past 12 months? How?

1 2 3 4 5 6 7 8 9 10

Do you constantly strive to improve?

1 2 3 4 5 6 7 8 9 10

Do you make decisions and manage your life with a knowledge of variation?

1 2 3 4 5 6 7 8 9 10

Do you make fact-based decisions?

1 2 3 4 5 6 7 8 9 10

Do you seek to achieve a quality experience in all that you do?

1 2 3 4 5 6 7 8 9 10

Do you define quality from the customer's point of view?

1 2 3 4 5 6 7 8 9 10

Do you seek out opportunities to learn and grow?

1 2 3 4 5 6 7 8 9 10

Can you see the common philosophy that runs through each man's quality approach?

1 2 3 4 5 6 7 8 9 10

Are you a role model for continuous improvement?

1 2 3 4 5 6 7 8 9 10

Chapter Questions

1. Why is an organizational philosophy focusing on delighting customers key to organizational success?

2. Using Dr. Feigenbaum's definition of quality as a guide, describe an experience you have had with a product or service.

3. Describe in your own words the two types of variation that Dr. Shewhart identified.

4. Which of Dr. Deming's 14 points do you find the most interesting? Why?

5. Describe an example from industry related to one of Dr. Deming's 14 points.

6. How do Dr. Deming's 14 points interact with each other?

7. How do the steering/diagnostic arms of Dr. Juran's program work together?

8. In your own words, describe the difference between big Q and little q. Use examples from your own experiences to back up your description.

9. People tend to make five erroneous assumptions about quality. What are two of these assumptions, and how would you argue against them?

10. Have you seen examples of Crosby's erroneous assumptions at your own work? Describe the incidents.

11. How are the teachings of each of the people in this chapter similar? Where do they agree?

12. How are the teachings of each of the people in this chapter different? Where do they disagree?

13. Briefly summarize the concept Dr. Taguchi is trying to get across with his loss function.

14. Describe Dr. Taguchi's loss function versus the traditional approach to quality.

3 Quality Systems

What standards and criteria exist to support effective quality management systems?

What is ISO 9000?

What are supplier certification requirements?

What is ISO 14000?

What is Six Sigma?

What is the Malcolm Baldrige National Quality Award?

Lessons Learned

Learning Opportunities

1. To become familiar with ISO 9000 and ISO 14000

2. To become familiar with TS 16949

3. To become familiar with the Six Sigma methodology

4. To become familiar with the requirements of the Malcolm Baldrige National Quality Award

In the 1970s Motorola learned about quality the hard way—by being consistently beaten in the competitive marketplace. When a Japanese firm took over a Motorola factory that manufactured television sets in the United States, it promptly set about making drastic operational changes. Under Japanese management, the factory was soon producing TV sets with 1/20th the number of defects they had produced under previous management. Then Motorola CEO Bob Galvin started the company on the quality path and became a business icon largely as a result of what he accomplished by improving quality at Motorola. In accepting the first ever Malcolm Baldrige National Quality Award at the White House in 1988, Bob Galvin briefly described the company's turnaround. He said it involved something called Six Sigma.

Paraphrased from:
"Six Sigma Is Primarily a Management Program,"
Quality Digest, *June 1999*

WHAT STANDARDS AND CRITERIA EXIST TO SUPPORT EFFECTIVE QUALITY MANAGEMENT SYSTEMS?

In order to best fulfill customer needs, requirements, and expectations, effective organizations create and utilize quality systems. *Within a quality management system the necessary ingredients exist to enable the organization's employees to identify, design, develop, produce, deliver, and support products or services that the customer wants.* Effective quality management systems are dynamic, able to adapt and change to meet the needs, requirements, and expectations of their customers. Leaders of effective organizations seek tools and insights that enable them to manage better, faster, and more effectively in the face of worldwide competition. Quality systems and methodologies provide these individuals with the guidance and direction they seek (Figure 3.1). Effective organizations use standards such as ISO 9000 and QS 9000, programs such as Six Sigma, and awards such as the Malcolm Baldrige National Quality Award to provide guidance for establishing their quality management system's structure, maintaining records, and using quality techniques to improve processes and systems.

WHAT IS ISO 9000?

Continued growth in international trade revealed a need for a set of quality standards to facilitate the relationship between suppliers and purchasers. The creation of the ISO 9000 series of international standards began in 1979 with the formation of a technical committee with participants from 20 countries. Named the International Organization for Standardization, this Geneva-based association developed and continues to revise and update the standards. The name "ISO 9000" has its origin in the Greek word *isos*, meaning "equal." The intent of the standards is to make comparisons between companies equal.

The purpose of the ISO standards is to facilitate the multinational exchange of products and services by providing a clear set of quality system requirements. Companies competing on a global basis find it necessary to adopt and adhere to these standards. The standards provide a baseline against which an organization's quality system can be judged. The foundation of this baseline is the achievement of customer satisfaction through multidisciplinary participation in quality improvement efforts, documentation of systems and procedures, and other basic structural elements necessary to quality systems. The generic nature of the standards allows interested companies to determine the specifics of how the standards apply to its organization. Many companies use ISO 9000 as the foundation for their continuous improvement efforts.

Accepted in 90 countries, ISO 9000 is applicable to nearly all organizations, including manufacturers of pieces, parts, assemblies, and finished goods; developers of software; producers of processed materials, including liquids, gases, solids, or combinations; municipalities; logistics providers (i.e., UPS, FedEx, DHL); hospitals; and service providers. The ISO organization estimates that, worldwide, 8 out of 10 cars contain parts or components designed or manufactured under the ISO 9001:2000 certification system. Since its inception, ISO 9000 has become an internationally accepted standard for quality in business-to-business dealings. As of 2007, more than 500,000 organizations worldwide have attained certification.

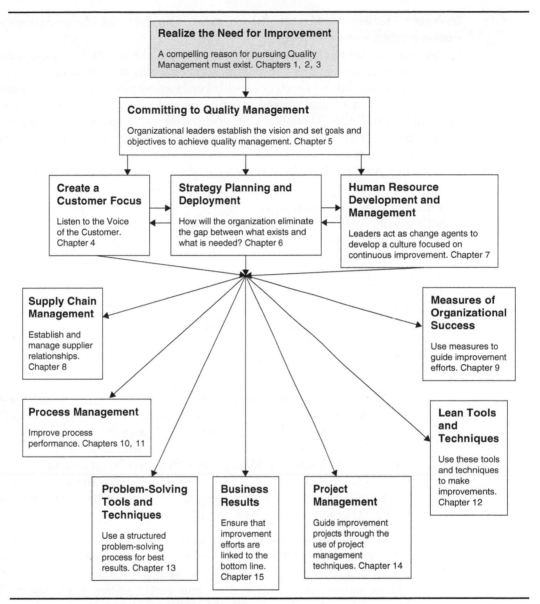

Figure 3.1   Achieving Organizational Success

In 2000, the ISO 9000 standards were revised significantly so that their structure more closely resembles the way organizations are managed. The wording of the standard was made more generic so that it is applicable to a wider variety of business sectors, including government services, business enterprises, e-business, manufacturing, and service industries. Effective organizations recognize that the real value of pursuing certification is internal.

In order to achieve certification, the organization must develop the discipline necessary to sustain performance excellence.

ISO 9001:2000 takes a process-oriented approach. The standard focuses on quality management systems and requires the identification of quality management processes as well as their sequence and their interactions with key business processes. A quality management system describes the organizational structure, procedures, and resources necessary to manage quality. The ISO 9000 requirements describe *what* a company must accomplish in order to meet customer expectations. However, *how* these goals are accomplished is left up to the particular company.

Eight key principles have been included in the ISO 9000:2000 standards:

- Customer-focused organization
- Leadership
- Involvement of people
- Process approach
- Systems approach to management
- Continual improvement
- Factual approach to decision making
- Mutually beneficial supplier relationships

ISO 9000:2000 is composed of three areas:

1. **ISO 9000:2000, Quality Management Systems: Fundamentals and Vocabulary,** provides information about the concepts and vocabulary used in the other two standards. This standard serves as a reference to support the interpretation of the ISO 9001:2000 requirements but contains no actual requirements.
2. **ISO 9001:2000, Quality Management Systems: Requirements,** provides the requirements that organizations must meet in order to achieve certification. ISO 9001 is designed to be used by all organizations, regardless of type, size, or industry sector (Figure 3.2). ISO 9001 consists of four main sections:

   Management Responsibility: This section focuses on how the analysis of data affects the performance of the organization's quality management system. Information is sought on how the organization's management establishes quality policies, makes plans, achieves objectives, and communicates customer requirements. These topics are covered in Chapters 5 and 6 of this text.

   Resource Management: The requirements in this section call for details on resource availability and deployment. Resources include information, facilities, communication, people, and work environment. Training effectiveness is also evaluated. These topics are covered in Chapters 6, 7, 8, and 9 of this text.

   Product and/or Service Realization: Product and/or service realization concentrates on how customer requirements and organizational self-assessments lead to continued improvement of processes and work methods. These topics are covered in Chapters 4 and 10 of this text.

ISO 9001:2000

Section 4: Quality Management System

4.1 General Requirements
4.2 Documentation Requirements

Section 5: Management Responsibility

5.1 Management Commitment
5.2 Customer Focus
5.3 Quality Policy
5.4 Planning
5.5 Responsibility, Authority, and Communication
5.6 Management Review

Section 6: Resource Management

6.1 Provision of Resources
6.2 Human Resources
6.3 Infrastructure
6.4 Work Environment

Section 7: Product and/or Service Realization

7.1 Planning of Product Realization
7.2 Customer-Related Processes
7.3 Design and Development
7.4 Purchasing
7.5 Product and Service Provision
7.6 Control of Monitoring and Measuring Devices

Section 8: Measurement, Analysis, and Improvement

8.1 General
8.2 Monitoring and Measurement
8.3 Control of Nonconforming Product
8.4 Analysis of Data
8.5 Improvement

Figure 3.2   ISO 9001:2000

Measurement, Analysis, and Improvement: This section examines the methods a company uses to measure its systems, processes, products, or services. This topic is covered in Chapter 9 of this text.

3. **ISO 9004:2000, Quality Management Systems: Guidelines for Performance Improvement,** provides guidelines for those companies wishing to go beyond ISO 9001:2000 and establish a quality management system. ISO 9004 provides a continuous improvement model focused on meeting customer requirements and improving performance. ISO 9004:2000 is not a requirement and does not lead to certification. Effective organizations often use ISO 9004 as a template for their continuous improvement efforts.

Documentation and record keeping are important aspects of ISO 9000. ISO 9000 requires records of many plant activities, including procedures, policies, and instructions; employee training records; process control charts and capability records; purchasing records; test and reliability data; audit records; incoming and final inspection records; and equipment calibration records. Companies following the ISO 9000 criteria need to keep records of any information that is useful in the operation or the organization. Evidence that established procedures, policies, and instructions are actually being followed on a day-to-day basis must also accompany these records.

The advantages of a fully documented quality management system are many. Documentation describes how work must be accomplished. Structured correctly, documentation will apply to a variety of situations, not just specific products. Documents serve as guides and ensure that work is performed consistently. Sound documentation can be used to determine and correct the causes of poor quality. Documentation defines existing work methods and provides a foundation for improvement.

In ISO 9000 a great deal of emphasis is placed on the need for excellent record keeping. In most cases, since the product has left the manufacturing facility or the service has been performed, only clearly kept records can serve as evidence of product or service quality. Sloppy or poorly maintained records give the impression of poor quality. High-quality records are easy to retrieve, legible, appropriate, accurate, and complete. Necessary records may originate internally or be produced externally. Customer or technical specifications and regulatory requirements are considered external records. Examples of internally produced records include forms, reports, drawings, meeting minutes, problem-solving documentation, and process control charts. A high-quality documentation control system will contain records that are easily identified and used in the decision-making process.

Companies seeking registration must have their compliance with the ISO 9001 standard judged by an independent ISO 9000 certified registrar. Figure 3.3 shows the flow of a typical registration process. Before the ISO 9000 governing body grants certification, a registrar conducts a thorough audit to verify that the company does indeed meet the requirements as set forth in ISO 9001. When the company desiring accreditation feels that it is ready, it invites an auditor to observe the company's operations and determine its level of compliance with the standards. Many firms find that conducting an internal audit before the actual registrar visit is more effective than a single audit. During this preaudit, deficiencies in the company's methods can be identified and corrected prior to the registrar's official visit. Once registration has been achieved, surveillance audits are conducted, often unannounced, approximately every six months. These audits are intended to ensure continued compliance. When preparing for a certification audit, a company must develop an implementation plan that identifies the people involved and defines their roles, responsibilities, deliverables, time lines, and budgets. Once created, the plan is managed through frequent review meetings to determine the progress. Once the system has been fully documented, the company can contact a registrar and plan the certification audit visit. Audits are discussed in greater detail in Chapter 15.

Decide to implement ISO 9000	Establish steering committee	Conduct internal audits	Continue internal audits	Implement and document procedures	Revise, improve, update, review, take corrective action based on initial visit	Begin preassessment procedures	Conduct registration assessment	Complete registration
Form management committee	Communicate intentions to entire corporation	Organize quality system	Begin documentation:	Initial visit made	Revise quality manual	Correct deficiencies		Strive for continuous improvement
Develop strategic plan	Select and train audit teams	Select registrar	Analyze processes		Conduct management review	Document and implement practices		Conduct surveillance audits
Begin ISO 9000 training	Continue ISO 9000 training	Continue to train audit teams	Write procedures					Continue management reviews
Educate workforce	Continue education	Continue ISO 9000 training	Continue to train audit teams					
Do cost assessment	Begin implementing standards	Continue education	Continue ISO 9000 training					
Do self-assessment		Define areas for improvement	Continue education					
			Create quality manual					
			Implement new procedures					

Figure 3.3 ISO 9000 Registration Cycle

Companies wishing to implement ISO 9000 should determine management's level of commitment. Management support can be gained by pointing out the benefits of compliance. An assessment of the company's current situation will reveal the present costs associated with poor quality as evidenced by scrap or rework or lost customers or other factors. Organizations who have successfully received ISO 9001 certification can list a wide variety of benefits. Three of these are universal. By achieving ISO 9001 certification, organizations establish the foundation of a good quality system. Maintaining certification means that systems and processes are reviewed regularly driving organizational discipline. As an internationally recognized standard, ISO 9001 provides access to markets and suppliers worldwide. Companies that have achieved certification cite increased revenue as a major benefit. Since ISO 9000 is recognized globally, certification allows these companies to expand their geographic markets. They are also able to service new customers who require ISO 9000 compliance from their suppliers. The usefulness of a common standard such as ISO 9000 has become particularly evident as companies purchasing their supplies worldwide have had to deal with inconsistent product quality. Effective organizations use these quality systems as a link between their suppliers and customers in order to create a continuous quality improvement chain. Additionally, existing customers benefit from ISO 9001 certification. Companies complying with ISO 9001 requirements have been able to improve their product and service quality and pass the benefits on to their customers. Internally, companies benefit from compliance. Reduced costs are evidenced through decreased scrap and rework, fewer warranty claims, improved customer satisfaction, reduced customer support costs, and improved productivity.

Obtaining ISO 9001 certification is a time-consuming and costly process. Depending on the current state of an organization's quality system, preparation for certification may take several thousand employee hours and cost thousands of dollars. Costs depend on the size of the company, the strength of the organization's existing quality system, and the number of plants within the company requesting certification. As with any major process improvement, the risk of failure exists. Attempts to incorporate ISO 9000 into the way a company does business may be hindered by a variety of forces, including insufficient management involvement in the process, inadequate resources, lack of an implementation plan, or lack of understanding about ISO 9000 and its benefits. This last force, a lack of understanding about ISO 9000, is crucial. ISO 9001 requires significant documentation. The additional burden of paperwork, without an understanding of how this newfound information can be used in decision-making and organizational improvement, leads to problems. It is important to realize that standardized procedures and organized information go a long way toward preventing errors that lead to poor-quality products and services. Organizations that are unaware of how to access the power that procedures and information provide may miss out on improvement opportunities. It is up to leadership to encourage the use of this information, thus gaining the maximum benefits from ISO 9000.

The high cost of certification is counterbalanced by the benefits an organization will receive by using the requirements as a guide to improve their processes. Quality becomes more consistent, and the percentage of "done right the first time" jobs increases. Enhanced

How Do We Know It's Working?

As a small job shop, JQOS had few formalized systems and processes. Under new leadership, changes focus on standardizing the way they do business. Having become familiar with quality systems like ISO 9000 and ISO 16949 at their previous positions, JQOS leadership understood the challenge before them. To achieve ISO 9001 certification, an organization must set up standard procedures and document their processes and practices. They feel that ISO 9001 certification will bring them several benefits. The standards will guide them as they establish the foundation of a good quality system. Maintaining certification involves reviewing and improving existing systems and processes. These activities will keep them in a proactive mode. The greatest benefit they see is that the highly recognized standard will provide them with access to a broader range of customers.

To begin their pursuit of ISO 9000 certification, leadership worked with an ISO 9000 consultant to lay out a three-year strategic plan. As a starting point, they identified their key processes as:

Quoting
Order processing
Product creation (machining)
Inspection
Shipping
Payroll

From the beginning, employees were involved in process analysis and improvement of these key processes. During the first year, the processes were recorded, studied, improved on, standardized, and finally computerized. Employee training took place as each new procedure came online. Other changes took place over time. In future chapters, several *"How Do We Know It's Working?"* features will detail many of the changes made.

procedures, upgraded record keeping, and the removal of redundant operations also dramatically improve an organization's effectiveness. ISO 9000 standards facilitate international trade. To learn more about ISO 9000, visit the ISO website at www.iso.org.

WHAT ARE SUPPLIER CERTIFICATION REQUIREMENTS?

Major corporations often purchase raw materials, parts, subassemblies, and assemblies from outside sources. To ensure quality products, the suppliers of these parts and materials are subjected to rigorous requirements. Purchasers establish these requirements and judge conformance to them by visiting the supplier's plant site and reviewing the supplier's quality system.

In certain business sectors, particularly motor vehicle manufacturers, although each purchaser had developed its own requirements, strong similarities existed in quality system and documentation requirements. Redundant requirements and multiple plant visits from purchasers placed a significant burden on suppliers. Conforming to several different, yet very similar, sets of requirements meant unnecessarily expended time, effort, and money. Recognizing the overlap in requirements, some major manufacturers have developed quality systems that often have as their foundation ISO 9000. These systems, tailored to a particular industry, eliminate redundant requirements while maintaining customer-specific, division-specific, or commodity-specific requirements. For instance, in 1999, ISO/TS 16949, Quality Management Systems: Automotive Suppliers—Particular Requirements for the Application of ISO 9001:2000 for Automotive Production and Relevant Service Part Organizations, was introduced. Developed by the International Automotive Oversight Bureau and submitted to ISO for approval and publication, ISO/TS 16949 defines automotive industry standards worldwide. ISO/TS 16949 is based on ISO 9001:2000, QS 9000 (United States), AVSQ (Italian), EAQF (French), and VDA6.1 (German) systems. ISO/TS 16949 allows automotive companies to retain individual control over more of the specific requirements. These customer-specific requirements may include the use of methods for Statistical Process Control (SPC), Production Part Approval Processes (PPAP), Failure Modes and Effects Analysis (FMEA), Measurement Systems Analysis (MSA), Advanced Product Quality Planning and Control Planning (APQP), and Quality System Assessment (QSA). Other additions to the ISO 9001:2000 document include terms and definitions specific to the automotive industry; requirements related to engineering specifications and records retention; process efficiency expectations; product design skills and training related to human resources management; product realization, acceptance, and change control requirements; and customer-designated special characteristics. For more information concerning TS 16949, contact the Automotive Industry Action Group at www.aiag.org.

Although ISO 9000 and TS 16949 are the best known of the quality system certifications, other quality management system certifications have been developed in a variety of industries. As the benefits of utilizing an organized quality management system become better known, an increasing number of business sectors have created appropriate standards. For example, in the telecommunications industry, TL 9000 has been developed to ensure better relationships between suppliers and service providers. In the medical devices and pharmaceutical industries, many organizations follow ISO 13485, which, in addition to product realization requirements, requires risk assessment planning be used in formulating products. The need for such a standard comes from the increasing number and variety of pharmaceutical and medicinal products, the globalization of pharmaceutical and medical devices industries, and the changes in biotechnology and pharmaceutical sciences. ISO/IEC 17025 focuses on calibration management. Calibration is no longer primarily a maintenance activity. Now it includes calibration, validation, and maintenance procedures that all must be documented, approved, and managed.

In the aerospace industry, the International Aerospace Quality Group (IAQG) works to establish commonality of quality standards and requirements, using AS9000 to encourage continuous improvement processes at suppliers, determine effective methods to share results, and respond to regulatory requirements. AS9000 emphasizes design control, process control, purchasing, inspection and testing, and control of nonconformances, adding 83 additional and specific requirements to the 20 elements of ISO 9001.

WHAT IS ISO 14000?

The overall objective of the ISO 14000 Environmental Management Standard is to encourage environmental protection and pollution prevention while taking into account the economic needs of society. The standards can be followed by any organization interested in limiting its negative impact on the environment. A company with an environmental management system like ISO 14000 is less likely to have environmental problems. Often, firms following ISO 14000 incur significant savings through better overall resource management and waste reduction.

ISO 14000 is divided into two main classifications: Organization/Process-Oriented Standards and Product-Oriented Standards. A company complying with these standards is monitoring its processes and products to determine their effect on the environment. Within the two classifications, six topic areas are covered: Environmental Management Systems, Environmental Performance Evaluation, Environmental Auditing, Life Cycle Assessment, Environmental Labeling, and Environmental Aspects in Product Standards. Launched in 1996, the ISO 14000 series of standards enables a company to improve environmental management voluntarily. The standards do not establish product or performance standards, establish mandates for emissions or pollutant levels, or specify test methods. The standards do not expand upon existing government regulations. ISO 14000 is a guide for environmentally conscious companies wishing to lessen their impact on the environment.

Figures 3.4 and 3.5 provide information about the general structure of the ISO 14000 process- and product-oriented series of standards. For each series, specific activities are required. For instance, within ISO 14001, Environmental Management Systems (Figure 3.6), companies must provide details concerning their environmental policy and the structure of their overall environmental management plan. Section 4.2, Environmental Policy, serves as the foundation for the environmental management plan. Section 4.3, Planning, provides details on what must be done to support the targets and objectives set by the policy. Section 4.4, Implementation and Operation, establishes procedures and processes that allow the plan to be implemented, including the structure of the communication and document control systems, as well as emergency preparedness and response systems. Section 4.5, Checking and Corrective Action, monitors the environmental management system. Companies seeking ISO 14000 certification are expected to have in place an internal auditing system supported by corrective action plans. Section 4.6, Management Review, ensures that the entire process is reviewed regularly so that opportunities for improvement are not overlooked.

As companies are becoming more environmentally conscious and more globally responsive, many are becoming ISO 14000 certified and mandating that their suppliers do so too. ISO 14001 certification becomes more popular each year. In 2003, over 60,000 companies worldwide were certified, over 13,800 in Japan, 5,000 in China, and 3,400 in the United States. As with ISO 9000, this standard is reviewed and updated regularly. For more complete information, visit the ISO website at www.iso.org.

■ ISO 14001
   Environmental Management Systems—Specification with
   Guidance for Use

■ ISO 14004
   Environmental Management Systems—General Guidelines
   on Principles, Systems, and Supporting Techniques

■ ISO 14010
   Guidelines for Environmental Auditing—General Principles on
   Environmental Auditing

■ ISO 14011
   Guidelines for Environmental Auditing—Audit Procedures—
   Auditing of Environmental Management Systems

■ ISO 14012
   Guidelines for Environmental Auditing—Qualification Criteria
   for Environmental Auditors

■ ISO 14014
   Initial Reviews

■ ISO 14015
   Environmental Site Assessments

■ ISO 14031
   Evaluation of Environmental Performance

■ ISO 14020
   Goals and Principles of All Environmental Labeling

■ ISO 14021
   Environmental Labels and Declarations—Self-Declaration
   Environmental Claims—Terms and Definitions

Figure 3.4   ISO 14000 Process-Oriented Document Requirements

How Do We Know It's Working?

JQOS's new leadership is environmentally conscious. Using ISO 14000 as a guide, they have taken many positive actions to reduce their organization's environmental impact. To improve the recycling of metal shavings and scrap, they quickly purchased a puck-making machine. This machine compresses scrap metal into a more easily reusable shape. Vacuums at each chip-making machine capture the chips directly into a barrel. When full, workers transfer chips and shavings to the "puck" machine, which compresses the chips into recyclable discs of metal. This enables nearly 100% of the shavings and scrap metal to be recycled.

A study of the shipping area resulted in significant changes to how the parts are shipped. The new methods greatly reduce part damage during shipping. These same changes reduced the amount of packing material by 40%. Made from recycled materials, the new packing is reusable or recyclable.

Grinders, lathes, mills, and drills all use coolant. In the past, this coolant was disposed of through a waste hauler. Under the new management, systems were put in place to clean and reuse coolant. In addition, a more environmentally friendly coolant was chosen.

Office paper usage has declined through improved processes and computerization. Scrap office paper is no longer thrown out but is shredded and recycled.

Policies about energy usage have been put in place. Lights, fans, computers, and other equipment must be turned off when not in use. The HVAC system has been updated and is now maintained regularly, allowing it to operate with greater efficiency.

Many of these changes paid for themselves through the selling of recyclable materials, lower energy consumption, and material reuse. Positive changes are happening all the time at JQOS. Through environmentally sound changes, the focus is on improving overall organizational effectiveness.

- ISO 14022
  Environmental Labels and Declarations—Symbols
- ISO 14023
  Environmental Labels and Declarations—Testing and Verifications
- ISO 14024
  Environmental Labels and Declarations—Environmental Labeling
  Type 1—Guiding Principles and Procedures
- ISO 1402X
  Type III Labeling
- ISO 14040
  Life Cycle Assessment—Principles and Framework
- ISO 14041
  Life Cycle Assessment—Life Cycle Inventory Analysis
- ISO 14042
  Life Cycle Assessment—Impact Assessment
- ISO 14043
  Life Cycle Assessment—Interpretation
- ISO 14050
  Terms and Definitions—Guide on the Principles of ISO/TC 207/SC6 Terminology Work
- ISO Guide 64
  Guide for the Inclusion of Environmental Aspects of Product Standards

Figure 3.5   ISO 14000 Product-Oriented Document Requirements

ISO 14000
Section 4: Environmental Management Systems Requirements

4.1 General Requirements
4.2 Environmental Policy
4.3 Planning
    4.3.1 Environmental Aspects
    4.3.2 Legal and Other Requirements
    4.3.3 Objectives and Targets
    4.3.4 Environmental Management Program(s)
4.4 Implementation and Operation
    4.4.1 Structure and Responsibility
    4.4.2 Training, Awareness, and Competence
    4.4.3 Communication
    4.4.4 Environmental Management System Documentation
    4.4.5 Document Control
    4.4.6 Operational Control
    4.4.7 Emergency Preparedness and Response
4.5 Checking and Corrective Action
    4.5.1 Monitoring and Measurement
    4.5.2 Nonconformance and Corrective and Preventive Action
    4.5.3 Records
    4.5.4 Environmental Management System Audit
4.6 Management Review

Figure 3.6   ISO 14000, Section 4: Environmental Management Systems Requirements

WHAT IS SIX SIGMA?

Six Sigma is a methodology that blends together many of the key elements of past quality initiatives while adding its own special approach to business management. Essentially, Six Sigma is about results, enhancing profitability through improved quality and efficiency. Six Sigma emphasizes the reduction of variation, a focus on doing the right things right, combining of customer knowledge with core process improvement efforts, and a subsequent improvement in company sales and revenue growth. Effective leaders focus Six Sigma efforts on processes and problems that relate to their specific market niche. Before approving projects, they ask tough questions about deliverables and project cost savings. The Six Sigma methodology encourages companies to take a customer focus in order to improve their business processes.

The Six Sigma concept was conceived by Bill Smith, a reliability engineer for Motorola Corporation. His research lead him to believe that the increasing complexity of systems and products used by consumers created higher-than-desired system failure rates. His reliability studies showed that to increase system reliability and reduce failure rates, the components utilized in complex systems and products have to have individual failure rates approaching zero. With this in mind, Smith took a holistic view of reliability and quality and developed a strategy for improving both. Smith worked with others to develop the Six

Quality Philosophies	Statistics
Performance Measures/Metrics	Data Collection: Data types and sampling
Problem Solving	techniques
Process Mapping	$\overline{X}$ and R charts
Check Sheets	Process Capability Analysis
Pareto Analysis	P, u, c charts
Cause-and-Effect Diagram Analysis	Root Cause Analysis
Scatter Diagrams	Variation Reduction
Frequency Diagrams	Six Sigma philosophy
Histograms	Green Belt Project

Figure 3.7   Training Typically Required for Green Belt Certification

Sigma Breakthrough Strategy which is essentially a highly focused system of problem solving. Six Sigma's goal is to reach 3.4 defects per million opportunities over the long term.

The Six Sigma methodology is based on knowledge. Practitioners need to know statistical process control techniques, data analysis methods, and project management techniques. Systematic training within Six Sigma organizations must take place. Motorola Corporation utilizes terminology from the martial art of karate to designate the experience and ability levels of Six Sigma project participants. Green Belts are individuals who have completed a designated number of hours of training in the Six Sigma methodology (Figure 3.7). To achieve Green Belt status, they must also complete a cost-savings project of a specified size, often $10,000, within a stipulated amount of time. Black Belts are individuals with extensive training in the Six Sigma methodology (Figure 3.8). To become a Black Belt, the individual must have completed a specified number of successful projects under the guidance and direction of Master Black Belts. Often companies expect the improvement projects overseen by a Black Belt to result in savings of $100,000 or more. Certification as a Black Belt is offered by the American Society for Quality. Master Black Belts are individuals with extensive training who have completed a large-scale improvement project, usually saving $1,000,000 or more for the company. Often before designating someone a Master Black Belt, a company will require a master's degree from an accredited university. Master Black Belts provide training and guide trainees during their projects. Master Black Belts and Black Belts serve in a number of roles. They are mentors because they encourage individuals to become involved in Six Sigma activities. They are teachers and coaches because they

Green Belt Requirements plus:

Variables Control Charts	Reliability
Attribute Control Charts	ISO 9000, MBNQA
Process Capability Analysis	Voice of the Customer: Quality Function Deployment
Hypothesis Testing	Regression Analysis
Design of Experiments	Black Belt Project
Gage R&R	

Figure 3.8   Training Typically Required for Black Belt Certification

Responsibility	Phase
Management	Recognize
Management/Master Black Belts	Define
Black Belts/Green Belts	Measure
Black Belts/Green Belts	Analyze
Black Belts/Green Belts	Improve
Black Belts/Green Belts	Control
Management	Standardize
Management	Integrate

Figure 3.9   Six Sigma Responsibility Matrix

provide training in new strategies and tools, both formal and one-on-one. They are explorers because they discover new application opportunities for Six Sigma and influence the direction of the organization. Figure 3.9 shows the responsibilities of project participants.

EXAMPLE 3.1   Becoming a Black Belt

Following graduation, Chris joined a multinational logistics organization. Early on, his manager encouraged him to further his education by becoming a Six Sigma Green Belt. Since his college program of study had included a four-course series in quality assurance, Chris felt confident that he was well prepared for this adventure. When he began his Green Belt certification process in the Fall, his classes included exposure to the quality tools listed in Figure 3.10. Since his training was really a review of his college courses,

Quality Philosophies
Performance Measures/Metrics
Problem-Solving Model
Process Mapping
Check Sheets
Pareto Analysis
Cause-and-Effect Diagram Analysis
Scatter Diagrams
Frequency Diagrams
Histograms
Statistics
Data Collection:
    Data types and sampling techniques
X and R charts
Process Capability Analysis
P, u, c charts
Root Cause Analysis
Variation Reduction
Six Sigma Philosophy

Green Belt Project

Figure 3.10   Training Typically Required for Green Belt Certification

Green Belt Requirements plus:

Variable Control Charts
Attributes Control Charts
Process Capability Analysis
Hypothesis Testing
Design of Experiments
Gage R&R
Reliability
ISO 9000, MBNQA
Voice of the Customer
Regression Analysis

Black Belt Project

Figure 3.11   Training Typically Required for Black Belt Certification

Chris was able to quickly grasp the concepts and tools and apply them to his project. The project, selected by his Master Black Belt Mentor, involved developing a corrective action procedure to identify root causes of process failures in a package sorting line. Following an in-depth review of his training and project, the Master Black Belt, an individual accredited through the Six Sigma Academy, Chris received certification early the next spring.

That summer, Chris began Black Belt training. Over the period of a year, he attended 160 hours of training in basic statistical techniques, advanced statistics, hypothesis testing, analysis of variation, regression analysis, control charts, and design of experiments (Figure 3.11). Black Belt trainees met once a month for training and then applied what they learned for the next 3 weeks to their everyday work. At each training session, Master Black Belts reviewed what he had applied from previous training sessions. Training also included the study of many projects completed by other Black Belts. Trainees were given the original project plans and asked to simulate the project and analyze the results. During his last class, Chris participated in a one-day project that included elements of design of experiments. For his Black Belt project, a logistics pick-pack-ship optimization project, Chris developed an experiment to study the optimal configuration for the pick-pack-ship line in his facility. He ran the experiment on Saturday, analyzed the data on Sunday, made appropriate changes to the line based on the experiment and then monitored the new setup for several weeks. Once his predicted results were confirmed, he wrote up his results, calculated financial benefits, and presented his report to management. Shortly thereafter, Chris was certified as a Six Sigma Black Belt by the Master Black Belt.

Though certification was granted by the Master Black Belt, Chris decided to formalize his certification by taking the Black Belt Certification test from the American Society for Quality. His score: 95%!

Companies using the Six Sigma methodology see an enhanced ability to provide value for their customers. Internally, they have a better understanding of their key business processes and these processes have undergone process flow improvements. Improved process flow means reduced cycle times, elimination of defects, and increased capacity and productivity rates. Accelerated improvement efforts reduce costs and waste while increasing product and service reliability. All of these changes result in improved value for the customer as well as improved financial performance for the company.

The Six Sigma methodology focuses on customer knowledge. By translating customer needs, wants, and expectations into areas for improvement, the Six Sigma methodology concentrates effort on improving critical business activities. Six Sigma focuses attention on these processes to make sure that they deliver value directly to the customer.

The backbone of Six Sigma efforts are improvement projects, chosen based on their ability to contribute to the bottom line on a company's income statement by being connected to the strategic objectives and goals of the organization. Since the Six Sigma methodology seeks to reduce the variability present in processes, Six Sigma projects are easy to identify. Project teams seek out sources of waste, such as overtime and warranty claims; investigate production backlogs or areas in need of more capacity; and focus on customer and environmental issues. Six Sigma projects have eight essential phases:

    recognize,
    define,
    measure,
    analyze,
    improve,
    control,
    standardize, and
    integrate.

This cycle is often expressed as DMAIC (define, measure, analyze, improve, and control). As Figure 3.12 shows, the generic steps for Six Sigma project implementation are similar to the Plan-Do-Study-Act problem-solving cycle espoused by Dr. W. Edwards Deming and Dr. Walter Shewhart, early pioneers of the quality field. Using DMAIC as a guideline, organizations seek opportunities to enhance their ability to do business. The improvement process begins by recognizing a problem exists that fundamentally affects operational issues. Steps are taken to define the problem clearly before beginning the activity of measuring the capability of the processes involved. The measurements are analyzed to see if there are any patterns or trends present. Following this, the source of variation is identified, and improvements are created to reduce or eliminate them. Control refers to the need to determine whether the changes worked. If they did, then the changes are adopted and standardized wherever applicable. The new methods and processes are integrated into the way the organization does business. Process improvement of any kind leads to benefits for the company, including the reduction of waste, costs, and lost opportunities. Ultimately, it is the customer who enjoys enhanced quality and reduced costs.

The tools utilized during a project include statistical process control techniques, customer input through Quality Function Deployment, Failure Modes and Effects Analysis, Design of Experiments, reliability, teamwork, and project management. Six Sigma also places a heavy reliance on metrics and graphical methods for analysis. Statistical and probabilistic methods, control charts, process capability analysis, process mapping, and cause

Define, Measure, Analyze (Plan)

1. Select appropriate metrics: key process output variables (KPOV). These are critical to quality characteristics.
2. Determine how these metrics will be tracked over time.
3. Determine current baseline performance of project/process. What is the process capability?
4. Determine the key process input variables (KPIV) that drive the key process output variables (KPOV).
5. Identify variation sources.
6. Determine what changes need to be made to the key process input variables in order to positively affect the key process output variables. Define performance objectives.

Improve (Do)

7. Make the changes.

Control (Study, Act)

8. Determine if the changes have positively affected the KPOVs. Determine new process capability.
9. If the changes made result in performance improvements, establish control of the KPIVs at the new levels. If the changes have not resulted in performance improvement, return to step 5 and make the appropriate changes.
10. Implement process controls.

Figure 3.12   Six Sigma Problem-Solving Steps

and effect diagrams are key components of Six Sigma. As with any methodology, a variety of Six Sigma acronyms exist (Figure 3.13).

The term Six Sigma describes a methodology, while the mathematical designation, 6σ (also referred to with lowercase letters as six sigma), is the value used to calculate process capability. Providers of products and services are very interested in whether their processes can meet the specifications as identified by the customer. The spread of a distribution of average process measurements, that is, what the process produces, can only be compared with the specifications set by the customer by using C_p, where

$$C_p = \frac{USL - LSL}{6\sigma}$$

When $6\sigma = USL - LSL$, process capability $C_p = 1$. When this happens, the process is considered to be operating at three sigma (3σ). This means that three standard deviations added to the average value will equal the upper specification limit and three standard deviations subtracted from the average value will equal the lower specification limit (Figure 3.14). When $C_p = 1$, the process is capable of producing products that conform to specifications provided that the variation present in the process does not increase and that the average value equals the target value. In other words, the average cannot shift. That is a lot to ask from a process, so those operating processes often reduce the amount of variation present in the process so that $6\sigma < USL - LSL$.

Some companies choose adding a design margin of 25% to allow for process shifts, requiring that the parts produced vary 25% less than the specifications allow. A 25% margin results in a $C_p = 1.33$. When $C_p = 1.33$, the process is considered to be operating at four sigma. Four standard deviations added to the average value will equal the upper specification

APQP	Advanced product quality planning
CTQ	Critical to quality
DFSS	Design for Six Sigma
DMAIC	Define, measure, analyze, improve, control
DPMO	Defects per million opportunities
DPU	Defect per unit
EVOP	Evolution operation
FMEA	Failure modes and effects analysis
KPIV	Key process input variable
KPOV	Key process output variable
Process Owners	The individuals ultimately responsible for the process and what it produces
Master Black Belts	Individuals with extensive training qualified to teach Black Belt training classes who have completed a large scale improvement project, often a master's degree is required
Black Belts	Individuals with extensive training in the Six Sigma methodology who have completed a number of improvement projects of significant size
Green Belts	Individuals trained in the Six Sigma methodology who have completed an improvement project of a specified size
Reliability	Measured as mean time to failure
Quality	Measured as process variability and defect rates

Figure 3.13   Six Sigma Acronyms and Definitions

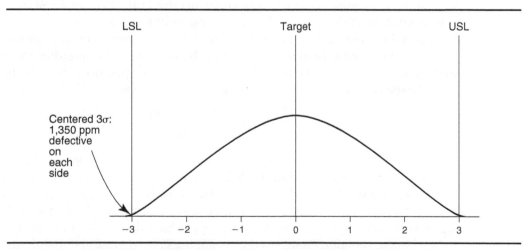

Figure 3.14   Three Sigma Occurs When 6σ = USL − LSL, Process Capability, C_p = 1

limit and four standard deviations subtracted from the average value will equal the lower specification limit. This concept can be repeated for five sigma and C_p = 1.66.

When C_p = 2.00, six sigma has been achieved. Six standard deviations added to the average value will equal the upper specification limit and six standard deviations subtracted from the average value will equal the lower specification limit (Figure 3.15). Those who

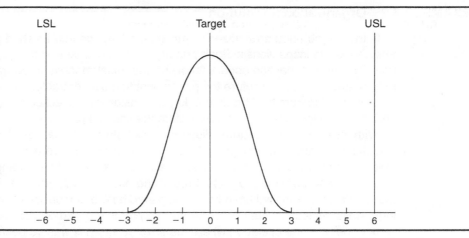

Figure 3.15 Six Sigma Occurs When 6σ < USL − LSL, Process Capability, C_p = 2

developed the Six Sigma methodology felt that a value of C_p = 2.00 provides adequate protection against the possibilities of a process mean shift or an increase in variation.

Operating at a Six Sigma level also enables a company's production to have virtually zero defects. Long-term expectations for the number of defects per million opportunities is 3.4. Compare this to a process that is operating at three sigma and centered. Such a process will have a number of defectives per million opportunities of 1,350 out on each side of the specification limits for a total of 2,700. If the process center were to shift 1.5 sigma, the total number of defects per million opportunities at the three sigma level would be 66,807. A process operating at four sigma will have 6,210 defects per million opportunities over the long term, while a process operating at the five sigma level will have 320 defects per million opportunities long term. Even if the cost to correct the defect is only $100, operating at the three sigma level while experiencing a process shift will cost a company $6,680,700 per million parts. Improving performance to four sigma reduces that amount to $621,000 per million parts produced. Six Sigma performance costs just $340 per million parts with a yield rate of 99.9997% (Figure 3.16). Each improvement in sigma level can result in up to a 10% increase in net income improvement.

Sigma	Defects per Million Opportunities	Yield (%)	Cost of Quality
1	690,000	30.90	Noncompetitive
2	308,000	69.20	Noncompetitive
3	66,807	93.30	25–40% of sales
4	6,210	99.40	15–25% of sales
5	320	99.98	5–15% of sales
6	3.4	99.9997	<1% of sales

Figure 3.16 Six Sigma Performance

EXAMPLE 3.2   Organizational Evolution

PLC Inc., a job shop specialized in machining forgings into finished products, utilizes three separate inspections as their primary method of ensuring the quality of their products. The first inspection occurs following the initial machining operations, (grinding, milling, and boring) and before the part is sent to a subcontractor for heat-treating. After the part returns from heat treatment, key dimensions are checked at the second inspection. A final inspection is conducted before the finished part leaves the plant.

With these three inspections, discrepancies between actual part dimensions and the specifications are found only after the part had completed several machining operations. This approach results in significant scrap and rework costs. If the forgings pass incoming inspection and begin to progress through the machining operations, for a typical part, four to six operations will have been completed before any errors are caught during the in-process inspection that occurs before the part is shipped out for heat treatment. The work done after the operation where the error occurred is wasted because each subsequent machining operation is performed on a faulty part.

A project team studied the inspection method and determined that process improvements could be made to improve incoming forging consistency. Operators at each step in the process are now responsible for checking actual part dimensions against specifications. The team also initiated a corrective action plan that required a root cause analysis and corrective action for each nonconformance to standards. Corrective actions plans prevent future errors.

As a result of their efforts, PLC, Inc. has seen an improvement of one sigma which translates to a 10-fold reduction in the number of defects. At three sigma, they could expect 66,807 defects per million. At an average cost of $10/piece to fix, the costs incurred were $668,007. Now that they are operating at four sigma, they expect 6,210 defects per million. At $10/piece to fix, their costs are $62,100. Further improvement activities are planned to improve performance to five sigma.

As with any process improvement methodology, there are issues that need to be examined carefully. One criticism is that Six Sigma methodology does not offer anything new. Comparisons have been made between the qualifications for a Black Belt and those for a Certified Quality Engineer (CQE). The similarities are striking (Table 3.1). Certifications are available through the American Society for Quality (www.asq.org). Comparisons have also been made between ISO 9000 and Six Sigma, the Baldrige Award and Six Sigma, and continuous improvement strategies and Six Sigma strategies, and significant similarities exist (Table 3.2).

The Six Sigma methodology is often adopted by large corporations with the financial resources to focus on training and projects. Because the methodology requires training and team participation, it is perceived to be costly to implement. Six Sigma should not be limited to large firms. Its ideas and concepts are viable regardless of the size or type of industry with costs being scaled accordingly.

Another criticism is the focus on defectives per million. Can an organization really call them defectives? The term itself brings to mind product liability issues. How does a customer

Table 3.1   Comparison of Body of Knowledge Required for CQE and Black Belt Certification

Category	ASQ Certified Quality Engineer (CQE) Certification Requirements	Black Belt Requirements
Leadership	Management and Leadership in Quality Engineering	Enterprise-Wide Deployment
Business Processes	Not covered	Business Process Management
Quality Systems	Quality Systems Development, Implementation, and Verification	Not covered
Quality Assurance	Planning, Controlling, and Assuring Product and Process Quality	Not covered
Reliability	Reliability and Risk Management	Not covered
Problem Solving	Problem Solving and Quality Improvement	Define-Measure-Analyze-Improve-Control
Quality Tools	Problem Solving and Quality Improvement	DMAIC
Project Management	Not covered	Project Management
Team Concepts	Not covered	Team Leadership
Statistical Methods	Probability and Statistics, Collecting and Summarizing Data	Probability and Statistics, Collecting and Summarizing Data
Design of Experiments	Designing Experiments	Design of Experiments
Process Capability	Analyzing Process Capability	Analyzing Process Capability
Statistical Process Control	Statistical Process Control	Statistical Process Control
Measurement Systems (metrology/calibration)	Measurement Systems Metrology	Measurement Systems Metrology
Lean Manufacturing	Not covered	Lean Enterprise
Other Techniques	FMEA, FMECA, FTA	FMEA, QFD Multivariate Studies

view a company that is focused on counting defectives? Should defect counts be seen as the focus or are companies really trying to focus on process improvement? Six Sigma organizations must be prepared to face these issues.

WHAT IS THE MALCOLM BALDRIGE NATIONAL QUALITY AWARD?

Organizations, regardless of what they provide, care about productivity, quality, safety, and cost. The Malcolm Baldrige National Quality Award (MBNQA) interrelated core values and concepts provide a foundation for excellence. Beginning with visionary leadership, these values and concepts improve an organization's agility in the marketplace. The criteria focus on customers, organizational and personal learning, valuing employees, management by fact, continuous improvement, and social and community responsibilities. The systems perspective of the criteria emphasizes results and value. Adhering to these core values and concepts enables organizations to direct their process improvement efforts in ways that

Table 3.2 Comparison of ISO 9000, the Malcolm Baldrige Award Criteria, and Continuous Improvement/Quality Management

	ISO 9000	*Baldrige Award*	*CI/QM*	*Six Sigma*
Scope	Quality management system	Quality of management	Quality management and corporate citizenship	Systematic reduction of process variability
	Continuous improvement		Continuous improvement	
Basis for defining quality	Features and characteristics of product or service	Consumer driven	Customer driven	Defects per million opportunities
Purpose	Clear quality management system requirements for international cooperation	Results-driven competitiveness through total quality management	Continuous improvement of customer service	Improve profitability by reducing process variation
	Improved record keeping			
Assessment	Requirements based	Performance based	Based on total organizational commitment to quality	Defects per million opportunities
Focus	International trade Quality links between suppliers and purchasers Record keeping	Customer satisfaction Competitive comparisons	Processes needed to satisfy internal and external customers	Locating and eliminating sources of process error

increase the value an organization provides to its customers and stakeholders. Their value system is shared by their employees, resulting in improvements in communication, collaboration, empowerment, commitment, and trust. Together, this approach improves the organization's public image with a commensurate increase in customer confidence.

The Malcolm Baldrige National Quality Award (MBNQA) was established in 1987 by the United States Congress. The MBNQA is named after former U.S. Secretary of Commerce (1981–1987) Malcolm Baldrige. Due to his personal interest in quality management and improvement, he was instrumental in the design and establishment of the award. Similar to Japan's Deming Prize, it sets a national standard for quality excellence. The award is open to companies in three areas: Business, Education, and Health Care. It is managed by the American Society for Quality (www.asq.org). As stated in the chapter introduction, Motorola was the first company to win the Malcolm Baldrige Award. Every year, this rigorous award attracts several dozen applicants in each category. A group of qualified examiners compares and contrasts each application with the criteria for up to 300 hours. Only a very select few reach the site-visit stage of the award process. By the completion of the on-site visits, a company may have been examined for as many as 1,000 hours. Since 1988, 71 organizations have received the United States' highest award for organizational effectiveness. Organizations pursuing the MBNQA focus on their approach to doing business and the deployment of their strategic resources. These organizations have one key metric for measuring their results: the stock price. It is interesting to note that the Baldrige award winners, when their stock performance is reviewed, outperform the stocks in the Standard and Poor's 500 Index by a ratio of about 3 to 1. For more information about the Baldrige stock study, visit the following website: www.nist.gov/public_affairs/factsheet/stockstudy.htm.

The award allows companies to make comparisons with other companies. This activity is called benchmarking. **Benchmarking** *is a continuous process of measuring products, services, and practices against competitors or industry leaders.* Covered in more detail in Chapter 10, benchmarking lets an organization know where they stand compared with others in their industry. Companies also use the award guidelines to determine a baseline. **Baselining** *is measuring the current level of quality in an organization.* Baselines are used to show where a company is, so that it knows where it should concentrate its improvement efforts.

Standards to be used as baselines and benchmarks for total quality management are set in seven areas (Figure 3.17). The MBNQA criteria present a complete package. The criteria seek to provide guidance for improving organizational performance practices, capabilities, and results. Since winners of the award share nonproprietary information with U.S. organizations of all kinds, best practices are communicated and shared. The criteria are a working tool for managing performance. They provide a guide for organizational excellence by encouraging organizational and personal learning resulting in improvement of overall organizational capabilities. Effective organizations sustain their excellence by delivering ever-improving value to customers and stakeholders.

The criteria are built on core values and concepts. The 2007 award criteria list those as:

- visionary leadership
- customer-driven excellence
- organizational and personal learning
- valuing employees and partners
- agility
- focus on the future
- managing for innovation
- management by fact
- social responsibility
- focus on results and creating value
- systems perspective

There are seven categories:

- Leadership
- Strategic Planning
- Customer and Market Focus
- Measurement, Analysis, and Knowledge Management
- Workforce Focus

1.0 Leadership

    1.1 Senior Leadership
    1.2 Governance and Social Responsibility

2.0 Strategic Planning

    2.1 Strategy Development
    2.2 Strategy Deployment

3.0 Customer and Market Focus

    3.1 Customer and Market Knowledge
    3.2 Customer Relationships and Satisfaction

4.0 Measurement, Analysis, and Knowledge Management

    4.1 Measurement, Analysis and Improvement of Organizational Performance
    4.2 Management of Information, Information Technology and Knowledge

5.0 Human Resource Focus

    5.1 Workforce Engagement
    5.2 Workforce Environment

6.0 Process Management

    6.1 Work Systems Design
    6.2 Work Process Management and Improvement

7.0 Results

    7.1 Product and Service Outcomes
    7.2 Customer-Focused Outcomes
    7.3 Financial and Market Outcomes
    7.4 Workforce-Focused Outcomes
    7.5 Process Effectiveness Outcomes
    7.6 Leadership Outcomes

Figure 3.17   Malcolm Baldrige National Quality Award Criteria Categories, 2007 (U.S. Department of Commerce, www.nist.gov)

- Process Management
- Results

The following descriptions are paraphrased from the criteria:

1.0 Leadership

The leadership section of the criteria focuses on two questions: *How do your senior leaders lead?* and *How do your senior leaders govern and address your social responsibilities?* This section questions an organization's values and vision. Investigators study how well senior leaders personally promote an environment that fosters, communicates, requires, and results in legal and ethical behavior. Senior leaders are also responsible for taking action to ensure accomplishment of the organization's goals and objectives toward improved performance. The governance section reviews accountability, both fiscal and physical, auditing procedures, legal and ethical behavior, and support of key communities. Leadership accounts for 120 of 1,000 points.

2.0 Strategic Planning

This section has a strong focus on innovation, strategic advantages, and resource needed to accomplish the strategic objectives. The criteria focus on two questions: *How do senior leaders develop your strategy?* and *How do senior leaders deploy your strategy?* Examiners study the organization's strategy development process, in other words, how organizations determine their strengths, weaknesses, opportunities, and threats. How an organization establishes their strategic objectives in relation to opportunities and threats is also key. Organizations do not move forward unless their strategies are implemented, so the criteria study organizations' action plan deployment and performance. This section receives 85 of 1,000 points.

3.0 Customer and Market Focus

Capturing the voice of the customer is critical to the success of any organization. The criteria focus on two questions: *How does the organization obtain and use customer and market knowledge?* and *How does the organization build relationships and grow customer satisfaction?* Organizations are judged based on their response to questions dealing with how they identify their customers, how they capture the voice of the customer, and how they use this information to make business improvements. Strong relationships with customers are encouraged. Organizations are expected to measure customer satisfaction. Customer and Market Focus receives 85 of 1,000 points.

4.0 Measurement, Analysis, and Knowledge Management

Effective organizations need information and the technology to support it. This section asks two questions: *How does the organization measure, analyze, and then improve organizational performance?* and *How does the organization manage its information, information technology, and organizational knowledge?* Effective organizations track their progress through the use of performance measures. The criteria study how these measures are selected, collected, analyzed, and reviewed. This information must align with organizational goals and objectives and be used to improve performance. Where the information comes from and how it is gathered is investigated. Examiners also check out how the organization ensures information accuracy, integrity, reliability, timeliness, security, and confidentiality. How this information is disseminated is also critical to this section. This section is worth 90 of 1,000 points.

5.0 Workforce Focus

In 2007, this section was completely redesigned in order to focus on workforce engagement and the workforce environment. It seeks to find answers to these questions: *How does the organization engage its workforce to achieve organizational and personal success?* and *How does the organization build an effective and supportive workforce environment?* The emphasis is on workforce enrichment and creating an organizational culture conducive to high performance. Effective communication and clear policies are critical to an effective organization. Successful organizations grow talent from within through good leadership and workforce development practices. Effective organizations are organizations that continue to learn and grow. Recruiting, hiring, placing, training, and retaining a capable workforce is crucial. Much of an organization's ability to attract and retain highly capable people is based on its work environment. The criteria study how an organization ensures and improves workplace health, safety, and security. A total of 85 of 1,000 points are awarded in this section.

6.0 Process Management

This section was also redesigned in 2007. It focuses on these questions: *How does the organization design its work systems?* and *How does the organization manage and improve its key organizational work processes?* Answering these questions means understanding the organization's work systems, core competencies, and work processes. Examiners study work process design, work process management, and work process improvement. They are interested in how the organization determines its core competencies and then designs systems to support them. These systems must be regularly studied for improvement possibilities. This section also covers emergency preparedness. In an emergency situation, an organization must know how to prevent disaster, manage operations and recover. Process Management accounts for 85 of 1,000 points.

7.0 Results

This section ties closely to the preceding six sections in order to ensure that the organization is correctly measuring and achieving appropriate results. The focus is on outcomes, not activities. Many questions are asked in this section, including *What are the organization's product and service performance results? What are the organization's customer-focused performance results? What are the organization's financial and marketplace performance results? What are the organization's workforce-focused performance results? What are the organization's process effectiveness results?* and *What are the organization's leadership results?* Positive results mean sustainable future operations. This is by far the most important section, worth 450 of 1,000 points.

The criteria for the Malcolm Baldrige Award are updated annually. Recipients of the award are from a variety of industries, including telecommunications, banking, automotive, hospitality industry, education, hospitals, building products, and manufacturing (Figure 3.18). For more information about the award criteria, contact the U.S. Commerce Department's National Institute of Standards and Technology at www.nist.gov. For a complete list of Malcolm Baldrige National Quality Award recipients, including organization profiles and contact data, visit www.quality.nist.gov/Award_Recipients.htm. The award criteria will serve as a foundation for future chapters in this text. At the close of each chapter, a section shows the link between chapter material and the award.

LESSONS LEARNED

Effective organizations need systems in place that ensure that their products or services achieve the highest levels of quality possible in the most cost-effective manner. As this chapter reveals, a wide variety of standards and award criteria provide insight and guidance into how to establish and implement a quality system. In future chapters the MBNQA criteria, supported by Six Sigma and ISO 9000, will be used as a guide to creating an effective

Manufacturing

Sunny Fresh Foods Inc.
The Bama Companies
Motorola Commercial, Government & Industrial Solutions Sector
Clarke American Checks, Inc.
Dana Corporation—Spicer Driveshaft Division
ST Microelectronics, Inc. Region Americas
Boeing Airlift and Tanker Programs
Solar Turbines Incorporated
3M Dental Product Division
Solectron Corporation
ADAC Laboratories
Armstrong World Industries, Inc., Building Products Operations
AT&T Network Systems Group Transmission Systems Business Unit
Texas Instruments Incorporated Defense Systems & Electronics Group
Eastman Chemical Company
Cadillac Motor Car Company
Westinghouse Electric Corporation Commercial Nuclear Fuel Division
Xerox Corporation, Business Products Systems
IBM Rochester
DynMcDermott Petroleum Operations Co.

Service

Premier Inc.
Caterpillar Financial Services Corporation—U.S.
Boeing Aerospace Support
Operations Management International, Inc.
Merrill Lynch Credit Corporation
Dana Commercial Credit Corporation
Verizon Information Services
AT&T Universal Card Services
The Ritz-Carlton Hotel Company
Federal Express Corporation

Small Business

MESA Products Inc.
Park Place Lexus
Branch-Smith Printing Division
Pal's Sudden Service
Wallace Co., Inc.
Globe Metallurgical, Inc.
Granite Rock Company
Ames Rubber Corporation
Texas Nameplate Company, Inc.
Los Alamos National Bank
Trident Precision Manufacturing, Inc.

Figure 3.18   Partial List of Malcolm Baldrige National Quality Award Winners since 1988
Source: www.quality.nist.gov/Award_Recipients.htm

Healthcare

North Mississippi Medical Center
Bronson Methodist Hospital
Baptist Hospital, Inc.
St. Luke's Hospital of Kansas City
SSM Health Care

Education

Jenks Public Schools
Richland College
Monfort College
Community Consolidated School District 15
Pearl River School District
University of Wisconsin-Stout
Chugach School District

Figure 3.18   (continued)

organization. To date, the MBNQA criteria are the most complete set of criteria available, covering all of the crucial aspects necessary to create an effective organization (Table 3.2). To help the reader better understand the concepts discussed in each chapter, following each chapter an ongoing case study will use the MBNQA criteria to evaluate the organizational effectiveness of an actual company. The case accompanying this chapter introduces the company. Use the information provided in Appendix 3 as guide for a MBNQA student project.

Many options are open to companies seeking to optimize their performance in all areas of their organization. Regardless of which one they choose to adopt, it is critical to remember that all are based on the concepts and values first clearly espoused by Dr. W. Edwards Deming. The first of his 14 points stressed maintaining a constancy of purpose through setting goals, training employees, and continuous improvement efforts. His work highlights the importance of quality and customers, both internal and external to the organization. He understood that the most effective organizations involved employees throughout the organization in the improvement of processes and quality. Whether an organization chooses quality management, Six Sigma, business process reengineering, certifications, or the Malcolm Baldrige National Quality Award as a guideline, the emphasis should be the same: a constancy of purpose toward achieving overall organizational excellence.

1. ISO 9000 is a set of quality standards used to facilitate the relationship between suppliers and purchasers.

2. ISO 9000 has three key benefits: it establishes a sound quality system, regular reviews drive organizational discipline and improvement, and as a widely recognized standard, ISO 9000 provides access to markets and suppliers worldwide.

3. A wide variety of supplier certification standards based on ISO 9000 exist, including TS 16949, TL 9000, IEC 17025, and AS9000.

4. ISO 14000 encourages environmental protection and pollution prevention while taking into account the economic needs of society.

5. ISO 14000 is voluntary.

6. Six Sigma is data driven and profit focused.

7. Six Sigma enhances an organization's ability to provide value for their customers.

8. Six Sigma enhances an organization's understanding of their key business processes.

9. The MBNQA criteria are built on core values and concepts like visionary leadership, customer-driven excellence, organizational and personal learning, valuing employees and partners, agility, management by fact, focus on results, and others.

10. The MBNQA criteria has seven categories: leadership, strategic planning, customer and market focus, measurement, analysis and knowledge management, workforce focus, process management, and results.

Regardless of which approach an organization takes, all quality management systems stress eight key principles:

Customer focus: identifying customer expectations
Process improvement based on customer expectations
Fact-based decision-making: measures critical to customer expectations
Systems approach to management: tracking measures
Leadership: commitment to excellence
Involvement of people
Continuous improvement
Mutually beneficial supplier relationships

Are You a Quality Management Person?

Do you have a system in your life that keeps you focused on what is key to your success? Check out the following questions and find out whether or not you are a quality management person. How do you answer them on a scale of 1 to 10?

Do you have standards that you follow in some of your day-to-day activities?

| 1 | 2 | 3 | 4 | 5 | 6 | 7 | 8 | 9 | 10 |

Do you compare your performance on key activities against standards?

| 1 | 2 | 3 | 4 | 5 | 6 | 7 | 8 | 9 | 10 |

Do you, in the face of competition, set goals and objectives to meet based on known standards?

| 1 | 2 | 3 | 4 | 5 | 6 | 7 | 8 | 9 | 10 |

Have you ever had your performance in an activity "certified" by a judge or competition?

1	2	3	4	5	6	7	8	9	10

Do you occasionally have standards you expect those you interact with to meet?

1	2	3	4	5	6	7	8	9	10

Do you understand the effect of your activities or lifestyle on the environment?

1	2	3	4	5	6	7	8	9	10

Have you changed some of your activities or your lifestyle to reflect a more environmentally conscious approach to life?

1	2	3	4	5	6	7	8	9	10

Do you make your improvement efforts based on the results you expect?

1	2	3	4	5	6	7	8	9	10

When you analyze a problem, do you do so in phases like define, measure, analyze, improve, and control?

1	2	3	4	5	6	7	8	9	10

Do you try to reduce the variation present in your activities or processes?

1	2	3	4	5	6	7	8	9	10

Do you make an effort to strive for personal learning and growth?

1	2	3	4	5	6	7	8	9	10

Do you focus on results and creating value?

1	2	3	4	5	6	7	8	9	10

Do you do any strategic planning?

1	2	3	4	5	6	7	8	9	10

Do you make changes to improve your effectiveness?

1	2	3	4	5	6	7	8	9	10

Chapter Questions

1. Describe what is meant by the term *quality system*.
2. Why is a quality system critical to providing a quality product or service?
3. What attributes would you expect to be present in a company that has a sound quality system?
4. Describe why ISO 9000 is important for an effective organization.
5. Why is record keeping important? What types of records do effective organizations keep?
6. Why is it important to document work methods and procedures?
7. Describe the ISO 9000 registration process.
8. What are the benefits of certification?
9. Find an article discussing either ISO 9000 or QS 9000. After reading the article, answer the following questions:
   a. Who needs to become certified?
   b. What steps does a company have to take to become certified?
   c. What are the reactions to ISO 9000 or QS 9000?
10. Describe ISO 14000 to someone who has not heard of it.
11. What do Six Sigma projects focus on? Why?
12. Describe the Six Sigma methodology to someone who has not heard of it.
13. How does Shewhart's Plan-Do-Study-Act cycle compare with Six Sigma's DMAIC cycle?
14. Describe the changes that occur to the spread of the process when the amount of variation in the process decreases.
15. What are the benefits of implementing the Six Sigma methodology?
16. Why would a company want to follow the Six Sigma methodology?
17. Describe what it takes to become a Green Belt.
18. What does a person need to do to become a Black Belt?

19. Describe the difference between a Black Belt and a Master Black Belt.

20. How do Green Belts, Black Belts, and Master Black Belts interact when working on a project?

21. How does the Six Sigma methodology compare with the continuous improvement methodology?

22. What is the Malcolm Baldrige National Quality Award?

23. Why would an organization apply for the Malcolm Baldrige National Quality Award?

24. Describe each of the Malcolm Baldrige National Quality Award criteria.

25. Why is the Malcolm Baldrige National Quality Award the most comprehensive guide to organization-wide improvement? What does it have that ISO 9000 and Six Sigma lack?

Organizational Effectiveness at Remodeling Designs, Inc., and Case Handyman Services

This case study is designed to provide readers of the text with an idea of how one company achieved organizational effectiveness. The case is divided into the seven sections of the Malcolm Baldrige National Quality Award (MBNQA) criteria, and each section is placed at the end of the appropriate chapter:

Leadership	Chapter 5	MBNQA criteria Section 1.0
Strategic Planning	Chapter 6	MBNQA criteria Section 2.0
Customer and Market Focus	Chapter 4	MBNQA criteria Section 3.0
Human Resource Focus	Chapter 7	MBNQA criteria Section 5.0
Measurement, Analysis, Knowledge Management	Chapter 9	MBNQA criteria Section 4.0
Process Management	Chapter 11	MBNQA criteria Section 6.0
Business Results	Chapter 15	MBNQA criteria Section 7.0

Beginning in Chapter 4, each end-of-chapter case includes three parts:

1. Section goals of the MBNQA criteria stated in terms specific to Remodeling Designs, Inc. and Case Handyman.
2. Questions derived from the MBNQA criteria of number 1.
3. Review/discussion of the effectiveness of Remodeling Designs, Inc. and Case Handyman.

For the current MBNQA criteria, visit www.nist.gov.

The company Remodeling Designs, Inc. was reviewed by University of Dayton students under the direction of Dr. Donna Summers during their Quality Management course in the fall of 2001. The information in the case is based on the MBNQA criteria for 2001. The review served as partial fulfillment of course requirements. This review is used in its entirety with permission from Remodeling Designs, Inc., Dayton, Ohio. Minor differences between the headings of this review and current MBNQA criteria are due to continuous improvements in the Malcolm Baldrige National Quality Award criteria. Each year, the MBNQA criteria are updated.

BACKGROUND

Remodeling Designs, Inc.

Remodeling Designs, Inc. was started in April 1990 as a part-time job for two graduates from the University of Dayton with a common hobby and a big dream. The company began as a partnership between the Eggers and the Cordonniers. It was incorporated in 1991. Each year the company continues to grow steadily and now operates with several production crews and a handyman division. Remodeling Designs, Inc. offers full-service project management for all types of remodeling jobs, specializing in residential projects such as kitchens, bathrooms, basements, and room additions. The company oversees all phases of the job, from design to completion.

Case Handyman Services

Case Handyman Services was launched by Case Design/Remodeling, Inc., one of the largest and most respected leaders in the home improvement industry. Case has provided quality service since 1961 and began franchising Case Handyman Services in 1998. Case Handyman Services provides Home Repair Specialists to take care of most home repair and maintenance needs.

The Mission and Purpose Statement for Remodeling Designs, Inc. and Case Handyman Services is provided in Figure 1.

Our Mission

To be the leading Design/Build remodeling and handyman companies in the Miami Valley as measured by client satisfaction, profitability, growth, and success.

Our Purpose

To improve our clients' lives by providing the highest quality Design/Build remodeling and handyman services. To enable our team members to flourish and prosper.

Core Values for
Case Handyman Services and Remodeling Designs, Inc.

As defined by Webster's dictionary, a "core value" is the most essential, vital part of some idea or experience accepted by a group or individual. In real life terms, "core values" are what we use to make everyday decisions on how we conduct business and ourselves. We are convinced the key to creating a truly outstanding organization is an intense focus on the values that guide its members' actions. These are Remodeling Designs and Case Handyman Services core values and beliefs. If we are not living up to them let us know!

Our companies consist of the highest quality team members who aim for excellence at all times.

Professionalism is evident in everything we do.

We take great pride in our work.

Client Satisfaction is a top priority.

Team members are empowered, respected, and appreciated.

We continually strive to improve our companies and ourselves.

Work is an important part of life, and it should be enjoyable.

We make commitments with care. We do what we say we are going to do.

We guard and conserve our companies' resources with the same vigilance that we would use to guard our own personal resources.

The success of our companies and our projects relies on constant communication at all levels.

Understanding our mission, purpose, values, and goals is critical to our success.

Figure 1   Mission Statement

4 Creating a Customer Focus

Why is a focus on customers so important?

How do customers define quality?

How do customers define value?

What is the difference between satisfaction and perceived value?

Why is it important to understand how the customer views the process?

How does an effective organization create an unwavering focus on customers' requirements, needs, and expectations?

How do effective organizations know what their customers want?

How do effective organizations capture the voice of the customer and turn that voice into actions that drive customer-perceived value?

What is Quality Function Deployment?

How can the Malcolm Baldrige National Quality Award criteria help an organization create a more effective customer focus?

Lessons Learned

Learning Opportunities

1. To understand the difference between satisfaction and perceived value
2. To understand how to create a customer focus
3. To understand how to capture the voice of the customer
4. To become familiar with how to perform a quality function deployment

From a customer's standpoint, neither quality, cost nor schedule always comes first. When customers evaluate the products and services they receive, they make trade-offs between all three key factors in order to maximize value. The challenge that suppliers face is to provide their customers with the maximum value, which often is a balancing act between quality, cost and schedule.

"The First among Equals," Quality Digest, *June 1999*

Quality is not a thing. It is an event.

Robert Pirsig, Zen and the Art of Motorcycle Maintenance

WHY IS A FOCUS ON CUSTOMERS SO IMPORTANT?

The current global business environment is extremely competitive. Quality is dynamic. Today's consumers are more than willing to switch from supplier to supplier in search of better service or courtesy, or better product availability or features—or for any other variety of reasons. To attract and retain customers, effective organizations need to focus on determining and then providing what their customers want and value. Advertising, market positioning, product/service imaging, discounting, crisis handling, and other methods of attracting the customer's attention are not enough. Effective organizations survive because they talk to customers, translate what their customers said into appropriate actions, and align their key business processes to support what their customers want. These critical activities enable effective organizations to meet their customers' needs, wants, and expectations the first time, every time.

HOW DO CUSTOMERS DEFINE QUALITY?

In today's world, people expect product quality as evidenced by performance, features, conformance to requirements, reliability, durability, and aesthetics. Effective organizations differentiate themselves through service: timely delivery, ease of serviceability, organizational responsiveness, employee competence, ease of access, courtesy, ease of communication, security, and understanding. Effective organizations place as much focus on the service side of their organization as the production side. These include finance, marketing, sales, distribution, human resources, field service, accounts receivable, and so on. Some organizations, like hospitals, banks, and package services, are totally service focused. Regardless of organization type, their goal is to effectively and efficiently service both internal and external customers. This means that transactions are smooth, efficient, and effective. Improvements come about through shortening response times and improving accuracy. Manufacturing is about product; service is about transactions. Both share one critical to quality characteristic: customers expect quality, reliability, and consistency. Due to the significant human interaction required in a service environment, it is easy to see why services are seen as inefficient, labor-intensive, time-consuming, costly, and variable.

How does a customer know that he or she has received a quality product or service? What does quality represent to a customer? Is there one simple definition for quality, or do customers define quality as "knowing it when they see it"? The American Society for Quality defines quality as "a subjective term for which each person has his or her own definition. In technical usage, quality can have two meanings: (1) the characteristics of a product or service that bear on its ability to satisfy stated or implied needs and (2) a product or service free of deficiencies." Dr. W. Edwards Deming defines quality as "non-faulty systems." Dr. Joseph Juran describes quality as "fitness for use." From Chapter 2, Armand Feigenbaum's definition states:

> quality is a customer determination which is based on the customer's actual experience with the product or service, measured against his or her requirements—stated or unstated, conscious or merely sensed, technically operational or entirely subjective— always representing a moving target in a competitive market.

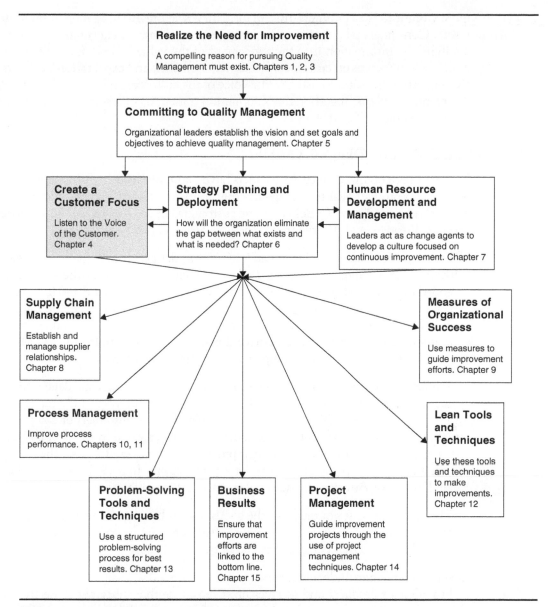

Realize the Need for Improvement

A compelling reason for pursuing Quality Management must exist. Chapters 1, 2, 3

Committing to Quality Management

Organizational leaders establish the vision and set goals and objectives to achieve quality management. Chapter 5

Create a Customer Focus

Listen to the Voice of the Customer. Chapter 4

Strategy Planning and Deployment

How will the organization eliminate the gap between what exists and what is needed? Chapter 6

Human Resource Development and Management

Leaders act as change agents to develop a culture focused on continuous improvement. Chapter 7

Supply Chain Management

Establish and manage supplier relationships. Chapter 8

Measures of Organizational Success

Use measures to guide improvement efforts. Chapter 9

Process Management

Improve process performance. Chapters 10, 11

Lean Tools and Techniques

Use these tools and techniques to make improvements. Chapter 12

Problem-Solving Tools and Techniques

Use a structured problem-solving process for best results. Chapter 13

Business Results

Ensure that improvement efforts are linked to the bottom line. Chapter 15

Project Management

Guide improvement projects through the use of project management techniques. Chapter 14

Figure 4.1   Achieving Organizational Success

Feigenbaum's definition points out that only a customer can determine if and how well a product or service meets his or her needs, requirements, and expectations. This determination is dependent on the customer's actual experience with the product or service. Dr. Feigenbaum's definition reminds effective organizations that experiences are memorable and that it is this emotional connection with a product or service that differentiates one organization from another. Selling the total experience makes an organization seem more vibrant to the customers; this, in turn, makes an organization more resilient to market forces. Consumers take their past experiences, requirements, expectations, and needs and combine the information into their judgment of the value of a product or service. Quality management focuses on customers' needs, requirements, and expectations. Leaders of effective organizations learn to listen to the voice of the customer and adapt what they hear into their strategic plans. Together, this information is used to guide a culture based on continuous improvement (Figure 4.1).

HOW DO CUSTOMERS DEFINE VALUE?

The world is a changed place. Consumers have extraordinarily high standards for customer service and product quality. Companies that want to retain their customers must continually strive to find ways to delight their customers. In an August 20, 2007, article titled "Beauty, Prestige, and Worry Lines," *The Wall Street Journal* began with:

> For more than 60 years, Estee Lauder Cos. has defined top-of-the-line cosmetics in the U.S. Its Clinique products led sales in upscale retailers over the past 12 months ended in June, while the eponymous Estee Lauder brand is second and its MAC brand is fourth, according to market data firm NPD Group. Even so, William Lauder can't afford to relax. As chief executive officer of the company founded by his grandmother, he faces growing competitive threats on a number of fronts. . . . Mr. Lauder must try to stay nimble and compete against these larger forces.

The article goes on to discuss Mr. Lauder's strategic plans to reach and delight his customers. These plans are being deployed, even as his company has held three of the top four sales spots for the past 12 months. Effective organizations never ignore customers or competition. They are always trying to figure out how to translate customer value in terms of customer needs, requirements, and expectations into their products and services. To do so, they must focus on the total customer experience. This is not a new concept. In a 1927 speech to the National Retail Dry Goods Association, industrial engineer Lillian Gilbreth stated,

> The customer must not only like the goods she has bought, the surroundings in which she has done the buying and the people from who she has done the buying, but she must also like herself better at the end of the transaction.

Effective companies understand this.

Providers of products and services must recognize that customers base their decisions about the quality of a product or service on the perceived value they receive. Value, *the attributed or relative worth or usefulness of a product or service*, is judged by a consumer each time he or she trades something of worth (usually money) in order to acquire the product or service. Consumers experience value when they benefit from the exchange. Because the value judgments of customers involve past experience, requirements, wants, needs, and expectations, the judgments are complex. As Feigenbaum's definition of quality cautions, the consumer may or may not be able to clearly define and state his or her requirements. Sometimes the consumer only senses an entirely subjective concept. When evaluating a product, the consumer might study the product's performance capabilities or its available features. Over the life of a product, a consumer has the opportunity to judge its serviceability and durability. When dealing with a service, a customer might be interested in the completeness, appropriateness, and timeliness of the service. Other aspects can be related to both products and services. Reliability, usually considered from the point of view of a product, may be applied to the reliability of a service person showing up at the appointed time. For both products and services, aesthetics may be key to customer satisfaction. The organization itself may be evaluated based on its credibility and reputation for responsiveness to customers, employee competence, ability to communicate, and courtesy. The customer's perception of value is what will ultimately make a difference between an unsuccessful customer and a successful one.

WHAT IS THE DIFFERENCE BETWEEN SATISFACTION AND PERCEIVED VALUE?

Satisfaction and perceived value are related but different concepts. Effective organizations recognize that they are offering product or service features to their customers, but what the customers are actually buying is the benefits those products or services offer. Perceived value is the customer's viewpoint of those benefits. Customer satisfaction, on the other hand, centers on how the customer felt the last time he or she bought a product or service from a company. It is a comparison between customer expectations and customer experience. Perceived value goes beyond customer satisfaction and concentrates on future transactions. Consumers' perception of the value they have received in a recent transaction will affect their decision to purchase the same product or service in the future. If they perceive their overall experience with the product or service as valuable, they will most likely purchase it again in the future; if they do not, they won't. Effective organizations realize that how the customer perceives the value of that transaction will determine whether he or she will buy from the same organization the next time.

Customer-perceived value is what enables a company to be successful in the future. To ensure the growth of their business, organizations need loyal customers. With such a wide variety of products and services available today, each time customers make a purchase, they compare the pros and cons of each alternative before purchasing the alternative they consider the most valuable. Loyalty, often described as customer retention, is in reality the absence of a better alternative. High customer-satisfaction ratings in the past do not necessarily equate to customer loyalty in the future, but high perceived-value ratings do. Effective organizations continually seek to increase their customers' perceived value of their products and services.

Information about customer-perceived value combined with information about customer satisfaction is powerful. With this information, effective organizations can make changes to the way that they do business in order to better serve existing customers while attracting future customers. Using one without the other has disadvantages too. For example, it is always useful to keep in mind that flying transcontinentally in a few hours is a better alternative than taking a ship for several days, thus equating to high

EXAMPLE 4.1   Alternatives

In a certain major metropolitan area, a number of grocery stores exist. Each week nearly all of the area grocery stores place ads advertising their specials in local newspapers and on radio and television. Yet one store does no advertising of any kind. Surprisingly, the store boasts a lengthy list of loyal customers and a healthy profit margin. In fact, once a customer experiences the services and products available at this store, he or she often prefers not to shop elsewhere. One disgruntled competitor even went so far as to say that it seemed as if this particular grocery store owned their customers. So how is it that one grocery store appears to be creating an atmosphere that enables customers to have a love affair with the products and services they offer? What magic is at work here? How has this effective organization created high perceived value with its customers?

It doesn't take long in a discussion with the owners of the small chain of stores to discover the truth. They willingly describe the efforts they make on the customers' behalf to constantly improve the products and services they provide. They use service to compete. Their customer service–oriented employees actively engage customers to find ways to improve. The owners are adept at using customer satisfaction, and they value information that helps them to change and improve their internal processes. Recent updates include creating customized coupons, altering aisle layouts, and modifying product offerings in order to better suit their clientele. They also engage in dialogue with their customers in order to determine the best ways to create value for them, for instance, healthy alternatives and ready-to-serve restaurant-quality meals.

Their overall focus on the total customer experience extends to the atmosphere of the store as well. The owners realize that customer perceptions of their shopping experience begin in the parking lot and end after the food has been consumed. For this reason, the owners have studied everything from parking lot layouts and lighting, to the cleanliness and location of their restrooms, to the shelf life of perishable foods. Their employees are trained to think of the customers and act accordingly. Is it any wonder that the store retains its existing customers while attracting new ones daily?

customer-perceived value. However, a study of customer satisfaction scores for airlines reveals that customers are far from satisfied with their experience. Effective organizations use all of the information available to them to focus on areas of concern for their customers.

WHY IS IT IMPORTANT TO UNDERSTAND HOW THE CUSTOMER VIEWS THE PROCESS?

Effective organizations take the time to be their own customers, to look at their processes as the customer would. They recognize the need to ensure that the process the customer sees is seamless, flawless, and easy to negotiate. Having hassle-free processes adds considerable value from the customer's viewpoint. These types of processes save money and time. Customers willingly participate in processes they can understand, which is essential in the service industries where customer input is vital to the success of the process. Information, whether about customer-perceived value or customer satisfaction, is more meaningful if it is obtained on the customers' terms and from their perspective. Connecting that information about how a customer views a process with internal business processes and action plans is the seller's responsibility.

EXAMPLE 4.2   Contact Yourself

One night, following a meeting of the e-Business Society, the president of an e-commerce firm received an interesting proposal from a dissatisfied customer who had attended the meeting: "If you think your company is so great, contact yourself." Feeling certain that the dissatisfied customer was exaggerating about the difficulties she had experienced, the president did just that, or at least tried to.

First he visited his company's website. There he selected the option labeled "Contact Us." When an e-mail box popped up on his computer screen, he typed in a question and sent it. Three days later he was still waiting for an answer. He repeated the procedure, sending another e-mail and waiting three more days. Still no response for either e-mail. Returning to the website, he searched for and finally located in very fine print a customer-service contact phone number. Dialing the number during normal business hours, he received the following rather abrupt message:

> You have reached EQ. Our regular business hours are 8 to 5 standard time.
> Please call back during regular business hours.

Since he was calling during regular business hours, the message surprised him. He tried circumventing the message by pushing various buttons, including 0 for the operator. Much to his surprise, he was not able to contact his own company. The enormous consequences of this problem left him breathless. Within moments he had called a meeting with his employees to remedy this situation.

Future plans include studying all of their processes from the customer's point of view. Intuit Inc. uses a "Follow Me Home" program to watch purchasers of their Quicken software while they install and use the program. Some supermarkets train new employees by giving them 10 minutes to find two dozen items. Watching your customers use your product is a great way to capture information overlooked by surveys and questionnaires.

HOW DOES AN EFFECTIVE ORGANIZATION CREATE AN UNWAVERING FOCUS ON CUSTOMERS' REQUIREMENTS, NEEDS, AND EXPECTATIONS?

The competitive business world isn't going to get any easier. So what can an organization do when they realize that they aren't doing very well? How do they improve? And, if they are doing well for the moment, how do they keep their competitors from gaining on them or surpassing them? How does an effective organization create an unwavering focus on customers' requirements, needs, and expectations?

Organizations practicing total quality management principles create a customer-focused management system and company culture that seeks to meet their customers' needs the first time and every time. Effective organizations analyze their customers' needs, wants, and expectations; translate them into technical specifics; and organize their key business operations accordingly. These organizations ensure that their leadership creates and implements strategic plans that focus on what is important to their customers and markets.

Effective organizations need an accurate understanding of what their customers expect. If they are going to properly target improvement activities, they also need to identify the gap between their current performance and customer requirements. They recognize the importance of studying both customer value perceptions and customer satisfaction. Customer-perceived value, the result of comparing purchasing alternatives, looks toward the future and is proactive. Knowledge of customer-perceived value allows an organization to change its future product or service offerings to better suit its customers. Customer satisfaction compares past experience or expectations to the realities experienced; it is reactive and retrospective. Organizations can use information about both customer-perceived value and customer satisfaction to help improve their existing processes.

EXAMPLE 4.3   What Does the Customer Expect?

At ROI, a distributor of construction materials, one of the principal strategies relates to business growth through acquiring customers who use ROI as their sole supplier. Recently, the sales team discussed a proposal provided by a customer. The customer would like to use ROI as a sole source for all building materials. In exchange for their commitment, they would like a reduction in brick prices, essentially purchasing each brick for one penny below ROI's cost.

The manager of brick sales is unwilling to agree to the price reduction because it would not optimize his department's overall figures. The sales team wants to agree to the deal because the resulting increase in sales would have a dramatic effect on the company's overall profit picture.

In order to convince the sales manager of bricks that this deal is a good one, the sales team gathered information about brick sales in general and specifically about the extras that accompany the sale of traditional brick. These extras include decorative finishing brick, arches, window sills, and other items that give a finished appearance to a brick building. They were able to show that the new business resulting from this deal would result in an overall increase in sales of these extra items that would more than compensate for the penny-a-brick loss. Once presented with these numbers, the manager of bricks realized that by remaining focused on customer needs as they relate to the overall strategy and health of the entire company, a better over-all business decision could be made.

HOW DO EFFECTIVE ORGANIZATIONS KNOW WHAT THEIR CUSTOMERS WANT?

They ask them.

Effective organizations talk with their customers. Effective organizations seek to understand every aspect of their customers' interaction with their company. They understand that this process begins when the customer first contacts the company and continues until the product has been consumed or the service completed. They realize what business they are really in, because they have asked that question from the point of view of the customer. Their customers have helped them define their business, including their principal strengths and weaknesses as well as what they need to do in the future in order to improve. Organizations know that they are supposed to talk to their customers. It is the effective organizations that ask, Are we doing too much talking and not enough listening? Effective organizations know they must listen; their business depends on it. They know better than to generalize about the behavior of all customers from just one example. They take steps to ensure that anecdotes in their organization that pertain to just one customer have not been told, retold, and altered just enough to be considered "common knowledge" about customer preferences. Effective organizations study the key issues by asking, probing, listening, digging, diagnosing, and changing. They translate what they have learned into actionable improvements. Then they do a very good job of communicating the value that they provide.

Listening to the customer has been referred to as trying to hug a cloud. Effective organizations pay close attention to the part of Feigenbaum's definition that cautions that quality is a moving target, that it can be subjective or technically operational, stated or unstated, conscious or merely sensed. Dr. Deming, in his text *Out of the Crisis*, said that it is important to measure customers' expectations through "consumer research . . . by which the manufacturer . . . is able to redesign his product to make it better as measured by the quality and uniformity that are best suited to the end users of the product and to the price that the consumer can pay." It is critical that effective organizations gain customer feedback by listening to the voice of the customer. Taking a reactive stance and responding to customers' complaints is one approach to interacting with customers. Effective organizations take a more proactive approach. They find ways to gather information to use to change the way they develop, innovate, and market their product or service. They recognize that providing products and services is not about selling them but rather about showing how the product or service can help a potential customer. They listen to their customers and find ways to give their customers what they want, not what they think their customers want. Every interaction with a customer is a chance to deepen the relationship, to solve their problems, or to make things easier for them.

How Do We Know It's Working?

Most machine shops rarely visit their customers. At JQOS, they take a proactive approach and regularly visit their customers. These visits focus on trying to understand what the customers want to do with their business, what business threats they face, the critical-to-quality aspects of their parts, how the parts are used, and the type of environment the customer's plant operates in. This knowledge helps JQOS create value for their customers. For instance, they changed the packaging of one customer's parts to support their receiving directly to the workstation environment. Another customer needed special shipping to support their just-in-time environment. Newer grinders were selected based on their ability to hold tolerances within 10 thousandths of an inch. Changes to the quoting and purchase order systems smoothed and speeded quoting, order placement, and order processing from the customers' point of view. Customers can now track the progress of their order online.

Customer visits provide a lot of information. As JQOS makes changes, their measure of performance is the quotes-to-orders turnover rate. Leadership at JQOS spends time tracking quotes. They expect to turn 40% of their quotes into firm orders. They know where they stand with their customers based on the number of purchase orders the customer has placed.

HOW DO EFFECTIVE ORGANIZATIONS CAPTURE THE VOICE OF THE CUSTOMER AND TURN THAT VOICE INTO ACTIONS THAT DRIVE CUSTOMER-PERCEIVED VALUE?

Translating customers' needs, wants, and expectations into actions that an organization can take is a necessary step toward creating organizational effectiveness. Effective organizations take the time to determine all of the reasons why a customer may contact their firm. They conduct research with past, present, and future customers to determine what those customers need and expect from each contact or purchase. They also determine what changes are needed in order to enhance their ability to provide a quality experience for a customer. Figure 4.2 provides an example of a translation of customer needs into measures of performance for a store.

Customer Question or Concern	Need
Does it fit in my car's cup holder?	Size, height
Will I have to worry about its breaking?	Does not break when dropped
Will it spill when it lies on its side?	Doesn't leak Opens/closes easily Has resealable lid
Is it lightweight? Is it too heavy?	Weight
Can I reuse it?	Reusable Recyclable
Can I chill it?	Stays cool Fits in most coolers

Figure 4.2   Translating Customer Needs into Measures of Performance

Quality Function Deployment

Quality Function Deployment (QFD) is a technique that seeks to bring the voice of the customer into the process of designing and developing a product or service. Using this information, effective organizations align their processes to meet their customers' needs the first time and every time. Companies use the voice-of-the-customer information obtained by a QFD to drive changes to the way they do business. Information taken directly from the customer is used to modify processes, products, and services to better conform to the needs identified by the customer.

Developed in Japan in the 1970s by Dr. Akao, QFD was first used in the United States in the 1980s. Essentially, QFD is a planning process for guiding the design or redesign of a product or service. The principal objective of a QFD is to enable a company to organize and analyze pertinent information associated with its product or service. A QFD can point out areas of strengths as well as weaknesses in both existing or new products.

Utilizing a matrix, information from the customer is organized and integrated into the product or process specifications. QFD allows for preventive action rather than a reactive action to customer demands. When a company uses the QFD format when designing a product or service, it stops developing products and services based solely on the company's own interpretation of what the customer wants. Instead, it utilizes actual customer information in the design and development process. Two of the main benefits of QFD are the reduced number of both engineering changes and production problems. QFD provides key action items for improving customer satisfaction and perceived value. A QFD can enable the launch of a new product or service to go more smoothly because customer issues and expectations have been dealt with in advance. Gathering and utilizing the voice of the customer is critical to the success of world-class companies.

A QFD has two principal parts. The horizontal component records information related to the customer, and the vertical component records the organization's technical response to these customer inputs. Essentially, a QFD matrix clearly shows what the customer wants and how the organization is going to fulfill those wants. The essential steps to a QFD are shown in Figures 4.3 and 4.4.

A QFD begins with the customer. Surveys and focus groups are used to gather information from the customers about their wants, needs, and expectations. Several key areas that should be investigated include performance, features, reliability, conformance, durability, serviceability, aesthetics, and perceived quality. Often, customer information—specifically, the way customers describe what they want—must be translated into appropriate, useful wording for the organization. That is, the wording needs to be translated into an action the organization can take; for example, when a customer says, "I can never find parking," this statement needs to be interpreted as "close, convenient parking readily available." In the first statement, the customer is expressing a need. The second statement turns that need into something the organization can act on.

Once this information has been organized into a matrix, the customers are contacted to rate the importance of each of the identified wants and needs. Information is also gathered

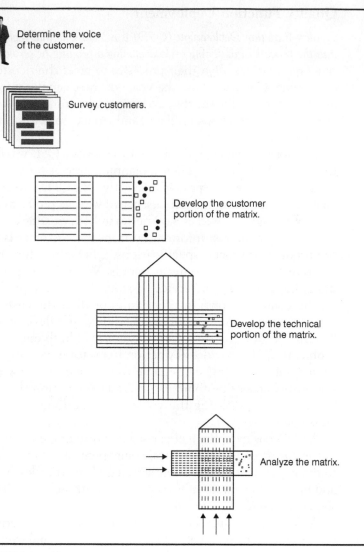

Figure 4.3   The QFD Process

about how customers rate the company's product or service against the competition. Following this input from the customers, technical requirements are developed. These technical aspects define how the customer needs, wants, and expectations will be met. Once the matrix is constructed, the areas that need to be emphasized in the design of the product or service will be apparent.

The following example shows the steps associated with building a QFD matrix.

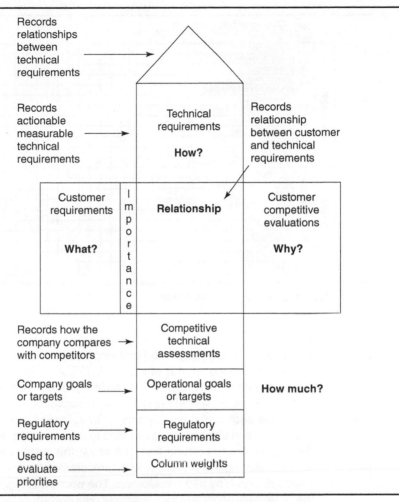

Figure 4.4   Summary of a QFD Matrix

EXAMPLE 4.4   Creating a QFD

AM Corporation sells sports drinks to the general public. They have always been very in tune with the health and nutritional needs of their customers. Recently their focus has turned to another aspect of their business: the containers their drinks are provided in. They have decided to utilize a QFD when redesigning their sports drink bottles.

1. *Determine the Voice of the Customer: What does the customer want?*
   The first step in creating the QFD involves a survey of their customer expectations, needs, and requirements associated with their sports drink bottles. AM Corp. met with several focus groups of their customers to capture their information. Following

Customer Requirements

		Customer Requirement	#
Container	Lids	Doesn't Leak	1
		Interchangeable Lids	2
		Freshness	3
		Open/Close Easily	4
		Sealed When Purchased	5
		Resealable Lid	6
	Shape	Doesn't Slip Out of Hands	7
		Fits In Cupholder	8
		Doesn't Tip Over	9
		Attractive	10
		Fits In Mini-Cooler	11
		Doesn't Spill When You Drink	12
Material	Characteristics	No Dents	13
		Doesn't Change Shape	14
		Is Not Heavy (Light)	15
		Does Not Break When Dropped	16
		Clear	17
		Reusable	18
		Recyclable	19
		Stays Cool	20
		No Sharp Edges	21
MISC	Cost	Inexpensive	22

Figure 4.5   Customer Requirements

these meetings, they organized and recorded the wants of the customers in the column located on the left side of the matrix (Figure 4.5).

2. *Have the customer rank the relative importance of his or her wants.*

   After AM Corp. organized the data, they reconvened the focus groups. At that time, they gave each of the participants an imaginary $100 to spend on the recorded wants. The participants were instructed to allocate more dollars to their more important wants. They recorded their values on the matrix next to the list of recorded wants. Following the meetings, AM Corp. created the final matrix by combining the values assigned by all the customers. The wants with higher dollar values are those the customers consider more desirable (Figure 4.6).

3. *Have the customer evaluate your company against competitors.*

   At the same meeting, the customers also evaluated AM Corp.'s competitors. In this step the participants divided $100 between AM Corp. and its competitors. The customers awarded more money to the company that they felt provided the best product or service for their recorded want. Following the meetings, AM Corp. created the final matrix by combining the values assigned by all the customers. Higher values represented where AM Corp. needed to focus their efforts (Figure 4.7).

4. *Determine how the wants will be met: How will the company fulfill the wants?*

   At this point AM Corp.'s efforts focused on determining how they were going to meet the wants identified by the customers. They spent many hours in meetings discussing the technical requirements necessary for satisfying the customers' recorded wants. These requirements were recorded at the tops of the columns in the matrix. AM Corp. made sure that the technical requirements, or "how's," were

Customer Requirements | Ranking

Container	Lids	Doesn't Leak	1	2
		Interchangeable Lids	2	6
		Freshness	3	6
		Open/Close Easily	4	3
		Sealed When Purchased	5	6
		Resealable Lid	6	8
	Shape	Doesn't Slip Out of Hands	7	4
		Fits In Cupholder	8	1
		Doesn't Tip Over	9	8
		Attractive	10	10
		Fits In Mini-Cooler	11	11
		Doesn't Spill When You Drink	12	9
Material	Characteristics	No Dents	13	11
		Doesn't Change Shape	14	8
		Is Not Heavy (Light)	15	7
		Does Not Break When Dropped	16	6
		Clear	17	10
		Reusable	18	6
		Recyclable	19	11
		Stays Cool	20	5
		No Sharp Edges	21	10
MISC	Cost	Inexpensive	22	5

Figure 4.6   Rankings of Desired Features

Customer Competitive Analysis

Customer Requirements | Ranking | | AM G C

Container	Lids	Doesn't Leak	1	2
		Interchangeable Lids	2	6
		Freshness	3	6
		Open/Close Easily	4	3
		Sealed When Purchased	5	6
		Resealable Lid	6	8
	Shape	Doesn't Slip Out of Hands	7	4
		Fits In Cupholder	8	1
		Doesn't Tip Over	9	8
		Attractive	10	10
		Fits In Mini-Cooler	11	11
		Doesn't Spill When You Drink	12	9
Material	Char.	No Dents	13	11
		Doesn't Change Shape	14	8
		Is Not Heavy (Light)	15	7
		Does Not Break When Dropped	16	6
		Clear	17	10
		Reusable	18	6
		Recyclable	19	11
		Stays Cool	20	5
		No Sharp Edges	21	10
MISC	Cost	Inexpensive	22	5

Figure 4.7   Customer Competitive Analysis

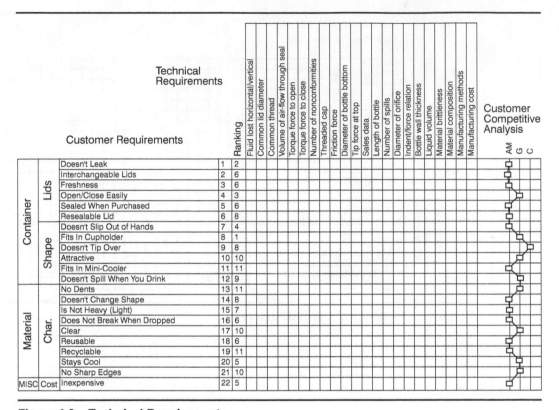

Figure 4.8   Technical Requirements

phrased in terms that were measurable and could be translated into organizational actions. Several of the wants needed two or more technical requirements to make them happen (Figure 4.8).

5. *Determine the direction of improvement for the technical requirements.*

   During the meetings discussing technical requirements, those involved also discussed the appropriate specifications for the technical requirements. They were able to identify how those technical requirements could be improved. For instance, for the comment "Fluid lost horizontal/vertical," the appropriate direction of improvement for this is "less," denoted by a downward arrow (Figure 4.9).

6. *Determine the operational goals for the technical requirements.*

   AM Corp. identified the operational goals that will enable them to meet the technical requirements (Figure 4.10).

7. *Determine the relationship between each of the customer wants and the technical requirements: How does action on (a change in) a technical requirement affect customer satisfaction with the recorded want?*

Figure 4.9 **Direction of Improvement**

The technical requirements (column headers) are:

- Fluid lost horizontal/vertical
- Common lid diameter
- Common thread
- Volume of air-flow through seal
- Torque force to open
- Torque force to close
- Number of nonconformities
- Threaded cap
- Friction force
- Diameter of bottle bottom
- Tip force at top
- Sales data
- Length of bottle
- Number of spills
- Diameter of orifice
- Indent/force relation
- Bottle wall thickness
- Liquid volume
- Material brittleness
- Material composition
- Manufacturing methods
- Manufacturing cost

Customer Requirements with Ranking (and a secondary value column):

		Customer Requirement	Ranking	
Container	Lids	Doesn't Leak	1	2
		Interchangeable Lids	2	6
		Freshness	3	6
		Open/Close Easily	4	3
		Sealed When Purchased	5	6
		Resealable Lid	6	8
	Shape	Doesn't Slip Out of Hands	7	4
		Fits In Cupholder	8	1
		Doesn't Tip Over	9	8
		Attractive	10	10
		Fits In Mini-Cooler	11	11
		Doesn't Spill When You Drink	12	9
Material	Char.	No Dents	13	11
		Doesn't Change Shape	14	8
		Is Not Heavy (Light)	15	7
		Does Not Break When Dropped	16	6
		Clear	17	10
		Reusable	18	6
		Recyclable	19	11
		Stays Cool	20	5
		No Sharp Edges	21	10
MISC	Cost	Inexpensive	22	5

Customer Competitive Analysis columns: AM, G, C

The team members at AM Corp. studied the relationship between the customer wants and the technical requirements (Figure 4.11). They used the following notations:

- A strong positive correlation is denoted by the value 9 or a filled-in circle.
- A positive correlation is denoted by the value 3 or an open circle.
- A weak correlation is denoted by the value 1 or a triangle.
- If no correlation exists, then the box remains empty.
- If there is a negative correlation, the box is marked with a minus sign.

8. *Determine the correlation between the technical requirements.*

   The team members recorded the correlation between the different technical requirements in the "roof" of the QFD "house" (Figure 4.12). This triangular table shows the relationship between each of the technical requirements. Once again, they used the same notations:

   - A strong positive correlation is denoted by the value 9 or a filled-in circle.
   - A positive correlation is denoted by the value 3 or an open circle.

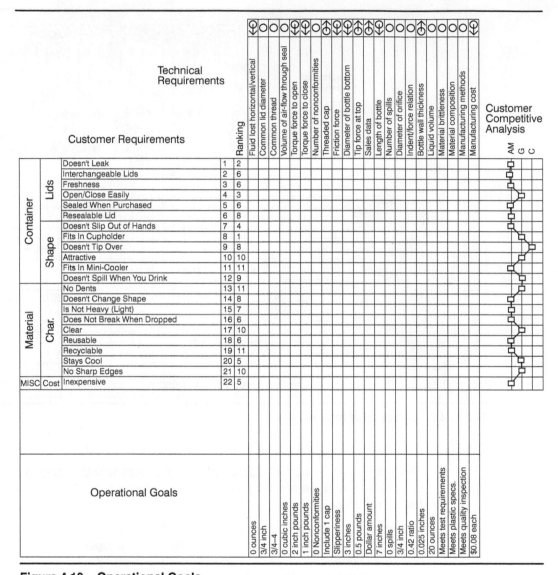

Figure 4.10   Operational Goals

- A weak correlation is denoted by the value 1 or a triangle.
- If no correlation exists, then the box remains empty.
- If there is a negative correlation, the box is marked with a minus sign.

9. *Compare the technical performance with that of competitors.*
   At this point AM Corp. compared their abilities to generate the technical requirements with the abilities of their competitors. On the matrix this information is shown in the technical competitive assessment (Figure 4.13)

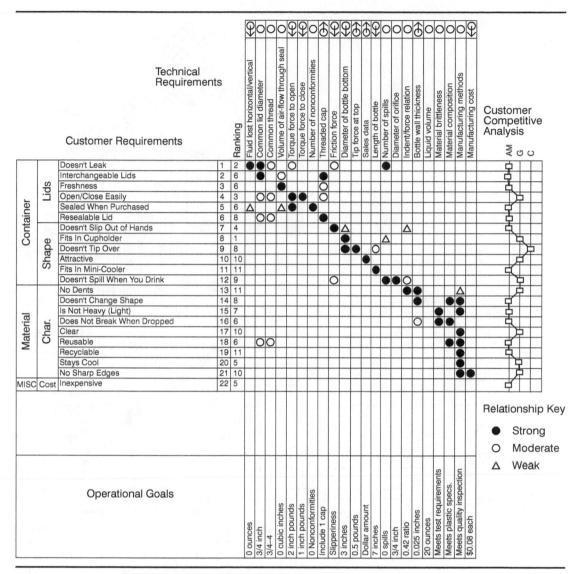

Figure 4.11   Correlation Matrix

10. *Determine the technical importance.*

At this point the matrix is nearly finished. In order to analyze the information presented, the correlation values for the wants and the how's are multiplied by the values from the rankings of the $100 test. For example, for the first column a ranking of 2 for "doesn't leak" is multiplied by a value of 9 (for "strong correlation"), for a total of 18. To this value the ranking of 6 (for "sealed when purchased") is multiplied by a value of 1 (for "weak correlation"). The grand total for the column is 24 (Figure 4.14).

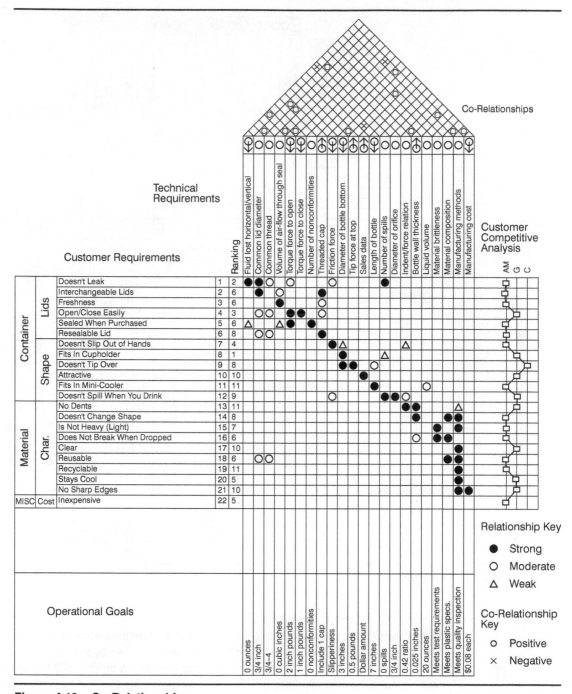

Figure 4.12   Co-Relationships

110

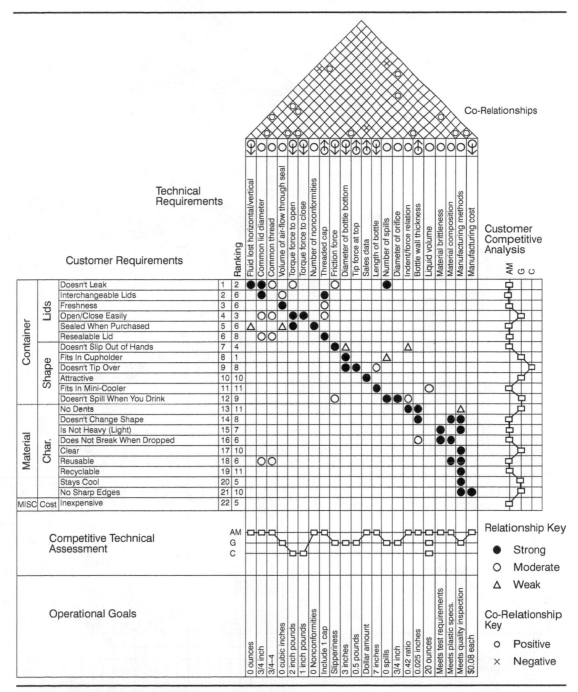

Figure 4.13 Competitive Technical Assessment

111

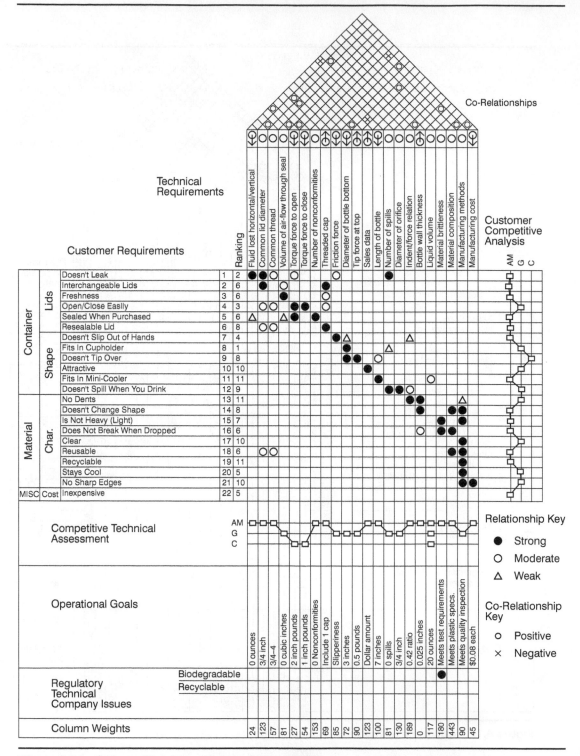

Figure 4.14　Column Weights and Regulatory Issues

Voice of Customer	Product or Service attributes	Process attributes	Process controls
Fits standard cup holder	Base must not exceed 3 inches in width	Design with attributes in mind	Design review
Does not leak at any time in any position	Lid diameter no larger than ¾ inch; 4 threads per cap; 1 in.-1b. force to close	Design with attributes in mind	Design review
Easy to open and close	No more than 2 in.-lb. of force to open; no more than 1 in.-lb. of force to close	Design with attributes in mind	Design review
Does not slip out of drinker's hand easily	Bottle diameter limited to 3 inches	Design with attributes in mind	Design review

Figure 4.15   QFD Results Matrix

11. *Add regulatory and/or internal requirements if necessary.*
    Here, any rules, regulations, or requirements set forth not by the customer but by some other agency or by government were identified and recorded (Figure 4.14).
12. *Analyze the QFD matrix.*
    What did the customer want? How is this supported by customer rankings and competitive comparisons? How well is the competition doing? How does our company compare? Where will our company's emphasis need to be?

AM Corp. studied the matrix they created and came to the following conclusions (Figure 4.15). In order to satisfy their customers and maintain a competitive advantage, they will have to focus their efforts on designing a sports drink bottle that:

- Fits into a standard cup holder in a vehicle (that is, the base must not exceed 3 inches).
- Does not leak at any time in any position. The technical requirements associated with this requirement are that (a) the lid diameter must be no larger than ¾ inch; (b) there must be 4 threads per cap; and (c) the cap must be reapplied at 1 in.-lb. of force maximum.
- Is easy to open and close, requiring no more than 2 in.-lb. of force to open and 1 in.-lb. of force to close.
- Does not slip out of the drinker's hand easily. For this reason, the bottle diameter should be no smaller than 3 inches, and the type of plastic utilized must have the appropriate coefficient of friction.

Asking customers what they want, need, and require is a time-consuming process. As was seen in the quality function deployment example (Example 4.4), translating customers' wants into an organization's how's is paramount to the success of any organization seeking to align its products and services with the processes that provide them with what the customer wants. Organizations that ignore the relationship between what a customer wants and how the organization is going to provide that want can never be effective. Leadership and strategic planning are crucial links in this alignment process. Leadership, covered in Chapter 5, sets the direction for the organization. Leaders point to the customer wants on which the organization is going to focus. Strategic planning, covered in Chapter 6, is the agenda in which leadership clearly lays out how the organization will meet their customers' wants. Strategic plans link the goals and objectives of projects and day-to-day activities in the organization with the customer wants. Because quality is such an essential dimension for a product or service, it is critical that an organization take the customer information they have gathered, translate it into organizational actions, and disseminate this information throughout the organization. Strategic plans, guided by good leadership, allow this to happen.

HOW CAN THE MALCOLM BALDRIGE NATIONAL QUALITY AWARD CRITERIA HELP AN ORGANIZATION CREATE A MORE EFFECTIVE CUSTOMER FOCUS?

The Malcolm Baldrige National Quality Award (MBNQA) criteria encourage organizations seeking to improve their effectiveness to expand their focus beyond the product itself or the moment the service takes place. The portion of the MBNQA criteria specifically devoted to studying the relationship an organization has with its customers is Section 3.0, Customer and Market Focus. This section examines how an organization determines the requirements and expectations of its customers and its marketplace. The section asks questions related to an organization's ability to gather information from the customer. In essence, the criteria ask organizations how well they know their customers. It focuses on the need to determine both short- and long-term requirements, expectations, and preferences of their customers and markets to ensure the relevance of their current and future products and services. Organizations must respond to questions related to how they determine the satisfaction level of their customers as well as how they plan to build and strengthen their relationships with customers in order to retain their current business and develop new opportunities for the future. Accompanying these questions, assessment guidelines provide guidance to firms seeking ways in which to improve their overall organizational effectiveness.

Creating an organization that has at its core the ability to sustain an unwavering focus on the customer is not a task for a single department within the organization. Effective organizations recognize and respond to the fact that the involvement of the entire firm is necessary. To create a customer-focused organization, the activities in a variety of areas must be integrated. When evaluating how effective an organization is at creating a customer focus, ask these questions based on the MBNQA criteria:

1.0 Leadership

- Has the organization's leadership listened to the voice of the customer and determined what the business needs to focus on?
- Do the organization's leaders support the focus through their actions?
- Have they converted policies to actions?
- Are the organization's leaders holding people accountable for actions that support the focus?

2.0 Strategic Planning

- Has the organization created and implemented a plan that supports the company's focus?

3.0 Customer and Market Focus

- Does the organization know what the customers' perceptions of value are?
- Has this information been shared throughout the organization?
- How has this information been put to use?

4.0 Measurement, Analysis, and Knowledge Management

- Is the organization gathering, analyzing, disseminating, and encouraging the use of information related to the customer?

5.0 Workforce Focus

- Does the organization's reward system reinforce customer-focused behavior?

6.0 Process Management

- Have the key processes supporting a customer focus been identified and improved?
- Has the organization selected improvement projects that support the needs of the customer?
- Are efforts focused on why the situation is undesirable from the customer's point of view?

7.0 Results

- Is the organization using business results to measure the gap between what they said they were going to do and what they actually did?
- Has the focus on what customers value resulted in improvements to the organization's business results?

LESSONS LEARNED

Creating an unwavering focus on the customer requires action in all aspects of an organization. Effective organizations talk with their customers in order to understand how their customers perceive value for their products and services. Then the organizations translate this information into technical specifications and align their business processes in order to provide the production service according to specifications. Effective organizations recognize that if they do not take care of the customer, someone else will.

1. Customers willingly change from supplier to supplier in search of better service or courtesy or better product availability or features.
2. Quality is a customer determination based on their experience.
3. Value is the attributed or relative worth or usefulness of a product or service.
4. Effective organizations find out what their customers want by asking them.
5. Quality function deployment is a technique used to capture the voice of the customer.
6. Effective organizations are proactive and use the information they gather from the customer to make changes to the way they do business.

Are You a Quality Management Person?

Quality management people see other people as their customers. Their interactions with them reflect their respect and concern for their customers' requirements, needs, and expectations. Check out the quality management person questions and find out whether or not you are a quality management person. How do you answer them on a scale of 1 to 10?

Have you asked your customers what they want?

| 1 | 2 | 3 | 4 | 5 | 6 | 7 | 8 | 9 | 10 |

Do you actively listen to your customers?

| 1 | 2 | 3 | 4 | 5 | 6 | 7 | 8 | 9 | 10 |

Do you know how your customers define quality?

| 1 | 2 | 3 | 4 | 5 | 6 | 7 | 8 | 9 | 10 |

Do you know how your customers define value?

| 1 | 2 | 3 | 4 | 5 | 6 | 7 | 8 | 9 | 10 |

Do you understand the difference between satisfaction and perceived value?

| 1 | 2 | 3 | 4 | 5 | 6 | 7 | 8 | 9 | 10 |

Do you understand how your customer views the process you interact with?

| 1 | 2 | 3 | 4 | 5 | 6 | 7 | 8 | 9 | 10 |

Have you ever been your own customer?

| 1 | 2 | 3 | 4 | 5 | 6 | 7 | 8 | 9 | 10 |

Do you treat all your customers, internal and external, well?

| 1 | 2 | 3 | 4 | 5 | 6 | 7 | 8 | 9 | 10 |

Do you practice good customer service?

1	2	3	4	5	6	7	8	9	10

Do you have an unwavering focus on your customers' requirements, needs, and expectations?

1	2	3	4	5	6	7	8	9	10

Chapter Questions

1. Why would an organization want to be effective at maintaining a customer focus?
2. What must an organization do to maintain a customer focus?
3. What are the benefits of maintaining a customer focus?
4. How do you, as a customer, define quality?
5. Describe a product or service experience you have had based on Feigenbaum's definition.
6. Describe the difference between satisfaction and perceived value.
7. Describe an experience you have had with a product or service based on customer satisfaction and perceived value.
8. Describe an organization you have witnessed creating an unwavering focus on the customer. How did they do it?
9. Describe the principal parts of a quality function deployment (QFD) matrix.
10. How is each of the principal parts of a QFD matrix created? What does each part hope to provide to the users?
11. Why would a company choose to use a QFD?
12. Describe how you would begin creating a QFD.

Section 3.0: Customer and Market Focus

The Customer and Market Focus section of the MBNQA criteria addresses how an organization determines requirements, expectations, and preferences of customers and markets. Also examined is how the organization builds relationships with customers and determines the key factors that lead to customer acquisition, satisfaction, and retention.

The key goals of this section are to:

1. Determine the primary markets and customers of Remodeling Designs, Inc. and Case Handyman Services.
2. Understand the methods that Remodeling Designs, Inc. and Case Handyman Services use to determine customers' critical success factors.
3. Determine how Remodeling Designs, Inc. and Case Handyman Services build and maintain relationships with customers.

This section was created by Rita Wendeln, Lisa Koebbe, Erich Eggers, Kelly Eggers, Karen Dillhoff, and Donna Summers.

QM Customer and Market Focus Questions

3.1 CUSTOMER AND MARKET KNOWLEDGE

1. How do you determine customer critical success factors and their relative importance to customers' purchasing decisions? How do you measure the effectiveness of this?
2. How do you target customers, customer groups, and/or market segments?
3. How does your marketing group define opportunities in the marketplace?
4. How do you keep your customer communication methods current with changes in technology?
5. How do you keep your customer communication methods current with business needs and direction?
6. How do you ensure that your strategic plans and leadership focus on what is important to the customer and the market to remain competitive? How is this information shared?
7. What changes have been made to the way you do business based on customer input?

3.2 CUSTOMER RELATIONSHIPS AND SATISFACTION

8. How do you build relationships with your customers to increase repeat business and positive referrals?

9. What is your complaint management process? How do you use this information to improve your organization?
10. How do you measure and track customers' critical success factors and use this information to predict future business from repeat customers and referrals?
11. How do you keep your approaches to determining customer satisfaction current with business needs and directions?
12. How do you follow up with customers on services to receive prompt feedback on which you can take action?

3.0 Customer and Market Focus

The Customer and Market Focus category examines how your organization seeks to understand the voices of customers and of the marketplace. Also examined is how your organization builds relationships with customers and determines key factors that lead to customer acquisition, satisfaction, and retention, and to business expansion.

3.1 CUSTOMER AND MARKET KNOWLEDGE

This item examines how your organization determines current and emerging customer requirements and expectations in order to remain competitive in the marketplace.

3.1a Customer and Market Knowledge

Remodeling Designs made the business decision to focus their efforts on acquiring high-end remodeling jobs. Once this decision was made, Remodeling Designs limited their advertisements to middle- to upper-class neighborhoods in Dayton. In contrast, Case Handyman services a variety of different areas in Dayton. In order to focus future marketing advertisements, both companies track where their job leads come from.

Remodeling Designs determines key product and service features by applying the Golden Rule. Employees at Remodeling Designs and Case Handyman complete a job with the same service that they would expect to receive. Customer satisfaction is measured by Customer Quality Audits, which are completed at the end of each job.

Remodeling Designs and Case Handyman distinguish themselves from competitors by going beyond customer expectations. The employees are required to wear uniforms and leave the work area tidy at the end of the day. They encourage constant communication with customers via e-mail, phone, and a notebook at the job site. Remodeling Designs provides a customer binder that contains the details of the job.

To stay competitive in a rapidly changing environment, members of the Remodeling Designs staff benchmark their peers in other states. Employees attend a Remodelers' Executive Roundtable twice a year. Different remodeling companies critique one another, share business strategies and marketing ideas, and establish goals. They also attend a remodelers' show and kitchen-and-bath show to stay current with new products and new ideas/designs. In addition, employees are required to attend seminars to keep up with continuing education.

3.2 CUSTOMER RELATIONSHIPS AND SATISFACTION

This item examines your organization's processes for building customer relationships and determining customer satisfaction, with the aim of acquiring new customers, retaining existing customers, and developing new opportunities.

3.2a Customer Relationships

Customer relationships begin with the initial customer inquiry. Following a phone conversation, an information folder is sent to the prospective customer. Kelly accompanies Mike on the first visit to the job site to talk to the customer and establish a relationship. The process map provided in Figure 1 describes the process for selling a new job.

Remodeling Designs maintains relationships with clients through mailings. Christmas cards are sent to previous and current customers. After the September 11 tragedies, American flag stickers were sent to customers along with a compassionate letter. Personalized gift baskets are sent to each customer during the remodeling of their home.

Remodeling Designs make themselves readily available to their customers. There are multiple means of communication available. Customers can call the office, send e-mail, or write concerns in a notebook; they can contact Remodeling Designs employees outside the office as well.

3.2b Customer Satisfaction Determination

Upon completion of a job done by Remodeling Designs and Case Handyman, customers fill out a Customer Quality Audit. Customers rate their satisfaction with the company's employees and the quality of the job done. Remodeling Designs have improved their method of gathering customer satisfaction measurements by changing their questionnaire from yes-no answers to actual numerical measures. Each audit is reviewed during a staff meeting, and problems, if any, are addressed at that time. Although the audits are filed in a binder, any problems written in the comments section are not tracked. Therefore, no data is kept on recurring problems.

Customer satisfaction is also measured by a company's record with the Better Business Bureau (BBB). Remodeling Designs and Case Handyman make sure that they are in good standing with the BBB.

Both Remodeling Designs and Case Handyman track warranty issues by a written Warranty Work Order. The Warranty Work Order is initiated by a customer complaint call. The warranty issue is filed in a Warranty Binder and recorded in the computer for each job. Although they keep a log for each job, they do not track how many warranty claims are made or how often a problem has recurred. This information could be used to prevent future use of bad products.

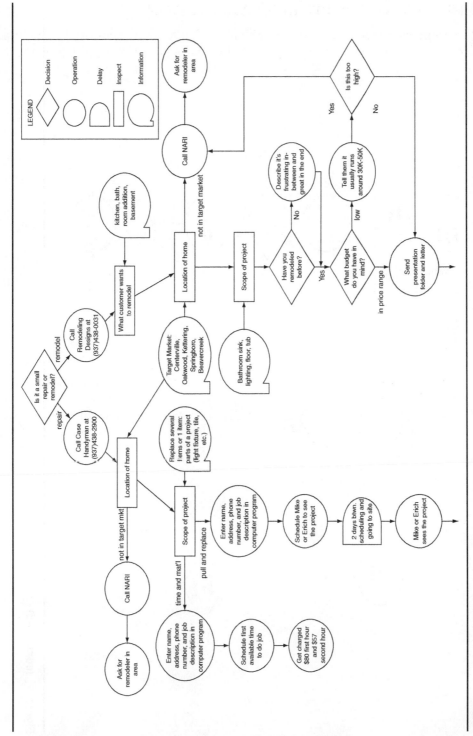

Figure 1   Process Map for Selling a New Job

121

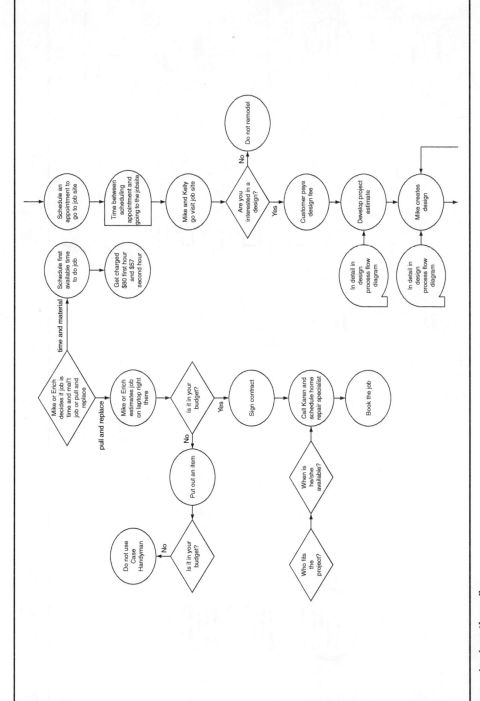

Figure 1 (continued)

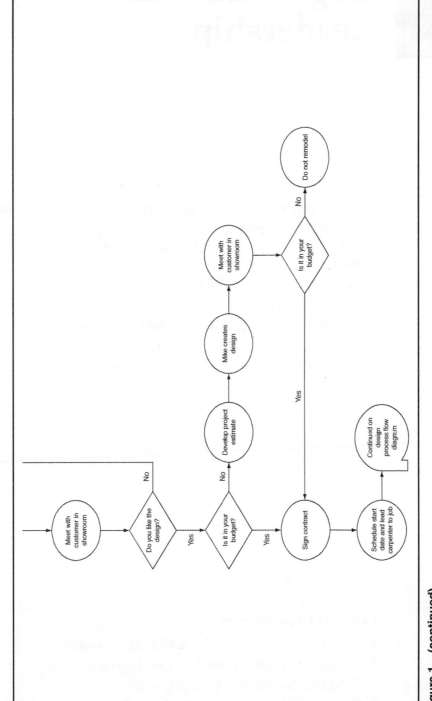

Figure 1 (continued)

5 Organizational Leadership

What is expected of leaders?

How do effective leaders create an organizational culture focused on creating value for their customers?

How do effective leaders translate the vision and mission into day-to-day activities?

What are the different leadership styles of effective leaders?

How do effective leaders manage by fact and with a knowledge of variation?

How do effective leaders practice ethical behavior and good citizenship?

How can the Malcolm Baldrige National Quality Award criteria help an organization's leadership become more effective?

Lessons Learned

Learning Opportunities

1. To understand what traits define effective leadership
2. To understand how leaders guide effective organizations
3. To become familiar with leadership styles
4. To understand how to manage by fact and with a knowledge of variation

POSITION TITLE   *President*

SPECIFIC DUTIES AND RESPONSIBILITIES   *The President will have full profit and loss responsibility with the goal of maintaining market leadership, increasing market share, and improving bottom line performance through the use of aggressive manufacturing management. Specific responsibilities include:*

- *Lead the development and implementation of strategic goals to maximize the business unit's growth and profitability.*
- *Develop an organizational culture with an aggressive mind-set toward product quality, continuous improvement, lean manufacturing, and bottom-line results.*
- *Ensure that the necessary financial resources and leadership are in place to provide accurate, timely, clear financial information for reporting and management decision purposes.*
- *Maintain market awareness of trends and expectations, including government regulations and policies, factoring this information into the company's competitive marketing plans. Play an active role in working with customers and distributors. Represent the company to suppliers, industry associations, governmental agencies, and the community.*
- *Develop and lead a management team of qualified professionals across all functional areas, working together toward a common goal. Ensure that the appropriate plans are in place to develop, motivate, and reward employees.*
- *Integrate strategic plans across all functional areas, including finance, marketing, sales, and manufacturing.*
- *Demonstrate strong communication skills with the ability to translate strategic concepts into clear, concise language for implementation at all levels. Practice a balanced management approach capable of leading change and driving an already successful company to its highest level of potential.*

WHAT IS EXPECTED OF LEADERS?

The job description on page 125 for the position of president is very explicit in its delineation of what is expected of leaders in their corporation. The list of requirements is daunting, to say the least. In his text *Out of the Crisis*, W. Edwards Deming states:

> The aim of leadership should be to improve the performance of man and machine, to improve quality, to increase output and simultaneously to bring pride of workmanship to people. Put in a negative way, the aim of leadership is not merely to find and record failures of men, but to remove the causes of failure: to help people to do a better job with less effort.

Leaders provide direction and purpose for an organization. Leaders are the people who realize that there is a compelling reason to change the way an organization operates. They are the ones who lead the way in the change process. The commitment to quality management can come only from leaders (Figure 5.1). Leaders outline the vision for the organization. They establish the goals and objectives based on this vision. From them flows the rest of the change process. It is leaders who provide the ultimate encouragement and reasoning behind why people would want to do their jobs. People perform on the job for a variety of reasons. They may perform for the financial incentives involved, the prestige, or the possibility of future rewards. Sometimes people even perform out of fear—fear of job loss, fear of retribution, fear of unpleasant consequences. The most effective reason people work is that they want to. According to Dr. Deming, leadership's overall aim should be to create a system in which everyone takes joy in his or her work. So how do leaders do that?

When an organization is committed to an ongoing focus of providing value for customers, its leadership must align this expectation at three levels: the overall organizational goals and objectives, the organization's processes, and the performance of individuals during their day-to-day activities. Leadership must define systems and standards that support the overall organizational goals and objectives. Under their guidance, employees work within the system to create value for the organization's customers.

HOW DO EFFECTIVE LEADERS CREATE AN ORGANIZATIONAL CULTURE FOCUSED ON CREATING VALUE FOR THEIR CUSTOMERS?

A *culture* is defined as "a pattern of shared beliefs and values that provides the members of an organization with rules of behavior or accepted norms for conducting operations." It is the philosophies, ideologies, values, assumptions, beliefs, expectations, attitudes, and norms that knit an organization together. In a unified organizational culture, the philosophies, ideologies, values, assumptions, beliefs, expectations, attitudes, and norms are shared by all employees. Effective leaders apply a missionary zeal to the job of creating an organizational culture focused on creating value for their customers. They use their boundless energy to put into practice the skills and techniques they have learned. When establishing and maintaining a culture, they realize that they must be committed to that culture and knowledgeable about what it will take to support the desired culture. Perhaps most importantly, effective leaders realize that they must visibly practice and support the desired

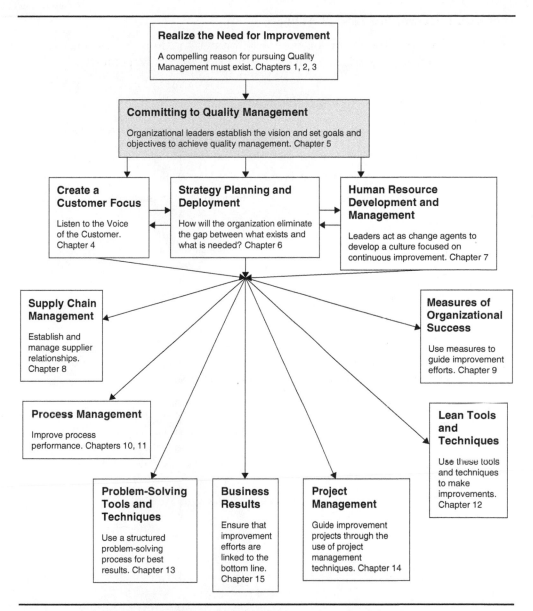

Figure 5.1   Achieving Organizational Success

culture on a daily basis. In order to encourage people to create value for their organization, effective leaders provide value-driven leadership. They ensure the alignment of the needs, wants, and expectations of the customer with the strategic objectives of the organization with the goals of the department with the goals of the jobs within the department with the goals of the individuals (Figure 5.2).

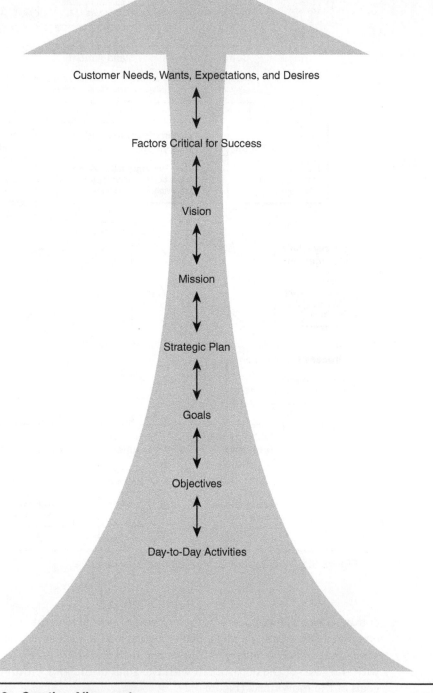

Customer Needs, Wants, Expectations, and Desires

Factors Critical for Success

Vision

Mission

Strategic Plan

Goals

Objectives

Day-to-Day Activities

Figure 5.2   Creating Alignment

How Do We Know It's Working?

Effective leaders understand how important it is to get to know employees. At JQOS, leadership makes it a point to talk personally and make eye contact with each employee on a regular basis. This means greeting employees while walking through the plant or offices, stopping to exchange a few words, and being present for both shifts. These meetings create an opportunity for interaction and enable leaders to hear employees' ideas and concerns. To get at employees' thoughts, he asks questions.

The owner sees his job as making sure that work, in terms of sales, gets into the shop and that his employees are responsible for the making of quality product. As a leader, he wants buy-in on his vision for the organization: grow the company. To do this, his employees will have to grow themselves, obtaining new skills and performing new jobs. These one-on-one meetings provide him with the opportunity to set performance expectations, provide feedback and encouragement, and track progress. He also has the opportunity to clarify any questions his employees may have. Lots of changes have been taking place over the past two years, and his one-on-one meetings have reduced the resistance to change.

The owner also serves as a decision maker. By walking through the plant and interacting with employees on a regular basis, he can see what is working and what is not. As he assesses the situation and analyzes problems, he can make adjustments to JQOS's strategy, goals, and objectives. He involves employees in designing solutions to the issues that arise. In many cases, he just asks the questions to get them started. If there are signs of frustration, he steps in and provides guidance and direction. This ensures that his employees feel good about the changes they helped put in place and the accomplishments they achieved. How does he know it's working? He has started to hear his own words coming back at him as his employees seek solutions to achieving the strategic plan.

Leaders are responsible for managing change in the organization. For this reason, leaders must learn to use their talents, training, and skill to serve two purposes: as a workforce motivator and as a decision maker (Figure 5.3). As a workforce motivator, leaders set performance expectations and clearly communicate them. While tracking the progress being made, they provide feedback. As befits a key decision maker, leaders also have an overall understanding of the situation. They assess the situation and analyze the associated problems. From this, they set a strategy that is aligned with the goals and objectives of the organization. This strategy guides them as they evaluate potential solutions and make effective decisions. They follow through with their decisions and deploy their strategy. Achieving these two purposes ensures that the goals of the individuals are aligned with the goals of the job, which are aligned with the goals of the department, which are aligned with the goals of the organization (Figure 5.4).

EXAMPLE 5.1   Responding to the Challenge

What follows is a job seeker's response to the job description that begins this chapter. Note how this job seeker's past accomplishments align with the needs of the organization.

Dear Sir,

Thank you for contacting me concerning the Division President position. As the very thorough job description shows, running a multi-million-dollar company is a complex assignment. I have made a careful study of the position description you provided and see many areas that correspond with the experience I have. I see the specific duties and responsibilities as follows:

Leadership:	Provide a focus and a cohesive motivation for the division's personnel.
Culture:	Create an organizational identity that results in enthusiasm and pride of workmanship.
Reporting:	Use performance measures to provide timely information on project status and financial data.
Market:	Assess, define, and support customer needs and market trends in order to determine future direction.
Market Presence:	Support distributors and demonstrate market leadership.
Employee Relations:	Ensure that the organization's most important resource, employees, are able to perform to the best of their abilities.
Communication:	Establish effective communication throughout the organization.

Having worked my way up through the ranks at my previous employment, I have had the opportunity to see how many of the key departments in our corporation operated. I believe that I have experience that matches well with this job description. For instance:

Leadership:	During my tenure as Vice President, and later President, through the use of performance measures, I aligned our organization's strategic objectives with departmental activities and held managers accountable for improving their operations long before the terms *lean manufacturing, Six Sigma,* and *value-added process mapping* became fashionable. I believe that it is very important for the organization to understand the company's strategic direction and goals. It is easy to become distracted by the daily operational activities. I pride myself on keeping the management team focused by showing them the order that exists within the chaos of daily events.
Market Awareness:	I view the company's strategic position and relevant goals in the market to be the focal point for the management team. Similarly, feedback from customers and intelligence about competitors are critical when implementing the strategic plan. During my 17 years with my former company, I was instrumental in guiding and directing its growth from a fledgling U.S. operation with 50 employees to a $100 million company with 390 employees and three U.S. facilities. As part of our expanding global influence, I often worked with customers throughout Europe, the United States, and Asia.
Employee Relations:	An organization's most powerful resource is its employees. Treating employees respectfully and clearly communicating expectations are critical to employee well-being. Employee involvement, through information sharing and problem-solving efforts, improves overall organization success.

I hope this letter provides some insight into why I am very interested in the President position. I look forward to discussing this opportunity in greater detail with you in the near future.

Sincerely,

Workforce Motivator

Set Performance Expectations
Set realistic goals
Set realistic standards
Establish checkpoints to measure progress
Establish measures of performance to measure progress
Encourage creativity

Communicate Effectively
Clearly communicate goals and objectives
Handle resistance to change
Listen and act upon employee concerns
Inspire cooperation and commitment
Encourage discussion of ideas

Provide Feedback
Create feedback system
Provide timely feedback
Provide constructive criticism
Manage conflict

Track Progress
Set priorities
Approve solutions
Encourage improvement
Manage differences
Provide recognition and rewards

Decision Maker

Assess Situation
Provide clarification
Ask questions
Simplify confusing situations

Analyze Problems
Formulate well-structured problem statements and descriptions
Isolate root causes
Look beyond the obvious
Avoid jumping to conclusions

Define Strategy
Create a strategy based on situation assessment and problem analysis
Clearly align strategy with goals and objectives of the organization

Evaluate Potential Solutions
Explore potential solutions
Think outside the box
Listen to others' ideas

Make Effective Decisions
State goals clearly
Align goals with strategy
Formulate alternatives
Involve others in decision making

Deploy Strategy
Follow through!
Take steps to make planned changes
Use measures of performance to monitor progress
Use checkpoints to track progress
Follow up with delegated tasks

Figure 5.3   Leadership Roles

132

CHAPTER 5

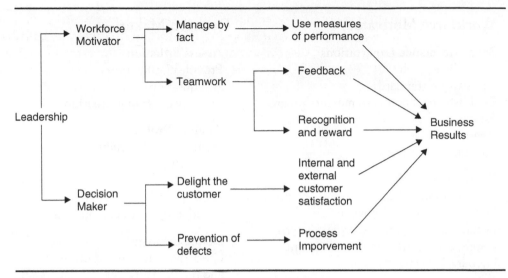

Figure 5.4   From Leadership to Business Results

Effective leaders "walk the talk." In other words, their actions display to other members of the organization what is expected. As Dwight Eisenhower so aptly stated, "They never listened to what I said; they always watch what I do." Effective leaders recognize that their employees closely watch their actions. Because employees focus on the activities of their leaders, leadership involvement in key activities is crucial when encouraging people to work in a manner that creates value for their firm.

How Do We Know It's Working?

JQOS leadership "walks the talk." They communicate with their employees verbally, in writing, and by their actions. Quarterly meetings are held with all employees to discuss key performance measures. Weekly data, such as sales numbers, number of returns, late shipments, internal quality issues, and number of good shafts out the door, are posted next to the employee break room. These numbers are important for employees to track since it affects their quarterly bonus. Through this bonus, JQOS communicates financially with their employees. Employees receive a bonus when quality parts shipped totals more than $1.5 million in a single quarter. The bonus is a percentage relationship based on their particular salary versus the total dollars spent on salaries during the quarter.

JQOS leadership shows that they mean business by their actions, too. During the first few weeks of new ownership, employees were surprised when they were ordered to turn off their machines under unsafe conditions. In the past, production needs meant ignoring unsafe conditions and actions. JQOS's new management dealt with safety issues first, returning guards to machines, opening up paths to fire exits, replacing/restoring fire extinguishers, reprimanding employees for unsafe acts, improving lighting, replacing hot fuses, redesigning transformers, mopping up spills and slick spots, making the break room and restrooms more hygienic, and getting live electrical cords out of coolant pans.

They made changes on the production side too. To improve quality, welds on shafts were no longer permitted. Fluorescent lighting was added to the existing mercury vapor lamps. This eliminated shadows on the parts and permitted employees to monitor their production more carefully, thus improving part quality. In the past, as machines wore out, employees developed all sorts of work-arounds. Over time, as money became available, the new leadership replaced several machines, starting with one machine that squeaked so badly that, when operating, the sound level in the plant approached OSHA's 90-decibel maximum permissible level. Grinders did not exhibit good repeatability, so money was spent to scrape the ways, replace the bearings, and replace switches and handles. New gauges have been purchased and setup boards created. Each chip-making machine now has a vacuum that sucks chips directly into a barrel. These chips are taken to a newly acquired "puck" machine that compresses the chips into recyclable discs of metal. A general cleanup of the plant is under way, too. Based on employee suggestions, lathes, once repaired, are painted blue and white; grinders, burnt orange; and mills, gray. New machine purchases ensured that more capable machines with newer CNC controls increased productivity and agility in the plant.

The shop isn't the only part of the company that is experiencing changes. The front office is involved in the improvement process, too. Customer quoting and order confirmation is now computerized and online. Basic services, such as payroll and accounting, have been either standardized procedurally or outsourced. By shopping around, better sources were found for everything from printer paper to shipping boxes and pallets. Establishing safety policies and improving plant conditions resulted in a $50,000 per year savings in insurance costs.

Leadership at JQOS is intent on creating a positive organizational culture focused on their vision of "Grow the Business." Their actions have spoken loudly to employees, telling them that there is a new way of doing business, one that respects employees and customers. Over time, as JQOS employees watched their leaders, they became involved in helping achieve the goals and objectives set in the strategic plan. By finding more and more ways to improve their business, they have benefited in the form of a better working environment and great bonuses.

Effective leaders can encourage people to create value by supporting key activities that drive organizational success. These key activities, whether they be providing customers with easy access to support information or developing a new manufacturing process, require investment of both financial resources and time. Effective leaders show their commitment and involvement by providing financial support, becoming involved in improvement efforts, and making sure that employees have time available in their workday to pursue the necessary activities. Effective leaders know what needs to be done, and they provide the tools to enable the workforce to do it.

EXAMPLE 5.2 Customer-Focused Leadership

Effective leaders establish a culture focused on customers by signaling to the organization what is important. To do this, effective leaders must be accessible to those customers. Direct customer contact provides insight into what the organization's customers expect on a day-to-day basis. Leaders must talk to their customers and then, through their personal actions, make sure that the customer information reaches and is utilized by the appropriate people within their organization.

The president and owner of the chain of grocery stores first mentioned in Example 4.1 understands the importance of customer service. His grocery stores have become known for their exemplary customer service and interesting array of product offerings. One of his customer-oriented policies requires that department managers hold no meetings during the busy shopping hours of 11:00 A.M. to 1:00 P.M. and 4:00 P.M. to 6:00 P.M. daily. During these times managers must be in the store, strolling the aisles, meeting and talking with their customers. During these times the president can also be seen strolling the aisles, greeting customers—often by name.

This policy is extended twice a year, during the three days before Thanksgiving and also the three days before Christmas. Every department manager is available and present in his or her department all day to aid customers during these busy shopping days. These managers are expected to ensure that their shelves remain stocked with critical items, pass out samples and tastes of products, help customers locate the items they seek, prevent aisles from being blocked by carts, and engage in other activities that create a more pleasant shopping experience for their customers. In keeping with his policies, the president and the stores' two vice presidents are also actively interacting with customers. Indeed, it is not unusual to see them bagging groceries and helping customers to their cars.

Effective leaders establish policies that support a focus on meeting customer needs and expectations the first time and every time. Effective leaders also understand the importance of living these policies.

EXAMPLE 5.3 Walking the Talk

While on a tour of a compressor manufacturing facility, a visitor was surprised by the cleanliness of the facility. The equipment, walls, and ceiling were painted white. The facility was brightly lit by overhead lighting as well as by windows in the walls. Although many machining operations occurred in the plant, the equipment itself was spotless, so clean that leaning against a machine would not result in any oil, grit, or grime coming off on your hands or clothing. The visitor wondered how this could be possible. Just the week before, he had toured a different facility that manufactured virtually the same product, and he found it to be dingy. Even the walls, painted gray like the machines, could not disguise the years of accumulated grime. He was further surprised to learn that both facilities were approximately the same age. He continued to wonder how all of this cleanliness could be possible, until his tour guide identified the man approaching their group as the president of the company. On his way toward them, the man stopped twice, once to pick up a stray piece of paper from the floor and deposit it in the nearby trash can and the other time to take a cloth from his pocket and wipe some dust off a work area. As employees watched, the president communicated through his own actions his own expectations for his employees' behavior.

HOW DO EFFECTIVE LEADERS TRANSLATE THE VISION AND MISSION INTO DAY-TO-DAY ACTIVITIES?

Effective leaders provide guidance, often in the physical "walk the talk" type of format. These leaders make sure that their actions correspond with a written vision or mission statement. *Vision statements describe where leadership sees the organization in the future.* Like the stars used as guides by ancient mariners, vision statements provide a star to chart a course by. *Mission statements are usually more specific and provide greater detail about the firm's objectives.* Figure 5.5 provides example wordings of a variety of vision and mission statements.

Effective leaders communicate the values of the organization to their employees by translating the vision and mission into day-to-day activities. To do this effectively, leaders talk with customers, identify the organization's critical success factors, and share this information about the things the organization absolutely must do well in order to attract and retain customers. Creating alignment, as shown in Figure 5.2, is essentially policy deployment, the step-by-step process of translating the organization's vision and mission into strategies supported by goals and objectives that in turn become work activities for the employees. Leaders ensure that the organization's vision, mission, strategies, goals, and daily activities remain focused on these critical factors.

In order to create alignment, effective leaders practice Dr. Deming's point: *implement leadership.* Alignment is not achieved by wishful thinking. Effective leaders are the first to ask:

Does the employee know what he/she is supposed to do?
Does the employee have the means to determine whether he/she is doing the job correctly?
Does the employee have the authority and the means to correct the process when something is wrong?

Effective leaders are also the first to realize that if any of these questions go unanswered, or if the answer to any of them is no, the fault probably lies with leadership, not with employees. To help create alignment, effective leaders carefully design employee policies supporting the links between the critical success factors identified by their customers and the employees' day-to-day activities. Effective leaders set realistic goals for their employees and give timely rewards to those who meet these goals. Leaders seek out employees' ideas and actively support the good ones. Very few people are truly good listeners, but effective

Hospital

As a major teaching institution, we will continue to be a leader in providing a full range of health care services. Working together with our medical staff, we will meet and exceed our patients' needs for high-quality health care given in an efficient and effective manner.

Grocery Store

The mission of our store is to maintain the highest standards of honesty, trust, and integrity toward our customers, associates, community, and suppliers. We will strive to provide quality merchandise consistent with market values.

Student Project

Our mission is to provide an interesting and accurate depiction of our researched company. Our project will describe their quality processes compared with the Malcolm Baldrige Award standards.

Pet Food Company

Our mission is to enhance the health and well-being of animals by providing quality pet foods.

Customer Service Center

Meet and exceed all our customers' needs and expectations through effective communication and intercompany cohesiveness.

Manufacturing

To produce a quality product and deliver it to the customer in a timely manner while improving quality and maintaining a safe and competitive workplace.

Manufacturing

To be a reliable supplier of the most efficiently produced, highest-quality automotive products. This will be done in a safe, clean work environment that promotes trust, involvement, and teamwork among our employees.

Figure 5.5   Examples of Mission Statements

leaders have learned to listen to their employees. It helps to keep in mind that the same letters spell the words "listen" and "silent." Leaders also place a high priority on expanding employee capabilities and responsibilities through education and training. Other employee-related activities of effective leadership are covered in Chapter 7.

Effective leaders have a very powerful tool that helps to link the activities of individuals within an organization with the customers served by the organization: the strategic plan. Strategic plans enable leaders to communicate effectively to all levels of the organization. Strategic plans are also the vehicle for translating the organization's vision, mission, and strategic objectives into deployable action plans. The strategic plan and its impact on creating organizational alignment will be covered in greater detail in Chapter 6.

WHAT ARE THE DIFFERENT LEADERSHIP STYLES OF EFFECTIVE LEADERS?

Four primary styles of leading exist: directing, consultative, participative, and delegating (Figure 5.6). Effective leaders realize that different situations call for different leadership styles. A leader must be skilled at recognizing the need for and then putting to use the appropriate style.

The *directing* style of leadership is an autocratic style. It is normally used when the leader must make a unilateral decision that must be followed without comment or question from the rank and file. The need to use the directing style of leadership may come about because the leader has more knowledge of the situation or because the decision affects the common good of the organization. This style of leadership style may be something as simple as a "no horseplay will be tolerated on the job" rule or as complex as "here are the customers the organization will focus on." Leaders using this style expect to be obeyed. Those who work for them have very little, if any, input.

The *consultative* style of leadership is used when a leader is seeking input from those working under him or her. It is considered a more developing style of leading because it encourages participation. Leaders may be facing a customer issue that requires the input of a specialist, such as a chemist; or they may be facing an employee issue that requires input from all employees, such as the selection of a health insurance carrier. When using this style of leadership, the leader seeks the advice, suggestions, and input of those around him or her but still remains the final decision maker.

When using the *participative* style of leadership, a leader assigns work to the employees, provides guidance during the work process, and makes a decision based on the conclusions of the employees working on the task. Unlike the consultative style of leadership, in this situation the leader is more likely to take the word or work of the employee(s) as the final decision in the matter.

Participative	Consultative
Provides guidance	Seeks input, advice, and suggestions
Gets involved only when necessary	Makes final decision
Accepts work and decisions of employees	Recognizes employees for their contributions
Helps others analyze and solve problems	
Recognizes employees for seeking support	

Delegating	Directing
Assigns responsibility	Engages in unilateral decision making
Assigns authority	Expects employees to follow orders
Provides minimal input	Gives information about what to do
Provides recognition	Gives information about how to do it
Verifies work	Gives information about why it should be done
Recognizes employees for accepting responsibility	Recognizes employees for following directions

Figure 5.6 Leadership Styles

The *delegating* style of leadership is the style in which the leader takes the smallest role. In this style of leadership, the leader essentially tells the employee or team what needs to be done, assigns the responsibility, and provides the individual or team with the authority to get the job done. The individual or team, having been given both the responsibility and the authority, completes the work with minimal input from the leader. In this style of leadership, the leader checks to verify the successful completion of the assignment and participates only if necessary.

Leaders must take care to wisely match their leadership style to the situation. Using a directing style when a developing style is called for may leave an employee feeling stifled and asking the question, "Why was I even asked to participate?" The directing style of leadership is the most heavy-handed. Applied incorrectly, it could stifle employee creativity and motivation. Conversely, employing a delegating style instead of a problem-solving style of leadership when more guidance is needed will leave the employee feeling stranded. The delegating style allows the employee the most freedom. However, if the needs of the project have not been clearly communicated, or if the employee is inadequately prepared to do the work, both the leader and the employee will be dissatisfied with the result. Leaders who try to apply the developing or problem-solving style in all situations may discover that employees make decisions contrary to key policies and procedures. Some circumstances require that established rules be followed under a directing style of leadership. Effective leaders feel comfortable adopting each leadership style as needed in appropriate situations.

HOW DO EFFECTIVE LEADERS MANAGE BY FACT AND WITH A KNOWLEDGE OF VARIATION?

Effective leaders know that people must be included in their organization's decision-making processes. They also know that sharing information is critical to making good decisions. Management by fact involves understanding the how's and why's of a situation before taking action. Management by fact requires an appreciation for and an understanding of the key systems of an organization. Effective leaders realize that management of systems requires knowledge of the interrelationships between all of the components within the system and the people who work in it. Information can sometimes be misleading. When leaders manage by fact, they use objective evidence to support their decisions. Objective evidence is not biased and is expressed as simply and clearly as possible. More importantly, it is traceable back to its origin, whether that be a customer, an order number, a product code, a machine, or an employee. Effective leaders remember to ask Dr. Deming's favorite question: "How do we know?" Knowing the answer to this question verifies the source of the information and its importance to and relationship with the issue at hand. Having this knowledge means having the facts available to support a plan of action.

Chapter 2 discussed the two sources of variation in a process as identified by Dr. Shewhart. Controlled variation, the variation present in a process due to the very nature of the process, can be removed from the process only by changing the process. Uncontrolled variation, on the other hand, comes from sources outside the process. Normally it is not part of the process and can be identified and isolated as the cause of a change in the behavior of the process. To manage with an understanding of variation means that a leader would recognize the type of variation present and respond accordingly. As Dr. Deming clarified, it would be a mistake to react to any fault, complaint, mistake, breakdown, accident, or shortage as if it came from a special cause when in fact there was nothing special at all—that is, it came from random variation due to common causes in the system. It is also a mistake to attribute to common causes any fault, complaint, mistake, breakdown, accident, or shortage when it actually came from a special cause. An understanding of variation enables leaders to make the right choices and deal appropriately with problems as they arise.

EXAMPLE 5.4   Managing by Fact and with a Knowledge of Variation

A sales manager has three salespersons, each covering different areas of the country. The performance of each salesperson is reviewed monthly. These reviews are often followed by praise for the highest achiever and a "pep talk" for the underachievers.

In response to upper management's request, the sales manager has graphed the performance of each salesperson versus the sales goal for the past 15 months. The resulting graphs (Figure 5.7) justified his suspicions concerning performance.

However, it is only when these graphs are combined with factual information and an understanding of variation that the true patterns of performance begin to emerge. Figure 5.8 shows how control limits were calculated based on the variation present in each individual's process. These limits show what each process is actually capable of, compared to the previous graph, which showed performance versus specifications (goals set). For more information on how these limits were calculated, please see the discussion on control charts in Chapter 13.

Note the cycle present in the Great Lakes Region sales. When the variation present in the process is studied, it appears that this salesperson expends efforts to get sales, but then his performance drops off until he gets too close to the established goal. Another fact to add to this analysis is that this firm specializes in automotive components and this sales territory encompasses the lucrative territories of Michigan, Ohio, and Indiana. Making sales in this region is relatively easy.

Note the Mid-Region (Kentucky, Tennessee, North and South Carolina) performance. A large amount of variation is present in the process. This region has been experiencing much growth in the automotive field. The large amount of variation present could signify a lackadaisical attitude on the part of this salesperson or a need for more training.

Note the performance of the Southern Region salesperson. The first graph, Figure 5.7, shows that his performance is consistently below the goal. Yet, if he were managed by facts with a knowledge of variation, it would become apparent that something good is going on here. Based on this graph, this salesperson has achieved high performance on several occasions. Further investigation reveals that although his region encompasses Florida, Georgia, and Alabama, areas not prime for automotive work, he has been able to generate significant sales. These sales, in the area of medical and computer devices, are helping the manufacturing firm diversify and move away from its dependence on automotive business. This salesperson should be congratulated for his efforts.

Using control charts and managing with a knowledge of variation provides an entirely different outlook for the performance of these individuals.

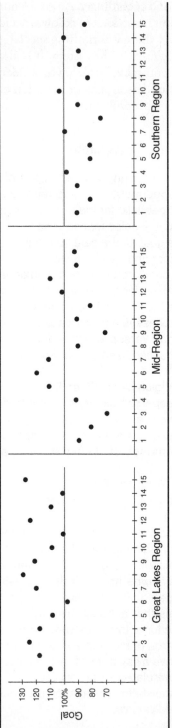

Figure 5.7 Performance Graphs of Each Salesperson for the Past 15 Months (Example 5.4)

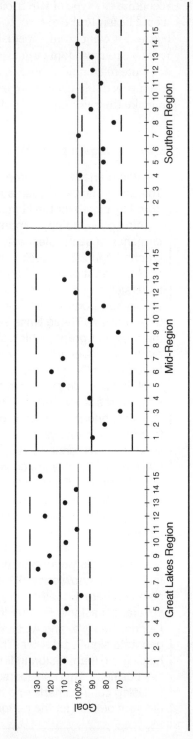

Figure 5.8 Graphs with Control Limits (Example 5.4)

HOW DO EFFECTIVE LEADERS PRACTICE ETHICAL BEHAVIOR AND GOOD CITIZENSHIP?

Leaders of effective organizations are expected to be role models for their employees. This role includes ethical behavior and good citizenship. Ethical behavior among leadership has taken a much greater emphasis since the collapse of Enron and Arthur Anderson. Leaders are expected to conduct themselves appropriately in all stakeholder transactions and interactions. Within their organization, they promote, enable, and ensure legal and ethical behavior. Leaders understand that they are accountable for their employees' behavior. They monitor and respond to any breaches in ethical behavior.

Organizations and their leaders have social responsibilities that include protection of public health and safety and the environment. Leaders should review activities conducted by the organization in order to determine their impact on society and the environment. Effective organizations address the adverse impacts on society and the environment of their products, services, and operations. They also anticipate future concerns. The strategic plan should document how the organization will lessen their impact and cite specific goals and objectives. Chapter 3 discusses ISO 14000 as a system to respond to environmental concerns and responsibilities. The strategic plan may include programs to improve the disposal of products or the reduction of waste created.

Effective leaders support their community. They are involved in its activities, and so is their workforce. They contribute to these communities in a variety of ways. Ideas include making safety or health care programs or information available to the public, strengthening community services, and collaborating on efforts to enhance the community. Employees, including leadership, should consider community service as part of their citizenship. Effective organizations go beyond mere compliance and create an organization that emphasizes ethical behavior and social responsibility.

HOW CAN THE MALCOLM BALDRIGE NATIONAL QUALITY AWARD CRITERIA HELP AN ORGANIZATION'S LEADERSHIP BECOME MORE EFFECTIVE?

The Malcolm Baldrige National Quality Award (MBNQA) criteria encourage organizations to examine how their leadership guides the organization. Senior leaders of the organization are asked how they direct their organization as well as how they review their organization's performance. Leaders must describe how they set and deploy organizational values, how they steer the organization toward its short-term and long-term goals, how they create value for their customers, and how they set and monitor performance expectations. Senior leaders are also expected to describe how they create an environment supporting organizational agility, employee empowerment, and employee learning. Leaders are reviewed in terms of their ability to translate priorities, strategies, and objectives into organizational improvement and opportunities for innovation. They are asked to show how they create alignment between the customers' value propositions and the organization's strategies, goals, objectives, and day-to-day activities. The criteria also address the organization's responsibilities to the public by focusing on how the leaders ensure that the organization understands and deals with the impact on society of their products, services, and operations. It asks leaders how they address their responsibilities to the public by practicing good citizenship. Employee involvement in the organization's key communities is also encouraged, as are ethical business practices.

The task of leadership is to create and maintain an effective organization with an unwavering focus on the customer. Leadership activities must not be confined to simply one single aspect of the organization. When evaluating how effective an organization's leadership has been, ask these questions based on the MBNQA criteria:

Leadership

- Does the organization have the understanding, commitment, and participation of its leadership?
- How will the organization's leadership maintain effective communication?
- How do the actions of the organization's leadership support an effective organization?
- Does leadership communicate priorities and hold people to them?
- Does leadership embrace change and act as agents of change?
- How do leaders establish the vision and values for the organization?
- How do leaders personally promote legal and ethical behavior?
- How do leaders create a sustainable organization?
- How do leaders govern the organization with an eye to accountability and protection of stakeholder interests?
- How do leaders ensure fiscal accountability?
- How do leaders encourage activities that support key communities?

Strategic Planning

- How do leaders communicate values, directions, and expectations throughout the organization?
- How will leadership communicate the strategic plan and its necessary actions to all levels of the organization?
- Has the organization's leadership converted the strategic objectives into deployable action plans?
- Does the organization's leadership use key performance measures to monitor the organization's action plans?
- Does the organization's leadership understand the need for alignment among the strategic objectives, goals, and action plans?
- How does the organization's leadership ensure alignment between the organization's strategic objectives, goals, and action plans?

Customer and Market Focus

- Are the leaders of the organization readily available to meet with customers?
- Do the leaders of the organization understand how vital it is to meet regularly with customers?

Measurement, Analysis, and Knowledge Management

- How do the organization's leaders analyze and use information?
- What measures of performance do the leaders use?

- How is feedback received concerning measures of performance?
- Does leadership share data and information openly at all levels of the organization?

Workforce Focus

- Do leaders support and implement policies and reward systems suited to the organization?
- How does leadership maintain effective communication with all levels of the organization?
- How does leadership create an environment supportive of innovation, agility, and organizational as well as employee learning?
- Does leadership recognize and reward employees who prevent or solve problems?
- Does leadership practice and support consensus decision making whenever possible?
- Does leadership search out employees' ideas and actively support the good ones while explaining why some ideas are rejected?
- Does leadership place a high priority on expanding employee capabilities and responsibilities?

Process Management

- How does leadership integrate process improvement into organizational activities?
- How do leaders use organizational performance findings to improve their own leadership effectiveness?
- Does leadership place a high priority on problem prevention?
- Does leadership focus on processes rather than actions?

Results

- How do leaders review organizational performance and capabilities to assess organizational success and competitive performance?
- How do leaders use measures of performance and key indicators established in the strategic plan to measure the success of the business in terms of its results?

LESSONS LEARNED

As the job description in the chapter introduction shows, in order to be a leader in today's global marketplace, an individual must be able to guide an organization by aligning the needs, wants, and expectations of the customer with the processes and day-to-day activities of its employees. A leader must be able to react to different situations with the appropriate leadership style in order to motivate employees. Effective leaders establish trust with their employees. They provide them with clear direction and guidance. They stay close to their employees, seeking information directly from those closest to the situation. They concentrate on the essential, establishing priorities and monitoring progress. A leader must have and utilize an understanding of managing by fact with a knowledge of variation when making decisions. A leader must create a customer-focused culture that works to create value for customers each and every day in all activities. With good leadership, positive changes can occur; without it, nothing is possible.

Effective leaders share key characteristics. They are optimistic and kind, with a preference for personal contact. While they display independent judgment, they are loyal team players, backing up their employees. Leaders display a characteristic calmness under stress. This trait enables them to face bad news squarely. They are decisive, able to combine a broad understanding of the whole picture and still see the detail. They define their jobs and the cultures of the organizations they work in.

1. According to Dr. Deming, the "aim of leadership should be to improve the performance of man and machine, to improve quality, to increase output and simultaneously bring pride of workmanship to people."
2. Leaders are change agents.
3. Leaders set the direction and the culture for an organization.
4. Leaders serve as workforce motivators and as key decision makers.
5. As a workforce motivator, leaders set performance expectations, communicate effectively, provide feedback, and track progress.
6. As a decision maker, leaders assess situations, analyze problems, define strategies, evaluate potential solutions, make effective decisions, and deploy strategy.
7. Effective leaders translate the vision and mission into day-to-day activities.
8. The four different styles of leadership are directing, consultative, participative, and delegating.
9. Effective leaders manage by fact and with a knowledge of variation.

Are You a Quality Management Person?

Each person has the potential to be an effective leader. Check out the quality management person questions and find out whether or not you are a quality management person. How do you answer them on a scale of 1 to 10?

Do you respect yourself and your abilities?

1	2	3	4	5	6	7	8	9	10

Do you know what you value?

1	2	3	4	5	6	7	8	9	10

Does what you value guide your life?

1	2	3	4	5	6	7	8	9	10

Do your values and your goals and objectives match so that what you accomplish in life is important to you?

1 2 3 4 5 6 7 8 9 10

Do you have a positive outlook on your life?

1 2 3 4 5 6 7 8 9 10

Do you take responsibility for your actions?

1 2 3 4 5 6 7 8 9 10

Do you treat other people with respect?

1 2 3 4 5 6 7 8 9 10

Do you treat other people fairly?

1 2 3 4 5 6 7 8 9 10

Have you done something in the past 12 months that you are proud of?

1 2 3 4 5 6 7 8 9 10

Has your satisfaction increased in the past 12 months?

1 2 3 4 5 6 7 8 9 10

Are you making better decisions than you were 12 months ago?

1 2 3 4 5 6 7 8 9 10

Do you take part in continuous learning in areas that interest you?

1 2 3 4 5 6 7 8 9 10

If someone asked you how you have changed since last year, would you like the answer?

1 2 3 4 5 6 7 8 9 10

Have you changed in an education or training manner in the past 12 months?

1	2	3	4	5	6	7	8	9	10

Do you live up to the standards and expectations you have set for yourself?

1	2	3	4	5	6	7	8	9	10

Chapter Questions

1. Why is it important to translate the vision and mission into day-to-day activities?

2. From your own experience, describe an example where the directing leadership style was appropriately used.

3. From your own experience, describe an example where the directing leadership style was inappropriately used.

4. From your own experience, describe an example where the developing leadership style was appropriately used.

5. From your own experience, describe an example where the developing leadership style was inappropriately used.

6. From your own experience, describe an example where the problem-solving leadership style was appropriately used.

7. From your own experience, describe an example where the problem-solving leadership style was inappropriately used.

8. From your own experience, describe an example where the delegating leadership style was appropriately used.

9. From your own experience, describe an example where the delegating leadership style was inappropriately used.

10. Your boss has just called a meeting to discuss lost work orders and customer order information. She has cited messy desks and a poorly used filing system as the root cause for the lost information. Since this information is critical for the firm, she has decreed that each desk and office will be cleaned and organized, starting today. As you leave the meeting, you notice that her office is very disorganized and that she has made no effort to clean it up. Will you make the change? Why? Why not? What would you do as an effective leader to make the change possible?

11. Your start time at work is 7:10 A.M. However, the boss doesn't come in until 7:30, even though he, too, has a start time of 7:10 A.M. It has come to his

attention that people are not arriving at 7:10. He wants to see the arrival time change to 7:10 in the future. Will you make the change? Why? Why not? What would you do as an effective leader to make the change possible?

12. Which of Dr. Deming's 14 points deal with leadership? Give an example of where you have seen these points applied (or where they need to be applied).

13. What does it mean to manage by fact and with a knowledge of variation?

14. When making a decision, what will you keep in mind to avoid the situation that happened in Example 5.4?

QM

Section 1.0: Leadership

The Leadership section of the MBNQA criteria addresses how senior leaders guide the organization in setting directions and seeking future opportunities. Primary attention is given to how senior leaders set and deploy clear values and high performance expectations that address the needs of all stakeholders, including the organization's responsibilities to the public and the organization's practice of good citizenship.

The key goals of this section are to:

1. Examine the key aspects of Remodeling Designs, Inc. and Case Handyman Services leadership and the roles of their senior leaders, with the aim of creating and sustaining a high-performance organization.
2. Examine how Remodeling Designs, Inc. and Case Handyman Services fulfill their public responsibilities and encourage, support, and practice good citizenship.

This section of the review was created by Joanie Zucal, Maria Dominique, Renee Cooper, Erich Eggers, Mike Cordonnier, and Donna Summers.

QM   Leadership Questions

1.1 ORGANIZATIONAL LEADERSHIP

1. What are some key organizational values for Remodeling Design/Case Handyman? How do you communicate those values to your employees? Can you give us an example that demonstrates this translation?
2. What specific action items and responsibilies have you developed in order to carry out your mission statement?
3. Who are your stakeholders? How were they identified?
4. How do you create and balance value for all stakeholders?
5. What are your key critical success factors? (What are the three to five things that you must absolutely do well in order for your business to prosper?)
6. Identify one long-term and one short-term goal pertaining to each critical success factor.
7. How do you communicate with employees and align the goals of the company with those of the employees to ensure that everyone is on the same page?
8. How do you reinforce the pursuit of these goals in the organization?
9. How do you know you are accomplishing your critical success factors? (What indicators have you established?)

10. How do you translate performance review findings into opportunities for improvement and innovation?
11. How do these findings trickle down to other parts of the organization, including suppliers? How do the findings affect your leadership skills?
12. From a leadership perspective, give an example of how you empower your employees and provide organizational and employee learning.
13. From a leadership point of view, how do you promote innovative actions? How agile is your organization in implementing changes? (Give an example.)
14. How does your culture support an environment for empowerment, innovation, organizational agility, and organizational and employee learning?
15. Are you as leaders willing to support the focus of the organization through your actions? (Give an example.)
16. How are you as leaders holding people accountable for actions that support and those that do not support the culture?
17. How have your leaders been able to focus the organization's efforts on converting policies into actions, which have in turn translated to the bottom line?
18. What are some of your key business processes?
19. Are you as a leader converting policies to actions relating to your key business processes?
20. Where do you gather information, and how do you use it for your decision-making process?

1.2 PUBLIC RESPONSIBILITY AND CITIZENSHIP

21. Are you knowledgeable about any present circumstances in the community that are affected by your company?
22. How do you anticipate public concerns about your current and future services?
23. How do you prepare for these concerns in a proactive manner?
24. Can you give us an example of how your company participates in community service?

1.0 Leadership

1.1 ORGANIZATIONAL LEADERSHIP

1.1a Senior Leadership Direction

At Remodeling Designs/Case Handyman Services the leaders encourage innovative ideas by setting a friendly environment. There is no "us versus them" attitude between leaders and employees. Both Mike and Erich deploy the values by "walking the talk." They lead by example. The leaders treat everyone (and expect all employees to do the same) using their main value, the Golden Rule: "Treat others as you would like to be treated." Another key value that they translate to their employees is to never second-guess the customer because the customers are always right. In the design process of any project, the leaders, lead carpenter, and customer all discuss what the customer's expectations are for the project, including materials to be used and design specifications. During this process Remodeling Designs photocopies or places samples of everything that the customer is choosing in a project book. They do this to protect themselves if the customer changes his or her mind halfway through the project or if any confusion arises as to what the customer chose. Before Remodeling Designs begin the remodeling, they discuss this book with the customer for a final verification of the materials and design. If during or after the remodeling the customer thinks he should be getting or should have gotten something more than what Remodeling Designs/Case Handyman has remodeled, and if the problem is due to a flaw in communication, the company simply gives the customer what he wants, within monetary limits. If the customer does not like the outcome of the remodeling at all and wants it all changed, Remodeling Designs will change it for an additional charge.

As leaders, Mike and Erich set their organizational values and performance expectations by aligning them with the organization's mission. Although this process is correct, it has one major flaw. The written mission statement, which is what the customers read, is not the same as the one that Mike and Erich translate to their employees. Remodeling Design's written mission statement is: "To improve our clients' lives by providing the highest quality Design/Build remodeling and handyman services. To enable our team members to flourish and prosper." In contrast, the translated verbal mission statement is "to make people happy and continue to be profitable." This translation does not specify what is required to make people "happy." Each employee may interpret "happy" in his or her own way, creating misalignment with the goals of the company. As leaders, Erich and Mike need to communicate the written statement to the employees.

The short- and longer-term directions are set in group meetings with the entire organization present. They hold this meeting not only to communicate with employees about goals and to make sure that the employees understand the goals. They also want to make all employees feel valued in a family-oriented company. At the end of every year, the leaders set monetary and budget goals for the next year. Every month these yearly long-term goals are broken down and revised into monthly short-term goals. In the monthly meetings they discuss projected cost, projected revenue, and any other possible issues that may occur that month. At the weekly meetings they assess their monthly goals to make sure they are heading in the right direction. They also communicate the progress details of the current projects.

To create a balanced value for their customers and other stakeholders, the leaders make sure that there is clear communication between the two. In order to accomplish clear communication with the customer, the lead carpenter becomes fully in charge of the project. Any problems, concerns, or questions are directed to the lead carpenter, who will answer them to the best of his abilities. If Remodeling Designs/Case Handyman Services must deal with suppliers or subcontractors, they must communicate with the lead carpenter only. This allows Remodeling Designs/Case Handyman Services to maintain control of the tone previously established with the customer. Furthermore, to maintain balance, Remodeling Designs/Case Handyman Services try to use only one or two suppliers or subcontractors for different aspects of the job. This helps establish a close-knit relationship between Remodeling Designs/ Case Handyman Services and their suppliers and subcontractors. Also, Remodeling Designs/Case Handyman Services try to give all pertinent information or necessary tools to suppliers or subcontractors to make the job run smoothly for all stakeholders.

As leaders, Mike and Erich have complete faith that their lead carpenters are going to do an extraordinary job. They allow them to make all of the decisions pertaining to the project on which they are currently working. The lead carpenter is the primary source of communication with the customer; anything the customer wants or needs goes by the lead carpenter.

Every year Remodeling Designs/Case Handyman Services send their employees to conventions to learn about new ideas for improvement and to become aware of what is changing in the remodeling world. This promotes employee learning and innovation. Furthermore, if the leaders or any employees are aware of any education or training that is relevant to their work, the leaders strongly consider sending those key employees to learn and gain these value-added skills. For example, one of the lead carpenters had heard about a man who made beautiful cabinetry; he thought it would be a great idea to learn how to make these cabinets and use them in remodeling jobs. The leaders sent the lead carpenters to learn this new trade, and they now use these skills in some of their remodeling jobs. If a carpenter would like to learn a more challenging skill, the leaders may place the carpenter with other employees who have mastered the skill, or they may send the carpenter to training outside the company. The leaders are also willing to send employees to school to learn grammar and technical writing skills.

1.1b Organizational Performance Review

The organizational review that the senior leaders of Remodeling Designs/Case Handyman Services perform includes not only how well their organization is currently performing, but also how well it is moving toward the future. Erich Eggers and Mike Cordonnier identified the following performance measures that they use to indicate how well they are conducting business: Valued Client Feedback Forms; repeat and referral rates; benchmarking; profitability, on individual jobs as well as overall; and employee reviews.

After the completion of each project, Remodeling Designs/Case Handyman Services send their client a survey called a Valued Client Feedback Form. Mike and Erich examine the returned feedback forms, and, depending on what they discover from them, they take the appropriate action. Regardless of whether the survey comes back with positive and/or negative results, Mike and Erich translate that feedback to everyone in the organization. These forms are the indicators that tell Mike and Erich how well their business is performing.

The second measure Mike and Erich have established is the number of repeat clients and the number of referral clients they have. This data shows them that they have performed exceptional service that has resulted in satisfied customers, because either their clients return for additional service or the clients recommend Remodeling Designs/Case Handyman Services to someone else.

The next performance measure that they use is benchmarking. By attending various conferences, most of which are national, Mike and Erich are able to compare their performance against different sets of standards or against the performances of best-in-class companies. With the information provided by the comparison, Mike and Erich can determine how to improve their business. This practice is important for them to continue, but it does not address how well Remodeling Designs/Case Handyman Services are competing locally. It is critical that the leaders be aware of what their competitors in the Miami Valley area are doing. Without this knowledge, Remodeling Designs/Case Handyman Services cannot remain up-to-date on their competitors' developments and changes.

Careful accounting, record keeping, and job tracking enable Remodeling Designs to study each job's profitability. Since the successful completion of each job translates to overall organizational profitability, Remodeling Designs and Case Handyman are careful to note whether the expectations and estimates for a job correctly predicted the actual costs associated with the job. When a discrepancy is uncovered, steps are taken to determine why it occurred and how it can be prevented in the future.

The final performance indicator that Mike and Erich use to judge how well their company is performing is yearly employee reviews. Before Erich and Mike conduct the review, they require the employee to answer eight questions regarding his own job. Then they provide the employee with three additional questions that will be discussion topics during the review. This allows the employee to prepare for the review. Once the review has been completed, Mike and Erich examine the written and verbal feedback they have received, and they take appropriate action. The employee review findings indicate to Mike and Erich what they need to change in order to help their employees perform their jobs better. Once this is brought to Mike's and Erich's attention, they make the necessary improvement and/or innovative changes. This in turn demonstrates to the employee(s) that Mike and Erich care deeply about them. It also shows that Mike and Erich understand that they need their employees as much as their employees need them.

As the performance reviews are completed, Mike and Erich translate their findings into opportunities for improvement and innovation. They communicate their results to all employees and then form groups to work on the projects. These improvement and innovative projects are part of Mike's and Erich's continuous improvement plan. By using the Plan-Do-Study-Act problem-solving technique, the groups implement permanent changes. Doing this has allowed them to remain technologically advanced in every process they perform, such as design, construction methods and tooling, billing, and communication with their customers and other stakeholders. If the change or improvement is minor and does not require a group effort, Mike and Erich work with the employee(s) whom the change involves; then they inform the entire company of this modification. Also, depending on the severity or complexity of the problem, Mike and Erich will let the specific employee(s) solve the problem on their own either by improving the existing idea or by developing an entirely new, innovative idea. In this way, Mike and Erich have demonstrated that they translate their review findings into action plans that are deployed throughout Remodeling Designs/Case Handyman Services to ensure organizational alignment.

1.2 PUBLIC RESPONSIBILITY AND CITIZENSHIP

1.2a Responsibilities to the Public

Remodeling Designs/Case Handyman Services have few services or operations that would have any impact on society in a positive or negative manner. As a remodeling firm and handyman service, they deal with few hazardous chemicals or waste. The one problem that was of some concern for Erich Eggers and Mike Cordonnier was the disposal of paint products. Their response to this issue was to take the leftover paint, paint thinner, or any other hazardous material back to the establishment where the material was purchased. These establishments dispose of the hazardous chemicals properly in accordance with the regulatory and legal requirements for disposal. The next waste issue was the construction waste produced by a remodeling or handyman service. Mike and Erich stated that they first try to reuse or recycle the waste in different ways instead of disposing of it in a landfill. If the material cannot be recycled or reused, it is disposed of in a landfill or dump site for construction products only.

One concern is that Mike and Erich are not anticipating public concerns with their services in a proactive manner. They are waiting for the issues to surface, and then they will tackle them. Remodeling Designs/Case Handyman Services have accomplished ethical business practices with all their stakeholders, who are customers, employees, subcontractors, and suppliers. It is very easy for these companies to apply ethical business practices when dealing with their employees because they are able to translate their core values to those employees. When dealing with customers, Remodeling Designs/Case Handyman Services set the tone for how they will run the service, and the majority of the customers follow the set tone. When it comes to subcontractors and suppliers, Remodeling Designs/ Case Handyman Services state that they try to use only two subcontractors or suppliers for each type of business needed in order to establish a relationship and set a tone that is conducive to ethical business practices.

1.2b Support of Key Communities

Remodeling Designs/Case Handyman Services are excellent in meeting the criteria of this section of the Malcolm Baldrige Award. Remodeling Designs/Case Handyman Services actively participate in Rehaborama, which is revitalization and rehabilitation of historical districts in the community. Their participation in Rehaborama is on a volunteer basis, and Remodeling Designs donates the materials used in the revitalization. Remodeling Designs has also designed projects for different aspects of the community free of charge. In addition, they participate in Rampathon, in which a ramp is built for someone with a disability who has no financial means for obtaining the ramp. Remodeling Designs/Case Handyman Services sponsor Centerville OM/Destination Imagination teams and girls' softball and soccer teams. Remodeling Designs/Case Handyman Services also make donations to various local school activities. Last, any cabinets or other materials removed from homes, or left-over scrap cabinets or materials that are in good condition, are sometimes donated to community shelters or the Boy Scouts.

We feel that overall, Remodeling Designs/Case Handyman Services are doing a great job of supporting and influencing key communities through their practice of good citizenship.

6 Strategic Planning

What is strategic planning?

How do strategic plans give an effective organization a competitive edge?

How do strategic plans support customer satisfaction and perceived value?

How are strategic plans created?

How do effective leaders create organizational alignment through strategy deployment?

How can the Malcolm Baldrige National Quality Award criteria help an organization's leadership create and deploy effective strategic plans?

Lessons Learned

Learning Opportunities

1. To become familiar with how strategic plans guide effective organizations
2. To understand how strategic plans support customer satisfaction and perceived value
3. To understand how to create strategic plans
4. To understand the importance of strategic plan deployment

"Cheshire-Puss," she began, rather timidly, as she did not at all know whether it would like the name: however, it only grinned a little wider. "Come, it's pleased so far," thought Alice, and she went on.

"Would you tell me, please, which way I ought to go from here?"

"That depends a good deal on where you want to get to," said the Cat.

"I don't much care where—" said Alice.

"Then it doesn't matter which way you go," said the Cat.

"—so long as to get somewhere," Alice added as an explanation.

"Oh, you're sure to do that," said the Cat, "if you only walk long enough."

Lewis Carroll, Alice's Adventures in Wonderland, *Chapter 6*

QM

WHAT IS STRATEGIC PLANNING?

What must a business do to survive? Grow? Beat the competition? Satisfy customers better? Be the low-cost producer? Provide higher quality? Furnish faster delivery? Design a higher-performance product? The answer to what a company must do in order to survive is different for every company and every market. Each company, in order to maximize its success, must decide what to emphasize and must allocate its resources accordingly. To do this, an effective organization must study the market and the competition and create a strategic plan for how it will compete in its industry or market. Without a strategic plan, companies would be a little like Alice in Wonderland: It will not really matter where they go so long as it is somewhere.

> The plan is nothing, but planning is everything.
>
>   Dwight David Eisenhower

As General and President Eisenhower said, a plan alone is of limited value, but the activity of creating a plan is everything. *Strategic planning is a process of involving everyone in matching the vision, mission, and core values of an organization with the current situation to focus tactical activities now and in the future.* Strategic plans set the direction and pace for the entire organization (Figure 6.1).

Unlike Alice in Wonderland, most businesses have an idea of what they provide and where they would like to go. Beginning with a vision statement that serves as the guiding star by which the course is set, the purpose of a strategic plan is to develop and achieve the organization's mission in a manner consistent with its values. The strategic plan spells out the specific goals and objectives that must be accomplished in order to reach the vision. A strategic plan seeks to align how the customers' needs are met through daily business activities with the values, mission, vision, and goals of the organization (Figure 6.2). Performance measures, integrated into the strategic plan, allow leaders to judge how the organization is progressing toward its goals and objectives and therefore its vision. Strategic quality planning takes a broader view of the planning process than traditional strategic planning (Figure 6.3).

Successful long-term strategic planning answers some basic questions:

- What business is the organization really in?
- What are the organization's principal strengths and weaknesses for competing in this market? What does it take to compete successfully?
- What does the organization wish to become in the future?

1. *What business is the organization really in?* Effective organizations answer this question from the customers' perspective and in broad terms. For instance, an amusement park owner may think of himself as being in the entertainment business. This broader view enables the park leadership to realize that they are competing for their customers' entertainment dollars, dollars that might be spent at a movie theatre or a mall. Knowing this, leaders can identify their strategies, goals, and objectives more clearly.

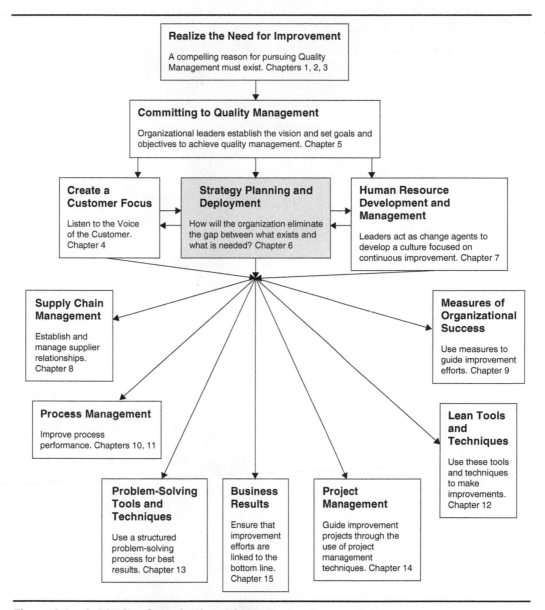

Figure 6.1   Achieving Organizational Success

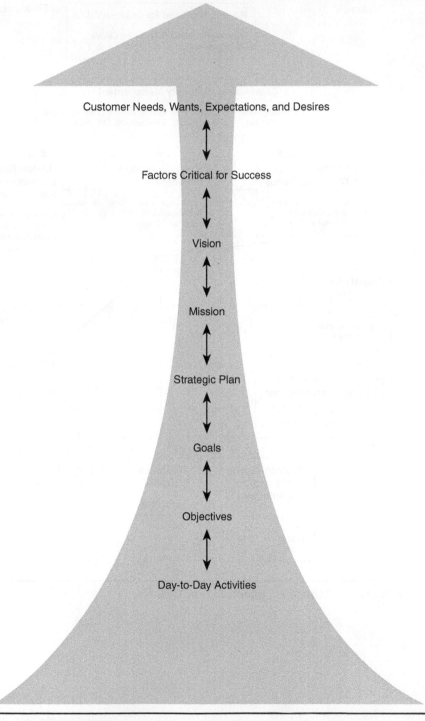

Customer Needs, Wants, Expectations, and Desires

Factors Critical for Success

Vision

Mission

Strategic Plan

Goals

Objectives

Day-to-Day Activities

Figure 6.2   Creating Alignment

Strategic Quality Planning	Traditional Strategic Planning
Focus is on customers.	Focus is not defined or is spread among many considerations.
Leaders determine critical success factors.	Leaders lack understanding of factors critical to success.
Goals and objectives are process and results oriented.	Goals and objectives are results oriented.
Goals and objectives are based on data and are driven by trend or pattern analysis.	Goals and objectives may be based on hunches or guesses.
Focus is on processes.	Focus is on products.
Alignment exists between critical success factors, mission, vision, goals, objectives, and day-to-day activities.	No alignment exists.
Everyone knows how his or her day-to-day activities align with critical success factors, mission, vision, goals, and objectives.	Few people know how their day-to-day activities fit in with the plan.
Improvement activities are focused on activities critical to success.	Improvement activities lack focus.
Improvement activities are both within and across functional areas.	Improvement activities are usually within functional areas.

Figure 6.3   Strategic Quality Planning versus Traditional Strategic Planning

EXAMPLE 6.1   What Business Is the Organization Really In?

Competing against national supermarket chains is difficult for owners of just one or two stores. The buying-power advantage of the larger chains makes competing on price nearly impossible. Savvy small-business owners recognize that although they can't compete on price, they can compete on service and quality. At Dorothy Lane Market (DLM) in Dayton, Ohio, the Mayne family manages its three stores by focusing on providing high-quality merchandise and exceptional service. They concentrate on the 20% of their customers who provide over 80% of their sales, maintaining regular contact with these people and ensuring that the products they seek are carried by their store.

DLM takes a broader view of who their customers are and what they want. In doing so, DLM has had to answer the question, Who is the real competition? Twenty years ago the average time spent creating an evening meal was 45 minutes. Today, the average time allotted to preparing an evening meal is just 15 minutes. How does a grocery store compete with "fast food"? Information provided by DLM's customers pointed to the need to have healthy alternatives to fast food. In response, DLM added a complete department that sells pre-pared foods from DLM's own kitchens. Shoppers can purchase a wide variety of healthy, preprepared foods, including chicken, salmon, lasagna, potato casseroles, salads, and so on. The menu varies on any given day according to the season, so a variety of healthy-alternative choices is always available.

2. *What are the organization's principal strengths and weaknesses for competing in this market? What does it take to compete successfully?* When answering this question, leadership must consider what it really takes to compete in their business. They need to know what their organization *absolutely must do well* in order to attract and retain customers. These factors critical to success are based on customer input. *Critical success factors are the vital activities that will have a proportionally greater impact on the orga-nization's ability to achieve their vision.* Business leaders must determine and then find ways to address their organizations' strengths, weaknesses, opportunities, and threats. A well-designed strategic plan will include ways to improve the strengths, remedy or minimize the weaknesses, take the best advantage of the opportunities present, and eliminate or neutralize the threats.

EXAMPLE 6.2 What Are the Organization's Principal Strengths and Weaknesses for Competing in This Market? What Does It Take to Compete Successfully?

Facing difficult competition from the wide variety of retail stores selling fashion clothing, the Gap has refocused its efforts on what guided the firm at its inception: selling inexpen-sive clothing that looks good. Paul Pressler, Gap's chief executive, was quoted in *Newsweek*, August 4, 2003: "This is what the Gap does well. It interprets fashion trends and finds new ways to present tried-and-true items." This back-to-the-basics move is the result of three years of poor financial performance. The Gap is essentially asking the question, What are the Gap's principal strengths for competing in this market compared to what it takes to compete successfully?

Other industries have been asking themselves the same question, resulting in a renewed focus on such business fundamentals as researching customer preferences and ensuring quality products and services. Companies that in the past surveyed customers only periodically are now doing so regularly. Studying the customers; defining target mar-kets; and providing processes, products, and services to support what the customers want are helping many companies redefine who they are and helping them reach new levels of success. Their successful long-term strategic planning answers the basic questions:

- What business is the organization really in?
- What are the organization's principal strengths and weaknesses for competing in this market? What does it take to compete successfully?
- What does the organization wish to become in the future?

3. *What does the organization wish to become in the future?* The vision of an organizational future is the star that guides the strategies that it follows. The answer to this question will enable an organization to determine what changes they need to make in order to be in the business of their future. Using a strategic plan, an effective organization is able to establish a clear path to closing the gap between what they would like to be in the future and where they currently are.

Creating a strategic plan is an iterative exercise, and the plans must be updated as time passes. Answering these questions requires a thorough understanding of customer demographics and the market, the competition, the trends, the economy, technological advancements, and the desires of the owners and stakeholders.

HOW DO STRATEGIC PLANS GIVE AN EFFECTIVE ORGANIZATION A COMPETITIVE EDGE?

Dr. Deming begins his 14 points with:

> Create a constancy of purpose toward improvement of product and service, with the aim to become competitive and to stay in business and to provide jobs.

With this statement Dr. Deming points out to leaders that their organization must be guided toward its ultimate goals. Strategic plans allow leadership to put down in writing in what direction the organization is heading and how it plans to get there. In a competitive business environment, an effective organization utilizes carefully designed strategic plans in order to create and sustain its competitive advantages and profit position. A well-structured strategic plan outlines the rationale for besting the competition in the market by exploiting market opportunities, maximizing organizational strengths, and playing off the competition's weaknesses. Efforts to address these issues typically result in a product or service that provides the customer with greater value, through improved quality, favorable economics, or enhanced service or performance.

In any organization, resources are finite. There simply isn't enough money, time, energy, people, machinery, and material available. Strategic planning improves the use of these precious resources. Leaders of effective organizations know that well-spent resources add value. They spend their resources in ways that create a worthwhile effect. The cost of that resource is not wasted. Wasted resources have two negative impacts: the waste of the original resource and the usefulness that might have come from wisely spending that resource. Strategic plans are the battle plans that enable an organization to accomplish its objectives. By implementing these plans, organizations are able to better place their products or services in the market. Strategic plans establish a direction for the organization. The results of implementing these plans are dependent upon the plans themselves, the individuals implementing them, and the forces at work in the market. Remember, other organizations are implementing their strategic plans too. A highly competitive market with mature products offering customers a wide variety of choices will require a very different sort of strategic plan from one for an entirely new product. In the mature market, if technology and resources are common among competitors, then the competitive advantage typically goes to the low-cost producer or the competitor with better distribution channels. In a new market, where innovation is the norm, product service will be a lower priority in the strategic plan compared to innovation of new products with expanded features. For example, manufacturers of personal digital assistants have focused their strategic plans on smaller units with increased functionality and smaller size.

Innovations in products or services are always possible, even in a mature market. The John Deere Company showed this in the 1970s by innovations in their services. John Deere noticed a marketwide weakness in replacement-part delivery time. The company implemented a strategic plan that supported their mission of having any part for any of their products available within 24 hours nationwide. A lawn-care specialist, construction manager, or farmer selecting a tractor, agricultural machinery, or other equipment knows that he will never have to delay work due to the unavailability of a $5 repair part. Choosing a John Deere is the right choice.

HOW DO STRATEGIC PLANS SUPPORT CUSTOMER SATISFACTION AND PERCEIVED VALUE?

By integrating customer information into the strategic planning process, effective organizations can identify the market segment in which they want to compete. Because of their understanding of their markets and customers, effective organizations are able to create and maintain a distinctive customer base. Customer needs, wants, and expectations translate directly into requirements for major design parameters to develop, produce, deliver, and service the product or service. A strategic plan uses this information and incorporates strategies for improving customer satisfaction by providing better products, services, economics, delivery, and quality.

EXAMPLE 6.3   How Do Strategic Plans Support Customer Satisfaction and Perceived Value?

As described in a previous chapter, for a certain distributor of construction materials, one of the chief strategies relates to business growth through acquiring customers who use them as their sole supplier. Recently, a salesperson and the sales manager discussed a proposal provided by a customer. The customer would like to use the distributor as a sole source for all building materials and in exchange for that would like a reduction in brick prices, essentially purchasing the brick at one penny below the distributor's cost. The manager of brick sales is unwilling to agree to the price reduction because it would not optimize his department's overall figures. The general sales manager wants to close the deal because the resulting increase in sales would have a dramatic effect on the company's overall profit picture.

In order to convince the manager of brick sales that this deal was a good one, the sales manager gathered information about brick sales in general and specifically about the extras that accompany the sale of traditional brick. These extras include decorative finishing brick, arches, window sills, and other items that give a finished appearance to a brick building. He was able to show that the new business resulting from this deal would result in an overall increase in sales of these extra items, which in turn would more than compensate for the penny-a-brick loss. They returned to the customer and agreed to work out a sole supplier agreement. Fact-based decision making enabled the organization to align customer-focused requirements with the overall business strategy.

HOW ARE STRATEGIC PLANS CREATED?

Strategic plans enable an organization's leadership to translate the organization's vision and mission into measurable activities to act upon. Developing a strategic plan requires taking a systematic look at the organization to see how each part of it interrelates to the whole. The strategic plan focuses on the key organizational goals that support the vision and mission. Strategic plans enable an organization to advance from wishful thinking—the customer is number 1—to taking action—changing corporate behavior and the actions of employees in order to support a focus on the customer.

A strategic plan defines the business the organization intends to be in; the kind of organization it wants to be; and the kind of economic and noneconomic contribution it will make to its stakeholders, employees, customers, and community. The plan spells out the organization's goals and objectives, including how the organization will achieve these goals and objectives. The strategic plan concentrates on the critical success factors (CSFs) for the organization, providing plans for closing the gaps between what the organization is currently capable of doing versus what it needs to be able to do. Using indicators or performance measures, the organization will monitor its progress toward meeting the short-term, mid-term, and ultimately long-term goals. These indicators of performance are critical because they enable an organization to determine whether they are on target toward reaching their goals. Effective organizations analyze the gap between what the current performance is and what the original strategic plan called for. In the case of a negative gap showing lower-than-expected performance, the organization must take corrective action to eliminate the root cause, improve performance toward the goal, and narrow the gap. In the case of a positive gap, showing better-than-expected performance, the organization may choose to take action to further enhance the gap. A good strategic plan also includes contingency plans in case some of the basic assumptions are in error or in case significant changes in the market occur.

In preparation for creating a strategic plan, the organization's leaders should determine:

1. The organization's business (what business are they really in?)
2. The principal findings from the internal and external assessments:
   a. Strengths and weaknesses
   b. Customer information
   c. Economic environment information
   d. Competition information
   e. Government requirements
   f. Technological environment

Figure 6.4   Elements Needed for an Effective Strategic Planning Process

Once these issues have been addressed, the creation of a strategic plan can begin. A strategic plan is essentially a framework that assists an organization in achieving its vision while allowing it the flexibility to deal with unforeseen changes in the business environment. The elements of a strategic plan are listed below; are shown in Figures 6.4, 6.5, 6.6, and 6.7; and are illustrated in Example 6.4.

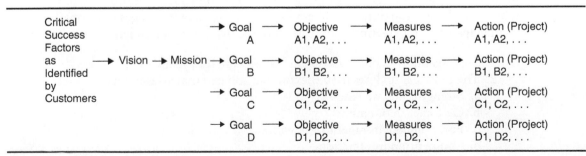

Figure 6.5   Strategic Planning

Preparation
1. The organization's business (what business are they really in?)
2. The principal findings from the internal and external assessments
   A. Strengths and weaknesses
   B. Customer information
   C. Economic environment information
   D. Competition information
   E. Government requirements
   F. Technological environment

Planning
1. *Vision:* the organization's strategic direction for the foreseeable future
2. *Mission:* the translation of the organization's vision into strategic actions
3. *Critical success factors as identified by customers:* the 3 to 10 things that a company absolutely must do well if the company is going to thrive
4. *Goals, objectives, and indicators of performance:*
   A. Goal 1
      Objective 1.1
         Indicator
      Objective 1.2
         Indicator
   B. Goal 2
      Objective 2.1
         Indicator
   C. Goal 3
      ⋮
5. Contingency plans

Figure 6.6   Generic Strategic Plan

1. **Vision:** the organization's strategic direction for the foreseeable future
2. **Mission:** the translation of the organization's vision into strategic actions
3. **Critical success factors:** the 3 to 10 things, as identified by customers, that a company absolutely must do well if the company is going to thrive
4. **Goals:** what must be achieved in order to support the CSFs
5. **Objectives:** the specific and quantitative actions that the organization must take in order to support the accomplishment of the goals and ultimately the mission and vision
6. **Indicators:** the performance measures that indicate whether the organization is moving toward meeting its objectives, goals, mission, and vision
7. **Contingency plans:** the plans in place that enable an organization to remain flexible in a complex, competitive environment

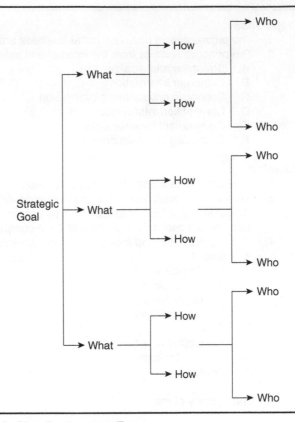

Figure 6.7   Strategic Plan Deployment Tree

EXAMPLE 6.4   PM Printing and Design: A Strategic Plan

Preparation

Leaders of PM Printing and Design recently met to clarify portions of their strategic plan. Using the plan, they hope to communicate to their employees the importance of creating and maintaining a customer-focused process orientation as they improve the way they do business.

Based on their meeting, the leaders have determined that their best market niche, or the business that they are really in, is concept to delivery. This includes designing, reproducing, and mailing customer brochures and literature. Market research has shown that no other full-service printing and design company exists in their market area, and customers are seeking an organization that can take their ideas and turn them into a finished product in their customers' mailboxes. This research has identified PM's strengths and weaknesses; provided customer information; defined their economic, technological, and competitive environments; and specified appropriate government requirements.

Planning

PM Printing and Design Vision

- PM Printing and Design will be recognized as the best source for printed and duplicated material through the recognition and implementation of customer-driven change in a service-focused environment.

PM Printing and Design Mission

- PM Printing and Design is a full-service design, reprographic, and mailing facility committed to serving the local community by producing the highest-quality product in the most cost-effective manner.

Critical Success Factors

- Provide a full-service reprographic facility
- Provide a full-service design process
- Provide a full-service mailing facility, including mailer creation and mailing processes
- Employ talented designers and skilled technicians
- Provide quality printed material using cost-effective services

Goals, Objectives, and Indicators of Performance

- Goal 1: Improve customers' knowledge of our services
  Objective 1.1: Advertise services communitywide
     Indicator: Number of customers
     Indicator: Number of repeat sales
  Objective 1.2: Increase market share
     Indicator: Number of customers
     Indicator: Number of repeat sales
     Indicator: Number of referrals
- Goal 2: Improve customers' perceptions of our services
  Objective 2.1: Reduce number of customer complaints
     Indicator: Number of complaints per month
     Indicator: Average time to resolve complaints
- Goal 3: Design and process orders rapidly while maintaining high quality
  Objective 3.1: Reduce time from order to completion
     Indicator: Average order turnaround time
- Goal 4: Increase customer value
  Objective 4.1: Improve quality and reduce cost
     Indicator: Cost per printed unit (impression)
     Indicator: Cost avoidance (work performed in-house versus contracting work)
  Objective 4.2: Remove non–value-added activities
     Indicator: Cost per printed unit (impression)
  Objective 4.3: Improve in-first-time through quality
     Indicator: Reduction in rework/scrap

- Goal 5: Provide a desirable work environment for PM employees
    - Objective 5.1: Expand employee opportunities for growth
        - Indicator: Progress toward cross-training goals for critical processes as identified by customers
    - Objective 5.2: Improve employee retention
        - Indicator: Length of employee service
        - Indicator: Number of employees with 1+ years of service

Contingency Plan

- Maintain good relationships with other area printers in order to have a source of extra production capability if needed for a rush job or in the event of an equipment malfunction.

Although this strategic plan is still in a state of development, it does provide a concrete example of how a strategic plan is aligned throughout. This alignment can be seen by goals, supported by objectives, and monitored by performance indicators.

Strategic planning, done correctly, takes time and effort. Well-constructed strategic plans support the idea of striving for improving how the organization conducts its business. Companies who are less than dedicated, or companies who don't understand the impact that strategic planning (or the lack of it) has on business operations, may choose to merely go through the motions of creating a strategic plan. These plans, often developed in isolation from the customer or from the day-to-day organizational operations, are little better than industrial wallpaper. When leadership exhibits a lack of interest in the planning process, the organization's future health is at risk. Without a vision supported by a strategic plan, these companies resemble Alice in Wonderland: It does not really matter which way they go.

Beyond a lack of understanding, commitment, and participation by leadership, a variety of pitfalls in strategic planning and its deployment exist. Sometimes an unbelievable vision is proposed, only to be followed by an inadequate definition of operating expectations. A strategic plan that lists objective after objective suffers from a lack of prioritizing. Strategic plans that fail to clearly assign responsibility for results are weak, as are those that fail to identify and utilize performance measures. The performance measures themselves may be a problem. Organizations do not benefit from strategic plans containing performance measures that are not connected with the activities proposed by the plan or that are vague or unclear. Strategic plans are also hampered when *what* the organization wants to accomplish is not supported by a corresponding *how* the *what* is going to be accomplished.

How Do We Know It's Working?

Every organization has "the way we do things," tried-and-true activities and methods developed to support the functioning of the company. But what happens when things change and new competitor or new technology or a new owner shakes up the status quo? How the organization adapts is crucial to its survival. Strategic plans can help guide the change process.

The leaders at JQOS took a strategic approach to business planning versus the former reactionary approach. Following a series of visits to the plant before the sale, the investors, led by Chuck, met to create a viable strategic business plan. They understood that success in any business venture is no accident and past success isn't any guarantee of future success. A sound strategic plan is critical to guiding their business.

"Grow the business" clearly articulates their vision. To support this, their strategic plan integrates a two-pronged approach: organic growth by enhancing their current specialties and acquisition growth by branching out into other manufacturing areas. The first year, the strategic goals focused on growing the business by reducing the amount of labor to make a shaft, improving maintenance, and replacing or repairing machines. Objectives to support these goals include establishing quality and lean manufacturing methods; creating formal hiring and training systems, instituting a productive maintenance program; improving and standardizing processes; partnering with vendors; enhancing technology usage; and dealing with emerging competition (Figure 6.8).

Each quarter, an off-site meeting is held to review the strategic plan. Discussions focus on measuring progress toward goals and objectives, setting new goals and objectives, and determining how much money to spend and where it should be spent. Having reached many of their strategic goals, at a recent meeting, JQOS leadership set a new strategic target of $12 million in sales and a consistent two-week order-to-delivery lead time. They are also studying potential acquisitions.

HOW DO EFFECTIVE LEADERS CREATE ORGANIZATIONAL ALIGNMENT THROUGH STRATEGY DEPLOYMENT?

To be effective, a strategic plan must be deployed. As living documents, strategic plans are not meant to sit on a shelf, to be touched only when it is time for an annual revision. Essentially, creating alignment is policy deployment. *Alignment means that if you push on one end, the other end will move in the direction you want.* Effective leaders enable members of the organization to make the transition between the strategic plan and daily business activities by translating *what* needs to be accomplished into *how* it will be accomplished. Effective leaders make sure that the day-to-day activities and the goals of the strategic plans of the organization are in harmony and are focused on what is critical to the success of the organization. The leaders want to ensure that if they push on the strategic plan, the actions of their employees will go in the desired direction. For this reason the strategy must be clearly communicated throughout the organization. Effective leaders make sure that the strategic plan contains clear objectives, provides and utilizes measures of performance, assigns responsibilities to specific individuals, and denotes timing. In order for a strategic plan to be effectively deployed, the organization's reward and recognition system must support plan deployment. This will be covered in greater detail in Chapter 7.

Vision: To Grow the Business

Mission: To manufacture shafts that meet our customer's expectations, the first time, every time.

Critical Success Factors: Order Processing Time, Quality, Delivery

Business Strategy 1: Organic growth through enhancing current capabilities and specialties.

Goal 1: Improve part quality

    Objective 1: Improve machining capability (see Goals 3 and 4)

        Indicator: # of machines able to hold tolerances

        Indicator: # of internal quality problems

        Indicator: # of external quality problems

    Objective 2: Improve operator capability (see Goal 1, Objectives 4 and 5)

        Indicator: # of internal quality problems

        Indicator: # of external quality problems

Goal 2: Reducing the amount of labor to make a shaft

    Objective 1: Improve and standardize procedures

        Indicator: # of processes with standardized procedures

    Objective 2: Establish quality system

        Indicator: # of internal quality problems

        Indicator: # of external quality problems

    Objective 3: Implement lean manufacturing methods

        Indicator: # of lean techniques implemented

    Objective 4: Create formal hiring system

        Indicator: employee retention

    Objective 5: Create formal training system

        Indicator: # of operators using standard procedures

        Indicator: # of quality problems

        Indicator: employee retention

Goal 3: Improving maintenance

    Objective 1: Hire skilled machine maintenance person

        Indicator: # of machines down during week

Goal 4: Institute productive maintenance system

    Objective 1: Replacing worn-out machines

        Indicator: # of new machines

    Objective 2: Repair machines unable to hold tolerance

        Indicator: # of grinders with new or reground ways

Goal 5: Enhance technology usage

    Objective 1: Update, improve, and computerize quoting and order processing system

        Indicator: # of computerized quotes

        Indicator: # of orders processed and tracked on computer

        Indicator: # of orders processed online

Business Strategy 2: Acquisition growth by branching out into other manufacturing areas.

Figure 6.8   JQOS Strategic Plan

EXAMPLE 6.5   PM Printing and Design: Assigning Responsibilities

Leaders of PM, the printing and design firm in Example 6.4, used the strategic planning information from Figures 6.6 and 6.7 as an example and created the responsibilities matrix shown in Figure 6.9. Note that specific activities have been created to support the goals and objectives. The matrix assigns responsibilities to specific individuals, shows the time frame for accomplishment, and establishes priorities for the objectives.

Goal 1: Improve customers' knowledge of our services

Objective 1.1: Advertise services communitywide
  Action: Contract with local radio, TV, and newspapers for advertisements
  Responsibility: QS, MF
  Due: September 15
  Priority: 1
    Indicator: Number of customers
    Indicator: Number of repeat sales
    Indicator: Number of referrals

Goal 4: Increase customer value

Objective 4.1: Improve quality and reduce cost
  Action: Compare and contrast reproducing machines available
  Action: Select improved machine if available
  Responsibility: RP
  Due: October 15
  Priority: 2
    Indicator: Cost per printed unit (impression)
    Indicator: Cost avoidance (work performed in-house versus
        contracting work)

Objective 4.2: Remove non–value-added activities
  Action: Create process maps for all critical processes
  Action: Set up one team per process to identify and remove non–value-added
        activities
  Responsibility: DS
  Due: December 1
  Priority: 1
    Indicator: Cost per printed unit (impression)

Figure 6.9   Example 6.5: PM Printing and Design Responsibilities Matrix

Effective organizations recognize what business they are in and what they absolutely must do well in order to compete in today's market. They have created strategic plans that support their vision of the future. These plans are filled with goals and objectives that identify, analyze, and close the gap between the current conditions and the vision. Most importantly, effective organizations are skilled at deploying their strategic plans and driving improvements.

How Do We Know It's Working?

The leaders of JQOS know that effective strategy deployment is added by an effective measurement system. This measurement system must be clearly constructed in order to allow all members of the organization to see and track progress. At JQOS, a bulletin board next to the employee break room is updated each Monday morning with the following figures:

Past weeks' sales
Year-to-date sales
Past weeks' shipments
Year-to-date shipments
Past weeks' returns
Year-to-date returns
Jobs past due
Quarterly results of all of the above
Quarterly targets
Progress toward quarterly targets
Safety performance

The figures are shown both numerically and graphically so that employees can track performance over time. Sales have risen from $4 million to $8 million a year. The number of jobs handled weekly by the plant has risen from 30 to 60 per week. Lead time has fallen from four to six weeks to nearly two weeks. First-pass quality has increased from 80% to 87%. Profit has risen from 2% to 8%, a figure that includes money spent on improvements. The number of employees has risen from 26 to 38, primarily because of adding a second shift to handle the increase in customer orders.

Instead of trying to make changes by describing attitudes, the leaders of JQOS have put in place strategies, policies, and procedures that influence the actions of their employees. To support these, they have carefully chosen measures that show the organization's progress toward its goals. They understand that what gets measured gets done.

HOW CAN THE MALCOLM BALDRIGE NATIONAL QUALITY AWARD CRITERIA HELP AN ORGANIZATION'S LEADERSHIP CREATE EFFECTIVE STRATEGIC PLANS?

The strategic planning section (section 2) of the Malcolm Baldrige National Quality Award (MBNQA) criteria examines how an organization sets strategic directions and how it creates key action plans to support the strategy. Applicants are asked to describe their organization's strategy development process. This process must address a variety of key factors, including customer and market needs, expectations, and opportunities; the competitive environment and the organization's reaction to it; and the organization's strengths and weaknesses, including technology or human resources. Recognizing that strategy development and strategy deployment go hand in hand, this section asks how the strategy developed by organizational leaders is deployed in order to be effective. Organizational strategy should address supplier or partner strengths and weaknesses. Organizations are also affected by existing economic, societal, and financial conditions, so these issues must also be addressed. Strategic plans must include objectives and the time frame for accomplishing those objectives. The award is interested in how an organization translates its strategic plan into actions, both short- and long-term. Emphasis is placed on the use of performance measures to track the organization's progress toward reaching the strategic goals and objectives outlined in the strategic plan.

Strategic planning enables an effective organization to create and maintain an unwavering focus on the customer. When evaluating the effectiveness of an organization's development and deployment of its strategic plan, ask these questions based on the MBNQA criteria:

Leadership

- Does the organization have the understanding, commitment, and participation of its leadership in order to ensure the successful implementation of the strategic plan?
- Has the leadership developed action plans to support the organization's key strategic objectives?
- Is the leadership deploying the action plans laid out in the strategic plan?
- How will the organization's leadership maintain effective communication?
- Does the strategic plan enable leaders to focus on the big picture as well as manage it?

Strategic Planning

- How does the organization develop its strategy?
- Has the organization converted the strategic objectives into deployable action plans?
- Are these action plans supported by key performance measures?
- Does the strategic plan address the following issues?
    - Customer and market needs, expectations, opportunities, and requirements
    - The competitive environment
    - The organization's capabilities
    - Technological changes
    - The organization's strengths, weaknesses, opportunities and threats
    - Supplier and/or partner strengths and weaknesses
    - Financial, environmental, societal, or other potential issues
- Is there alignment between the strategic objectives, goals, and action plans?
- How will the strategic plan and its necessary actions be communicated to all levels of the organization?
- Does the strategic plan communicate priorities to the organization?
- How is the organization's strategy deployed?
- What is the timetable for strategy deployment?

Customer and Market Focus

- Are the strategic objectives based on critical success factors as identified by customers?
- How is this information integrated into the strategic plan?

Measurement, Analysis, and Knowledge Management

- How do the analysis and the use of information support the strategic planning process?
- What measures of performance are used?
- How is feedback received concerning measures of performance?

Workforce Focus

- How do the organizational policies and reward systems support the strategic plan?
- How will leadership maintain effective communication with all levels of the organization?
- Do the action plans include a human resources component?
- How is feedback utilized to adjust actions?

Process Management

- What is the organization's overall strategic planning process?
- What are the key steps of the process? Who are the key participants? What is the planning horizon?
- How is this process improved from cycle to cycle?

Results

- How do measures of performance and key indicators established in the strategic plan help measure the success of the organization in terms of business results?
- Does the organization's skill at managing key processes include utilizing employees' skills, knowledge, and capabilities?
- Have the changes made to employee policies improved business results?
- Have employee policies, strategic planning, and customer and market information been successfully integrated? Do the business results reflect this?

LESSONS LEARNED

Organization leaders must focus on the big picture, using their knowledge and access to information to determine the direction the business will take. Having developed a strategic plan, senior leadership must stay alert to changes and shifts in their customer base, and they must be prepared to make the necessary changes to the business strategy. Strategy deployment is as crucial as strategy development. Operational leaders should be prepared to take the overall strategy to the day-to-day activity implementation level. Alignment of goals and objectives with the mission and vision and critical success factors make this task feasible. Feedback, through the use of performance measures, between all levels helps maintain the necessary alignment.

If good strategic planning is not practiced:

Goals are not known throughout the company.
Goals change too often.
Goals are not achieved.
Goals are achieved without real improvement.
Progress is not sustained.
Organizational frustration exists.

Without good strategic planning and deployment, there is short-term achievement at the expense of long-term organizational health. Leaders must create strategic plans that align actions with the critical success factors as identified by the customers. They must work toward total system alignment, aligning the organization's vision, mission, strategic objectives, processes, and day-to-day activities with their customers' needs, wants, and expectations. Effective organizations must change faster than their competitors and the external market environment in order to gain and sustain a competitive advantage. Effective leaders use the strategic plan to provide clarity. In the complexity of day-to-day operations, it is easy for members of the organization to become distracted and lose sight of the strategic goals and objectives. Effective leaders continually re-emphasize the ultimate targets established in the strategic plan.

1. Strategic plans support customer satisfaction and perceived value.
2. Strategic plans guide an organization as it tries to eliminate the gap between the current conditions and where it would like to be.
3. Strategic plans create a "constancy of purpose toward improvement of products and services."
4. A strategic plan contains the vision, mission, critical success factors, goals, objectives, and indicators of performance that guide the actions an effective organization takes.
5. A vision guides the formation of the strategic plan.
6. Goals and objectives link the vision with day-to-day activities.
7. Strategic plan deployment is as key as strategic plan development.

Are You a Quality Management Person?

Strategic planning sounds like something big businesses do, not people. But consider this: You are the CEO of your life. You are the one who reacts and acts on the events in your life. Check out the quality management person questions and find out whether or not you are a quality management person. How do you answer them on a scale of 1 to 10?

Do you have long-term goals for your life?

1	2	3	4	5	6	7	8	9	10

Do you have short-term goals for your life?

1	2	3	4	5	6	7	8	9	10

Are your long- and short-term goals aligned with your values and aligned with your day-to-day activities?

1	2	3	4	5	6	7	8	9	10

Do you have well-thought-out, realistic goals and objectives with achievable targets established for yourself?

| 1 | 2 | 3 | 4 | 5 | 6 | 7 | 8 | 9 | 10 |

Do you plan your daily activities around your short- and long-term goals?

| 1 | 2 | 3 | 4 | 5 | 6 | 7 | 8 | 9 | 10 |

Do you regularly prioritize important tasks that you need to get done?

| 1 | 2 | 3 | 4 | 5 | 6 | 7 | 8 | 9 | 10 |

Do you regularly prioritize routine tasks that you need to get done?

| 1 | 2 | 3 | 4 | 5 | 6 | 7 | 8 | 9 | 10 |

Do you regularly allocate the appropriate amount of time needed to accomplish what you have planned?

| 1 | 2 | 3 | 4 | 5 | 6 | 7 | 8 | 9 | 10 |

Do you create realistic plans and follow through with them?

| 1 | 2 | 3 | 4 | 5 | 6 | 7 | 8 | 9 | 10 |

As events change in your life, do you take the time to change your strategic plans?

| 1 | 2 | 3 | 4 | 5 | 6 | 7 | 8 | 9 | 10 |

Do you make changes to your activities in order to better align your life with your strategic plans?

| 1 | 2 | 3 | 4 | 5 | 6 | 7 | 8 | 9 | 10 |

Chapter Questions

1. How is quality strategic planning different from traditional strategic planning?
2. Why does an effective organization need a strategic plan?
3. What are the benefits of a strategic plan?

4. Describe each of the elements needed for the strategic planning process.

5. Describe the steps necessary to create a strategic plan.

6. What does an organization need to know about itself before creating a strategic plan? Why is this information important?

7. Why is strategy deployment as important as strategic planning?

8. How is alignment created in a strategic plan?

9. Ask to see the strategic plan of a company for which you work or have worked. How does it compare with what you have learned about the strategic planning process?

10. Select an organization. Using what you have learned in this chapter and the examples as a guide, create a strategic plan for the organization.

Section 2.0: Strategic Planning

The strategic planning section of the MBNQA criteria addresses strategic planning and deployment. This section stresses that customer-driven quality and operational performance excellence are key strategic issues that need to be integral parts of the organization's overall planning.

The key goals of this section are to:

1. Examine how Remodeling Designs and Case Handyman Services leadership set strategic directions and develop their strategic objectives, with the aim of strengthening their overall performance and competitiveness.
2. Examine how Remodeling Designs and Case Handyman Services leadership translate their strategic objectives into action plans to accomplish the objectives and to enable assessment of progress relative to their action plans. The aim is to ensure that their strategies are deployed for goal achievement.

This section of the review was created by Joanie Zucal, Maria Dominique, Renee Cooper, Erich Eggers, Mike Cordonnier, and Donna Summers.

QM   Strategic Planning Questions

1. Do you have a strategic plan? Who are the key participants? What steps are taken to develop that plan? (Ask for the document.)
2. How do you define the business you are really in?
3. Does your strategic plan address what business are you really in?
4. Has your strategic plan identified the key processes that you absolutely must perform well in order to stay in business?
5. What are your principal strengths and weaknesses for competing in the remodeling and home repair business?
6. How does your strategic plan reflect what you wish to become in the future?
7. When creating your strategic plan, did you consider all of your stakeholders?
8. How have you, as a leader, created a strategic plan that has translated into positive results on the bottom line?
9. Has the strategic plan mapped out a strategy for improving each of your key processes?
10. How does your strategic plan react and change in response to new information that affects the way you run your business?

2.0 Strategic Planning

2.1 STRATEGY DEVELOPMENT

2.1a Strategy Development Process

The overall strategic planning process that the leaders of Remodeling Designs/Case Handyman developed is mapped in Figure 1. The owners and key participants of this process are Mike Cordonnier and Erich Eggers. The five main outcomes that they completed were the following: a primary aim or vision of the company, a strategic objective, an organizational strategy, a management strategy, and an implementation strategy. Three of these main outcomes have also supporting outcomes. However, after this document was completed, it was filed and never used.

In creating the strategy development process, Remodeling Designs/Case Handyman included the key factors to ensure that the development of the strategic plan linked their mission statement to the bottom line. Since Remodeling Designs/Case Handyman is a service-oriented company, it focuses mainly on customer and market needs, expectations, and opportunities. Remodeling Designs Inc. (RDI) discussed the need for a better system to complete remodeling projects as efficiently as possible. To create this effective and efficient system, RDI highlighted a number of customer-based improvements. One improvement was high-level customer sales service, which would include a checklist or other document that explained to the customer what RDI's policies are regarding written quotes, choice of material, and so on. RDI also expressed the need for creating a profile of the perfect customer, which included assigning values describing a perfect customer based on the characteristics of previous good customers. This profile would help RDI better allocate their time spent on selling because they could spend more time servicing already existing, paying customers. In order to better serve their customers, RDI mapped out their sales process from the first phone call to the preconstruction meeting.

Remodeling Designs also examined the competitive environment in their strategy development process, including what they would have to do in the future to remain competitive. In the year 2000, RDI realized that the world was at a turning point and that everything around them was changing, including the remodeling business. "More and more remodeling companies were catching on to our professional angle. We needed to continue to be the progressive company that we had been in the first ten years. We recognized the fact that we needed people to still say, 'Well, let's just do it the way RDI does it, because they are always right.'" RDI also examined potential new changes in the way they do business in the present and key changes implemented in the past. Key changes were made to the internal operations that strongly affected servicing the customer. Previously, RDI "suffered from a lack of organization, a lack of information exchange, lack of detailed drawings, lack of material acquisition procedures, lack of material staging procedures, [lack of] material check-in procedures, among many other internal operations." With all of these defects, RDI was not able to provide high-quality customer service. The company leaders, the frontline carpenters, and the production manager all created subsystems that dealt with each of

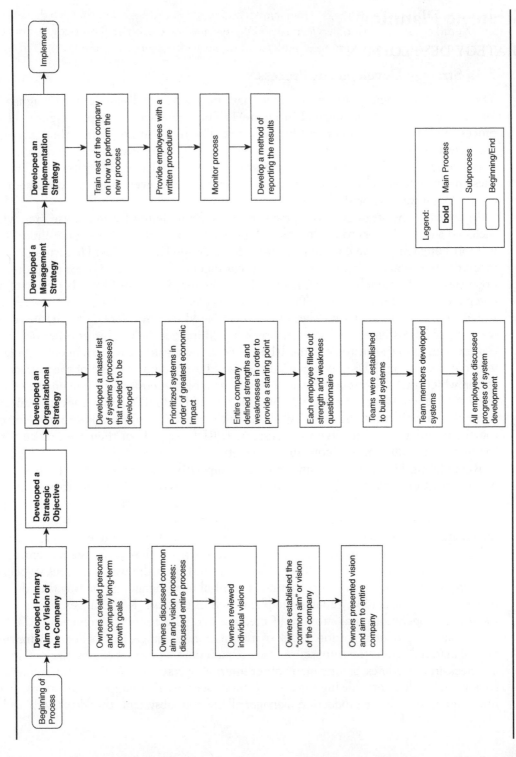

Figure 1 Process Map for How Remodeling Designs/Case Handyman Created Their Strategic Plan

the cases individually. A major change/addition was to create a book that would improve the overall communication between the RDI crew and the customer. This book contains all of the information necessary to complete a project without any personnel other than the lead carpenter.

After examining their existing process, RDI decided to improve on the already well-established processes by describing and laying them out more efficiently and in greater detail. In doing so, they also created new processes. In mapping the process, they found some major problems that they were able to correct along the way. After completely reorganizing the sales process, they moved on to restructure the marketing, finances, and accounting areas. As a result of these changes, RDI began to experience an improvement in flow, their net profit increased, jobs ran more smoothly, jobs were completed on time more often, and employees were much happier because they did not have to give excuses to the customers about finishing late.

RDI wanted to improve the process for the person taking the lead information for a project. This person had to be trained in obtaining the correct information and had to be able to provide rough estimates ("ballpark" figures) for jobs. RDI expressed the need to increase the duties of the lead carpenter to those of a project manager. The lead carpenter must see the big picture. RDI also discussed the need for the lead carpenter to have a job audit in which he or she could discuss what went right and what went wrong with the completed job. Some of the possible new processes and systems that RDI discussed were material requisition, acquisition, and receiving; the material disbursement process; job-costing strategies; closing out a job; ensuring quality standards; keeping and organizing job information and paperwork; and subcontractor management. The system for subcontractor management would include developing standards for subcontractors, developing ways to rate the subcontractors, developing billing/payment terms, and ensuring that insurance certificates are on file and up-to-date. The last key factor that RDI reviewed was possible financial risk. One way to address financial risk was to examine who or what was directly responsible for the issue(s) that caused their repeat business and referral business rate to drop more than 60%, to around 25%.

2.1b Strategic Objectives

Although RDI did not have a formal list of strategic objectives, possible strategic objectives could be found in the document "Remodeling Designs, Inc. Growth Charter and Long-Term Company Goals":

> Remodeling Designs, Inc. wants to provide the highest level of service to discerning customers throughout the Miami Valley, stress the importance of communication, total customer satisfaction, balance business/work and personal life, and company growth: increased sales, increased efficiency, and/or increased profits.

Remodeling Designs, Inc. established yearly goals up to the year 2010. Examples of yearly goals listed are amount of sales in millions, net profit after taxes, necessary systems in place for departments, positions that need to be filled, company reorganization steps completed, and company growth and developments (for example, building purchases).

Although these strategic goals and objectives are a good start, they do not address the following key factors: technological and other key changes that may affect their services, and supplier and subcontractor strengths and weaknesses.

2.2 STRATEGY DEPLOYMENT

2.2a Action Plan Development and Deployment

2.2b Performance Projection

Since Remodeling Designs/Case Handyman did not convert their strategic objectives into action plans, Section 2.2, Strategy Deployment, is devoted to explaining the purpose of a strategic plan, describing the correct format of a strategic plan, and suggesting improvements.

The main objective of creating a strategic plan is to provide a broad and long-range view of RDI's and Case Handyman's position and potential within their evolving competitive environment. Once this broad and long-range context is better understood, the leaders of RDI and Case Handyman can more effectively determine what fundamental actions they are going to initiate on a year-to-year basis in order to compete successfully within their business sector. In order to focus internally, the leaders of RDI/Case Handyman must ask themselves the following basic questions:

- What business are we really in?
- What are our principal strengths and weaknesses for competing in this business?
- What does it take to compete successfully?
- What do we wish to become in the future?

The answers to these questions provide a foundation for performing an internal assessment of the company.

In addition to this internal assessment, an external assessment must be completed as well. This external evaluation should focus on the following key areas:

- Consumer demographics and psychographics
- The emerging economic environment
- The global competitive business environment
- The political and legislative environment
- Advances in technology

After a thorough analysis of the information developed as a result of the internal and external assessments, the leaders must develop a series of assumptions on which plans and actions will be based. Then, based on their assessments and assumptions, they can make the transition from strategic planning to business planning. Business planning involves the following:

Determine the critical success factors, which are the 3 to 10 things RDI/Case Handyman must do extremely well in order for their business to survive and prosper

Exploit strengths and remedy or minimize weaknesses

Develop contingency plans

Set mid-term goals for a 3- to 5-year time horizon and identify annual objectives in order to progress toward those goals

Set performance measures
Determine how to fund the activities in order to pursue this plan

Ideally, RDI/Case Handyman need to strive for the above description of creating a strategic plan. However, because they fell short of completing the entire process, the following improvements should be implemented. The first major area would be to actually complete the strategic plan. Most importantly, action plans need to be developed because planning means nothing without implementing actions derived from those plans. As plans are implemented, performance must be measured in order to determine whether the plan is working. Performance measures expose gaps between what is happening and what is supposed to happen. The leaders must analyze this gap to determine the root cause; then they must take corrective action to eliminate it. It is also essential that this document becomes living and breathing instead of being filed and never used.

Another area of improvement would be to reorganize the format of the document. The existing one is in narrative form, which is a good start. However, it is not to the point. Also, it is difficult to determine what the broad and long-range view of their organization is. Preferably, the strategic plan should follow a concise format in which the mission or vision is stated at the beginning, followed by the critical success factors. Then, based on the critical success factors, goals must be developed, and for each goal objectives must be identified. Finally, the leaders must develop and set performance indicators for each objective.

7 Human Resource Development and Management

How do employees create an effective organization?

How do effective employees enable an organization to create value for their customers?

How do leaders in effective organizations motivate employees?

How do leaders in effective organizations manage change?

What modifications do effective organizations make to their reward system to support the desired culture?

What types of education and training do effective organizations provide for their employees in order to remain competitive?

How do effective organizations use teams?

How do individual personalities affect team performance?

What needs to happen in order to have effective meetings?

How can the Malcolm Baldrige National Quality Award criteria help an organization's human resources become more effective?

Lessons Learned

Learning Opportunities

1. To become familiar with the role that employees play in creating an effective organization
2. To become familiar with the role that employees play in creating value for customers
3. To understand how leaders motivate employees
4. To understand how leaders affect change
5. To understand how reward systems are used to support organizational culture
6. To understand the importance of education and training
7. To understand the role of teams
8. To understand how to run effective meetings

Much has been written about the "two pillars" of the Toyota production system—just-in-time and autonomation—designed by the late Taiichi Ohno. Although this remarkable combination played a large role in creating the efficiency and success of the Toyota Motor Company, these technological breakthroughs aren't the whole story. The fine-tuning that made the Toyota production system really work came not from upper management, nor from the engineers, but from the shop floor in the form of employee suggestions—over 20 million ideas in the last 40 years.

Yuzo Yasuda, 40 Years, 20 Million Ideas: The Toyota Suggestion System

The whole employee involvement process springs from asking all your workers the simple question, What do you think?

Donald Peterson, Former Chairman of Ford Motor Co.

HOW DO EMPLOYEES CREATE AN EFFECTIVE ORGANIZATION?

Great products and services are generated through great employees. Effective organizations seek out and employ effective people. These employees are the ones who understand how their jobs fit into the overall scheme of providing products and services to customers. Their knowledge, skills, and efforts are invaluable to the firm because successful organizations evolve on a daily basis in lots of small ways. Every day, effective employees think about how to improve their products and services and discover better ways to do their jobs.

Effective organizations tap into the knowledge and skills of their employees to drive improvements (Figure 7.1). Employees have ideas that can be put to use enriching a customer's experience and value perception of the organization's product or service. Effective leaders recognize the power that employee interaction on a day-to-day basis has on organizational performance. They go to the source of ideas and get to know their employees on a one-on-one basis. They select and implement the employee ideas that have the greatest potential to enhance their customers' value perceptions of their products and services. For instance, one grocery store, after having installed a coffee/snack bar area, acted on an unusual suggestion from an employee. The employee had noticed that people who stopped for coffee or a bite to eat often lamented that they had forgotten to purchase an item or two. They were reluctant to return to the store, pick up the item, and then stand in line again. The employee suggested designating a runner for the coffee area who would go and get the item(s). The customer could then pay at the coffee bar register. This idea turned out to be quite profitable. It also delighted their customers. Effective organizations are careful to hire, train, interact with,

---

EXAMPLE 7.1   Knowledge Is Power

On July 3 of 2003, *The Wall Street Journal* ran an article entitled "On the Factory Floors, Top Workers Hide Secrets to Success." In the article the author, Timothy Aeppel, reported that businesses striving to make improvements are often unable to do so because workers recognize that knowledge is power. Seldom do these workers reveal their secrets, even to other workers, fearing that management would use these secrets to change processes, speed up production, or eliminate jobs. Many of the jobs these workers perform require precision and skill. These workers are able to perform their work faster and at a higher level of quality than others in the job. Their abilities are key because minor alterations, mistakes, or errors could make the product worthless.

As organizations strive to remain competitive, they would like to incorporate employee knowledge into their processes. By tapping into their employees' accumulated expertise and the innovations they have incorporated into their jobs, they hope to make process improvements, remove non–value-added activities, increase productivity, and enhance customer satisfaction and perceived value. Unfortunately, some employees resist becoming involved in the effort, because doing so may create a threat to one of their few sources of power—accumulated knowledge and expertise.

Source: Timothy Aeppel, "On the Factory Floors, Top Workers Hide Secrets to Success," *The Wall Street Journal*, July 3, 2003.

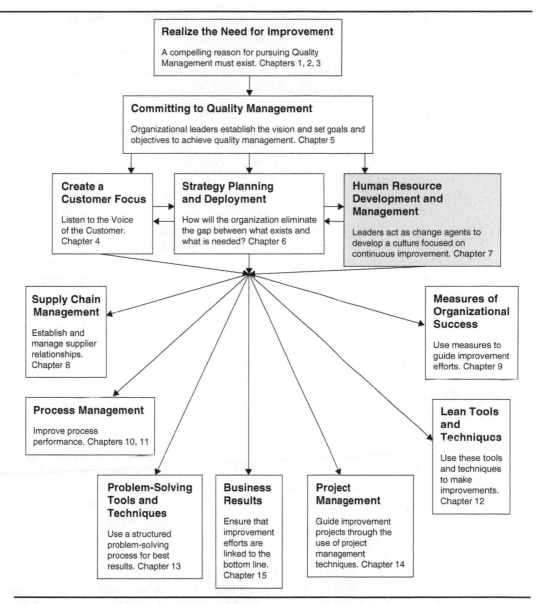

Figure 7.1   Achieving Organizational Success

support, and promote employees in a way that encourages continuous improvement. Loyal, flexible and talented employees are appreciated. As the following example shows, however, not all organizations are able to utilize their employees' expertise to improve.

If you contrast this example with the activities at Toyota, it becomes obvious that the culture and atmosphere of some organizations enable cooperation, information sharing, and communication, while others do not. How did a company like Toyota encourage its employees to provide 20 million improvement ideas? This chapter explores what it takes to create motivated employees who want to participate.

HOW DO EFFECTIVE EMPLOYEES ENABLE AN ORGANIZATION TO CREATE VALUE FOR THEIR CUSTOMERS?

When a craftsman creates an individual product for a customer, the direct connection of the craftsman with the customer leaves little doubt about the satisfaction level of the customer. However, modern manufacturing methods and the complexities of today's wide variety of services prevent many employees from having direct contact with the ultimate end user of their organization's product or service. For this reason, employees cannot meet the needs and expectations of their ultimate customers without support from all areas of the organization. Valuable customer information must be made available to employees in a format that is useful to them. Employees must be able to see the correlation between their day-to-day activities and the value of their product or service as perceived by the ultimate customer. In order to create customer value, employees must be familiar with how meeting the quality, cost, and schedule requirements of their job impacts customer value perception, company profits, and, ultimately, their future employment. Regular communication must take place between leadership and all employees, and this communication should go well beyond the quarterly summary provided by many companies. Employees at all levels must be involved in decisions that affect their jobs. If they cannot judge how their activities impact the customer, they cannot be effective in increasing customer satisfaction and value perception.

HOW DO LEADERS IN EFFECTIVE ORGANIZATIONS MOTIVATE EMPLOYEES?

A significant component in human resources effectiveness is the quality of the leadership. How leaders interact with their people is crucial to creating an effective work environment for employees. Effective leaders realize that the better they treat their employees, the better they treat their customers. Effective leaders have a relationship with their employees. Their employees trust them because they provide clear direction and establish clear expectations and priorities. These leaders also support their people and avoid situations like those in Figure 7.2. They aren't quick to jump to conclusions or lay blame. Effective leaders seek information and then make decisions.

Employees, regardless of their type of job, come to work with certain self-perceptions and needs. Most employees desire a feeling of achievement, and most want to take pride in their work. They want a job that allows them to use their knowledge, skills, and abilities. Many prefer challenging assignments and seek responsibility. Effective leaders recognize these traits in their employees and build on them to motivate their people. Creating and maintaining non-faulty systems and clear communication are two goals to keep in mind when trying to motivate employees.

In order to create and sustain motivated employees, leaders must design non-faulty systems. People don't come to work to do a bad job. They don't get up in the morning and say, "I think I will perform badly today." Still, things go wrong, and products are produced and services provided that are unsatisfactory. Usually, faulty systems are at the root of the

Figure 7.2   Awesome Odds

problem. As Dr. Deming showed in his red bead experiment discussed in Chapter 2, if employees must utilize faulty systems, they will not be able to help performing below expectations, even when they know the ultimate goal. Faulty systems do not allow people to have pride in their workmanship. It is leadership's responsibility to design and maintain nonfaulty systems, thus allowing employees to work at their greatest level of productivity. Faulty systems are unmotivating. As one of Dr. Deming's 14 points says, *leadership must remove barriers that rob people of their right to pride of workmanship*. In performing their jobs, employees often face difficulties that can be removed only by leadership.

Communication is a vital part of motivating employees. No leader can handle conflict or negotiate successfully without being a good communicator. *Communication occurs when information is imparted either by audible, written, and/or visual (postures/actions) means*. Communication is more than just presenting information. Effective communication seeks to both exchange information and provide understanding. How people perceive information is critical. Noise in the form of personal agendas, needs, values, fears, and defenses affects communication too. Understanding can be enhanced by communicating in the recipients' language and from their point of view. People must be able to recognize that the information affects them and that they are to act on that information. Good communication is evidenced by whether or not the message got through and got acted on. To ensure that everyone gets the same meaning from the message, good communicators ask questions to clarify the message. They paraphrase and restate what they think they heard or read. Effective communicators take the time to practice their skills and learn to express themselves well. They also listen. Practicing active listening involves taking time to listen not only for message content but also for feelings, responses, and cues sent by the other person.

Effective leaders realize that they must listen to their employees in order to determine what motivates them. Effective leaders do not rely on hunches or guesses. It is impossible to know what motivates an individual without asking him or her and then truly listening to the response. While listening, a leader learns about and comes to understand an employee's attitudes toward work, pay, benefits, supervision, appraisals, and other work-related topics. This information can be used to place employees in a work situation they find motivating.

EXAMPLE 7.2 Making the Most of a Situation

Over the years the owner of a mill that makes specialty restoration woodworking products has learned that the skills of a master carpenter are critical to attracting and retaining customers. He has also learned that he cannot expect to hire a master carpenter who is exceptionally talented, very efficient, and also very neat. He has been able to hire individuals with two of those three qualifications, but not all three.

Rather than being defeated by this quandary or proclaiming, "You just can't hire good people anymore," the owner asked himself, "What type of master carpenter is best for my business?" Through discussions with customers, he came to the realization that customers value a carpenter who can take their ideas and change them into reality in a timely manner (talent and efficiency). Since then, the owner has been hiring carpenters who meet those expectations. To achieve neatness, which is critical to the efficiency of the master carpenter as well as to the safety of shop personnel, he has also hired part-time individuals to clean up and organize the shop. A part-time person can also serve as an apprentice to the master carpenter if interested.

Rather than trying to change the worker to fit his mold, the owner focused on the elements crucial to the customer and adapted to the situation.

Employees spend a good portion of their working life trying to determine what is expected of them. Sometimes they act incorrectly because they do not understand what their leaders want. Unclear communication is usually the problem, leading to frustration of everyone involved. Effective leaders clearly communicate what is acceptable and what is unacceptable. They recognize that until an employee understands what behaviors and actions are expected of them, the leader has no basis for making corrections. Leaders also recognize that if they allow incorrect behaviors and actions, they have essentially trained employees that they consider those incorrect behaviors and actions acceptable. Effective leaders follow through and do something about incorrect behaviors and activities. They deal with situations in a timely manner knowing that people don't mind the word "no," but they do mind not knowing why. Employees can act on the basis of clear limits and expectations about behaviors and actions. Leaders can reinforce through their coaching, feedback, and rewards until the changes have been effected.

When effective leaders provide feedback to employees, suppliers, and customers, they do so constructively by following these suggestions:

- Give feedback directly to the person involved, not through channels
- Provide feedback only about things that you know for certain
- Cite specifics
- Give feedback when the person can do something about it
- Give feedback at the appropriate time
- Give feedback in small doses

How Do We Know It's Working?

Employees need to know where they stand. Annual reviews are one way of communicating expectations and providing feedback. Prior to taking ownership, at JQOS, no one received reviews of any sort. The new owner established a system where all employees, regardless of job title, are reviewed annually based on their hire date. Reviews may occur more frequently. Employees in their first two years of employment are reviewed every 90 days. Raises and bonuses, besides the quarterly bonus based on profitability, may occur at any time. He also made it clear that the reviews and any subsequent raises would focus on two issues:

How have you helped improve the productivity for JQOS?
How have you improved something about yourself?

The emphasis is on what the employee is doing to help JQOS get good product out the door on a timely basis. It is the answers to these questions that determine whether an employee receives a raise. Employees are encouraged to show productivity improvements by discussing how they trained another employee on their workstation, how they were able to use their own skills to move within the plant on an as-needed basis when staffing issues arose, how they discovered and implemented a productivity improvement, or how they worked to solve a quality, safety, customer, or productivity issue. Self-improvements include learning a new skill, learning how to operate another machine or work another job, or taking training off-site to increase their job knowledge. These questions encourage workers to look beyond being a good employee, practicing safe work methods, and being at work when scheduled. They reflect Dr. Deming's 14 points to constantly and forever improve the process of providing a product or service. They also reflect Dr. Deming's desire to encourage self-improvement. The management at JQOS wants employees to succeed. They provide many opportunities for training, skills assessment, mentoring, and apprenticeship.

HOW DO LEADERS IN EFFECTIVE ORGANIZATIONS MANAGE CHANGE?

Changing conditions in the marketplace often place pressure on organizations to modify the way they do business. Leadership must communicate to the employees the desired change and motivate individuals to make the change. Dr. Deming once said, "Nothing will happen without change. Your job as a leader is to manage the change necessary."

In an organization, two types of change can occur: strategic change and individual change. Strategic changes set the direction for the entire organization. These are larger, more dramatic changes related to the organization's vision, policies, and procedures. Strategic change may require changes to the organization's infrastructure or basic culture. Change at the individual level relates to how individual employees respond and adapt to strategic change. To drive individual change, effective leaders recognize that employees need support as they replace older, established ways of doing things with new, unfamiliar methods (Figure 7.3). Employees' reactions to change relates to their span of control. They can change only things within their power to change. Leaders motivate individual change by coaching, feedback, reinforcement, rewards, and recognition.

Change at either level is a cycle that requires momentum and clear direction (Figure 7.4). Through input from their customers and insights into the marketplace, leaders determine that a need to change exists. They create a vision to guide the change. Effective leaders always change with a plan in mind, which results in the least amount of stress on their human, equipment, and material resources. They determine where they want to go and where they want those who work for them to go. To involve employees in the change, leaders clearly communicate their goals and objectives. Deploying the communication strategy alerts employees to the new goals and objectives. Measures provide the feedback necessary to ensure that the change process is working. Leaders use measures to monitor progress and provide the necessary coaching, feedback, reinforcement, rewards, and recognition to achieve the change.

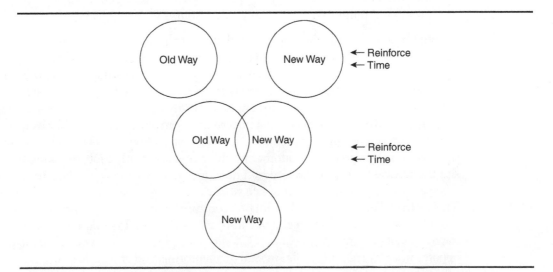

Figure 7.3   Making a Change

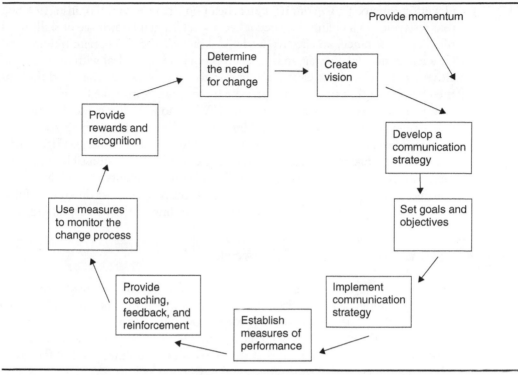

Figure 7.4   The Change Cycle

Change is rarely easy. People resist change because humans are control oriented, and when their environments are disrupted, they perceive that they have lost the ability to control their lives. Some people are more resistant to change than others. Resistance can be based on an individual's frame of reference—on his or her individual values, emotions, knowledge, and behavior. Organizational resistance goes beyond individual resistance and usually takes the form of a key respected individual who does not support change. Interference or resistance can take the form of lack of support, for instance, logistically or economically. Resistance is sometimes very visible and sometimes very subtle.

Change happens when people are motivated. Effective leaders ask, "What motivates people?" They recognize that without passion, nothing is fun. While not every employee can or wants to become passionate about their jobs, Dr. Deming tells us that everyone comes to work to do a good job. For this reason, when making changes, effective leaders establish reachable goals. Though they constantly raise the performance bar for their employees, they are careful to exploit organizational and, therefore, people successes. They use success stories to illustrate how much was accomplished in a similar situation. These stories help break the resistance based on habit and history exhibited by most people.

Leaders need to understand that resistance to change is a natural human reaction. Effective leaders know that in order for change to occur, resistance to those changes must be expected and planned for. Essentially, they manage change by managing resistance. They may take a preventive approach, meeting resistance head-on before it starts. For instance, they may realize that resistance to a change may occur because additional knowledge or skills will be required. Managing resistance, an effective leader will provide the appropriate training and education. Or a leader may take a wait-and-see approach, planning to deal with resistance as it arises. In either case, leaders must clearly communicate the new expectations and the reasons behind the changes so that people understand why they are being asked to change. Employees are looking to leaders and asking the question, "What do you want me to do tomorrow that is different from what I am doing today?" Effective leaders provide training and time to make the change. They also structure their reward system to support the change (Figure 7.5).

Making change in any organization can be difficult, if not impossible, without the involvement and cooperation of the employees. It takes only moments to proclaim a new culture or a new method, but it takes a great deal longer to get people to act differently. Effective leaders realize that employees mold their behavior according to their interpretations of the signals

How Do We Know It's Working?

What makes JQOS's change process so successful? JQOS's new leadership recognized the need for change. They realized that they should begin by making positive changes that directly affected the employees in order to gain their trust and interest. They conducted meetings with small groups of employees to discuss their issues and concerns. After just a few meetings, it became clear that the employees in the shop did not feel that they were valued as much as the front-office employees.

To employees, improving the shop environment expressed leadership's concern for shop employees. At a later meeting, they prioritized the changes they felt necessary. After reviewing the list, JQOS leadership explained which changes would be focused on and which ones would have to wait. The list was posted in the employee break room, and as each item was completed, it received a checkmark before the next item on the list was tackled. Cleaning and painting the break room, adding a refrigerator and microwave oven, improving the restroom facilities, and increasing the lighting in the shop took top priority. As employees began to see positive changes in their work areas, they became more open-minded to other changes as well.

Many of those other changes related to production improvements. With these, they started each change process with key employees, listening carefully to their questions, concerns, and issues. For instance, new policies, product standards, and procedures resulted in a more complete and timely work order packet being sent to the floor from the front office. Once streamlined, these procedures were computerized. Another change was the installation of computer workstations on key equipment on the shop floor, enabling information to flow smoothly throughout the production process. Improving productivity was crucial to the company's survival, so saying no to the changes wasn't an option. To support the changes, JQOS leadership recognized the need to link behavior change with rewards. As the changes progressed, they spent time coaching, providing feedback, and reinforcing the new methods.

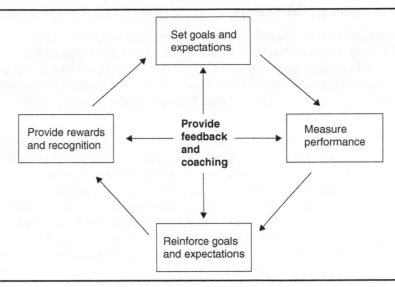

Figure 7.5   Supporting Change

leadership sends them. These signals may come as policies, requests, or edicts, or from the day-to-day actions taken by leadership. Effective leaders recognize that actions speak louder than words. Communicating through leadership actions and examples is paramount to changing behavior. Effective leaders recognize that the tools need to be in place to support the desired change. To maximize the change process, effective leaders ensure that the worker/machine/computer interface, as well as the worker-to-worker interface, is compatible with the needs, capabilities, and limitations of the worker. Further, effective leaders ensure that the reward system matches the desired expectations in order to change behavior. Alignment must exist between rewards, expectations, leadership actions, and customer needs. Both rewards and punishments must reinforce behavior expectations. Interestingly enough, many people do not know what is involved in making a change. Figure 7.6 shows a matrix for making a change, and Example 7.3 provides a model.

Desired Result	Supporting Actions	Timing	Motivation	Verification

Figure 7.6   Change Matrix

EXAMPLE 7.3   Making the Change

Effective leaders realize how difficult it is to change. They take the time to try to change something about themselves before asking someone else to change.

It is easy to tell someone else he has to change. If change is that easy, try to change one thing about yourself. Try changing. Consider exercising more, eating more healthfully, or breaking an undesirable habit. To help with the change process, consider the following points and questions:

What is the desired end result? Can you picture it? To change, a person must understand what the ultimate outcome of that change will be.

What actions will you take to make the change? To change, a person takes a series of actions which produces a result that moves the individual toward the desired outcome. This result may be seen and interpreted as either positive or negative. Based on the interpretation of the result, the person modifies his or her attitude and acts differently the next time in order to produce a more appropriate result.

What is the time frame for the change? Change takes time, yet if no timetable exists, a person is likely to say, "I'll get to that tomorrow." Effective change takes place when the person works toward the change according to a schedule.

How will you stay motivated? To change, it is necessary to stay motivated. Often, it is not easy to maintain the motivation and direction required to bring about change.

How will you know you have changed? What will your indicators be? A person needs to have some sort of feedback that enables her to understand how she is progressing toward the desired outcome. Otherwise, she may not recognize when the change is complete.

In the future, try using these questions and the matrix provided in Figure 7.6 to guide you when making a change.

WHAT MODIFICATIONS DO EFFECTIVE ORGANIZATIONS MAKE TO THEIR REWARD SYSTEM TO SUPPORT THE DESIRED CULTURE?

Too often, management leaders try to change the culture by describing the new attitudes they seek, but they fail to take the steps needed to influence the action employees take. Actions, not words, are the foundation of both a change and a culture. An effective manager makes sure that employees have the answers to the following questions:

How does the reward system reinforce behavior?

Are expectations aligned with results?

Are expectations and their alignment with organizational goals and objectives clearly communicated?

Regardless of the organization, in order to be effective, people need to know what their goals and objectives are. They need to know how to proceed toward those goals and objectives. They also need to have the appropriate tools, training, and skills to move toward those goals. During their movement toward their goals, they need to have a means of knowing how well they are doing, and they need to have a means for adjusting their work-in-process. All of this requires action and communication from leadership. Since customer needs and perceptions must drive the activities in an effective organization, internal employee cooperation and coordination must prevail. It is critical that corporate policies and programs reinforce the desired culture. When creating or modifying reward systems, consideration should be given to:

Customer expectations
Job descriptions based on identified customer expectations
Expected results
Performance expectations, including behavioral skills

How Do We Know It's Working?

Absences and lates directly impact the ability of an organization to meet their production demands. Schedules must be adjusted, overtime worked, replacements found, and work delayed. All of this results in higher costs, poorer deliveries, and lower productivity. Having employees ready to work when needed is crucial for organizational effectiveness. One of the many changes made at JQOS involved how absences and late arrivals were handled. Under the former management, unexcused absences and lates were tracked, but action was rarely taken. If punishment occurred at all, it was subjectively and inconsistently applied.

Before making any changes, the new owner did a thorough review of the previous two years' absences and lates. He discovered that the best worker missed just eight hours per year, while the worst worker missed 212 hours per year. These hours were compiled using just unexcused absences, where the worker never called in; late arrivals, again with no alerting phone call; and leaving early without telling anyone. This data did not include scheduled doctor's appointments, leaving during the day with supervisor's permission, or days missed when the person called at the very beginning of their shift. These numbers reflected workers who failed to show up, who failed to report in, or who left early. Over 80% of the workers missed very few hours, and just a few missed quite a lot.

Over several weeks, the owner worked with employees who volunteered to study the situation and develop an equitable absenteeism policy. Their formal recommendation included establishing a rolling 6-point system. Employees receive ½ point for each time they are late, 1 point for an absence, and 3 points when they are absent without calling in. At 3 points, they receive a verbal reminder. At 4 points, they lose their bonus and receive a written warning. At 5 points, they receive another written warning. Employees accumulating 6 points during a six-month period would be terminated. In a rolling system, as one month passes, the points gained in the oldest month drop off. This system does not include absences that are scheduled in advance or call-ins.

Following the volunteer employee committee's approval, the system was announced. The owner immediately received a long line of anxious employees waiting to see him in his office. Interestingly enough, each employee who came to see him had excellent attendance. The ones he was concerned about never said a word.

Employees tested the new system during the first year. Every once in a while, someone forfeited their bonus. The shop supervisor was careful to track this figure and encouraged workers to do the same. He administered pep talks when employees reached 3 points. Two employees have been terminated. One employee had a serious attendance problem to begin with. The other employee was a bit tougher to part with. Despite management's best efforts to help this employee get to work, he eventually accumulated 6 points. He had come close many times. Much discussion occurred when letting this employee go. Though frequently absent or late, he was a skilled machine operator. Because of his skill, most employees did not think management would follow through and terminate him. Many saw this as a true test of the system: would it be applied consistently? Though it was difficult, management understood the importance of fairly applying the system to everyone. No one doubts the system or management's intentions now. Overall, attendance has improved significantly, and shop production has increased.

Reward systems should recognize achievement. As employees enable an organization to reach its goals and objectives, the employees should be rewarded for their efforts. Since they are the direct recipients of rewards, employees should play an active role in creating or modifying a reward system. To ensure fairness, the reward system should employ a rating scale that accurately reflects actual employee performance, whether in a team or individually.

EXAMPLE 7.4   Clearly Set Expectations

Remember Example 5.3 where a visitor took a tour of a pristine plant in which the president of the manufacturing company "walked the talk"? The visitor encountered a similar situation on a different tour. This time, the plant he was touring was separated into two halves by a stout brick wall. Only two large openings existed in the wall, allowing room for forklifts and people to pass. When he went from one side of the wall to the other, it was as if two different worlds existed, as if the plant were actually two separate entities. On one side, the machining centers were dirty and disorganized. On the other side, the machining centers, aisles, and in-process storage were pristine. When investigating why the dramatic difference between the two areas existed, he was told that each area was under the responsibility of a different individual. He asked to be introduced to each of the plant managers. The plant manager of the clean and organized side of the facility was easy to find. The manager was out on the shop floor talking with operators about their processes.

Understanding that the visitor had many questions to ask, the plant manager invited him along while he went on his rounds. During their discussion, it became apparent that the plant manager established clear expectations about the cleanliness and organization level of the facility. It was clear that employees were rewarded for their efforts to maintain their equipment. They also received the appropriate training to do their jobs right. In order to support employees in their cleanliness efforts, at the end of each shift, 15 minutes were set aside to clean and tidy work areas and, more importantly, to jot down in log books comments about machine performance or maintenance needs. These log books were inspected each evening by a preventive maintenance team who quickly made repairs or adjustments based on the operators' comments. This plant manager focused on improving processes, removing the barriers that rob people of pride in workmanship, providing time and training to meet the expectations set, and using the reward system to reinforce behavior.

WHAT TYPES OF EDUCATION AND TRAINING DO EFFECTIVE ORGANIZATIONS PROVIDE FOR THEIR EMPLOYEES IN ORDER TO REMAIN COMPETITIVE?

Effective organizations avoid organizational myopia. They find ways to teach their employees new skills. They expose their employees to new situations and other organizations. As previous chapters have discussed, effective organizations seek to create alignment between the needs, wants, and expectations of their customers; the organization's strategies, goals, and objectives; and the day-to-day activities of their employees (Figure 7.7). Part of creating alignment is ensuring that employees have the appropriate education and training that enables them to perform their job requirements at a level that supports the ultimate customer's needs, wants, and expectations. Part of this education and training should include information about how quality, cost, schedule, and profit expectations for their jobs affect customer satisfaction. Effective employees understand the impact that their job has on each of these concepts. They are involved in the decisions that affect their jobs. Effective organizations provide their employees with the knowledge and skills necessary to excel in their jobs, as well as regular feedback so that they can monitor their impact.

Two of Dr. Deming's 14 points—"Institute training on the job" and "Institute a vigorous program of education and self-improvement"—focus on education and training. *Training* refers to job-related skill training and is usually a combination of on-the-job training and classroom-type instruction. Effective employees are provided with the appropriate training to give them the set of skills and knowledge needed to excel in their jobs. Job-related skill training prepares workers for the daily activities involved with their job. Such training should also include information that helps employees deal with experiences they may encounter only rarely. If training does not address problems that may arise infrequently, then workers will be required to handle those situations as they arise using only their own problem-solving and decision-making skills. If these infrequent situations are unsafe in any way, an accident or injury could result. Follow-up or refresher training is also key to skill acquisition. This type of training enables employees to maintain higher skill and performance levels by refamiliarizing the employees with the best practices and eliminating poor habits.

Compared to training, education is more broad-based; it provides individuals with a broader base of knowledge that helps them look at a situation from other dimensions. The education an individual receives may not be immediately applicable to the activities he or she is currently performing.

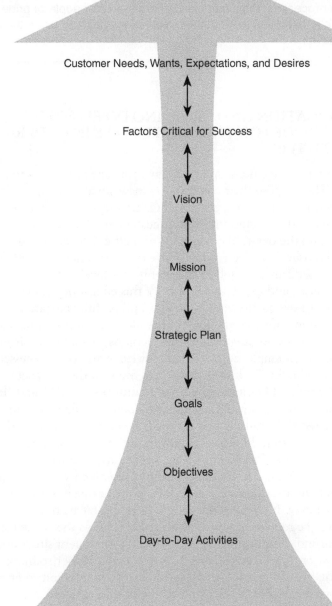

Customer Needs, Wants, Expectations, and Desires

↕

Factors Critical for Success

↕

Vision

↕

Mission

↕

Strategic Plan

↕

Goals

↕

Objectives

↕

Day-to-Day Activities

Figure 7.7   Creating Alignment

One key skill that effective organizations ensure their employees have is problem solving. Effective organizations train each and every one of their employees to isolate the root causes of problems, utilize quality tools to gain improvements, and lock in the gains they achieve to make the improvements permanent. Chapter 11 covers basic problem solving and introduces many quality tools for root cause analysis.

HOW DO EFFECTIVE ORGANIZATIONS USE TEAMS?

Effective organizations build high-performance teams. These teams contain people with complementary skills who work together to achieve their goals. As team members, they hold themselves responsible and accountable for goal accomplishment. Many types of teams exist, including:

> Management teams composed of heads of departments to do strategic planning
> Cross-functional teams with representatives from a large variety of areas for the design or development of complex systems
> Self-directed work teams made up of employees grouped by complementary skills in order to carry out production processes
> Project or problem-solving teams, which are often temporary groups of individuals from the appropriate functional areas with the necessary skills to work on a specific task

Teams are often given the task of investigating, analyzing, and finding solutions to problems. Project teams consist of people who have been given a mandate to focus on a particular process, area, or problem. Generally, this team is composed of those closest to the problem as well as a few individuals from middle management who have the power to effect change. The team may consist of people from a variety of departments depending on the problem facing them. The team may even include an outside vendor or a representative from the customer base. Upon resolution of the problem, the team will be disbanded or reorganized to focus on another project.

Henry Ford, founder of Ford Motor Company, said:

> Coming together is the beginning. Keeping together is progress. Working together is success.

Effective organizations recognize that teams are crucial to solving issues and problems facing the organization. Like Henry Ford, they also realize that teams do not always coalesce into highly functional entities without help. There are several stages of team development. Recognizing that teams experience growth throughout their existence helps leaders guide and direct team activities.

During the first stage the team forms. This *formation* stage is usually experienced in the first few meetings. During this time the team establishes its goals and objectives. It also determines the ground rules for team performance. For teams to work well, leadership must set clear goals that are aligned with the mission and strategic direction of the firm. When leadership sets the direction, the team is much more focused and tends not to get bogged down in the problem-selection process. The team must know the scope and boundaries within which it must work. Leadership must communicate how the team's progress and performance will be measured.

Oftentimes, following their initial formation, teams experience a rocky period during which team members work out their individual differences. This is the time that the team gets acquainted with one another's idiosyncrasies and the demands of the project. During the *stormy* stage, the goals and scope of the team may be questioned. Since a team is composed of a group of individuals who are united by a common goal, the best teamwork will occur when the individuals focus on the team's objectives rather than personal motives. While working together, team members must understand and agree on the goals of the team. They must establish and adhere to team ground rules for behavior and performance expectations. To ensure harmony in the team, all members must participate, and responsibilities and duties must be fairly distributed. Each team member must understand his or her role in the completion of the project. Knowledge of how internal or external constraints affect the project is also helpful. Team members must possess a variety of skills, including problem-solving skills, planning skills, facilitation and communication skills, and feedback and conflict management skills.

The third stage of team development occurs when the team starts to work together smoothly. It is during this *performing* stage that things get accomplished. To be successful, teams need the appropriate skills in a supportive organizational culture, and they need the authority to do the job that they have been asked to do. Leadership can do a lot to rid the team of the barriers that inhibit its performance. These barriers include inadequate release time, territorial behavior from involved functional areas, lack of training, inadequate support systems, lack of guidance or direction, and lack of recognition. Senior leadership's sincere interest in and support of the resolution of the problem is evidenced by their willingness to commit money and time for training in problem solving and facilitation. In any case, senior leadership must monitor and encourage their teams to solve problems. The teams will quickly become unmotivated if the solutions they propose are consistently turned down or ignored. Leadership support will be obvious in management's visibility, diagnostic support, recognition, and limited interference.

As the team finishes its project, the final stage—*concluding*—occurs. During this phase team members draw the project to its conclusion, verify the results, and disband the team. Several key events take place during this time. The team, having taken action, perhaps by implementing a solution to a problem, must verify that what they planned to do got done and what they did actually worked. Teams are not finished when they have proposed a plan of action. Teamwork is finished when the plans have been acted on and the results judged effective. Until then, the team cannot be disbanded.

EXAMPLE 7.5 Providing Leadership: Setting Goals and Objectives

At Perfect Choices, a mail-order catalog clothing company, the 1-800 number service had bogged down. Callers wishing to place an order were put on hold for an unconscionable amount of time. Once placed on hold, some callers merely activated their speaker phones and went about their business, picking up when the speaker phone signaled that a connection had been made. This high number of callers on hold for long periods of time was extremely costly not only in terms of toll-free charges incurred by Perfect Choices, but also in terms of business lost as disgruntled shoppers hung up.

When the team originally formed to solve this problem, it was given the brief, not very specific charter: *improve the 1-800 number.* After several meetings discussing what this meant, the real issue surfaced. Everyone on the team was aware of the company vice president's tight-fisted behavior when it came to spending money. The team members wanted to know how much they were allowed to spend making the 1-800 number better before working further on the problem. After all, adding more operators or increasing the number of 1-800 lines could be costly solutions.

Finally, the leader of the team was brave enough to invite the vice president to a meeting. At this meeting, after much hemming and hawing, the question of budget was laid before the man. His response was straightforward. He provided figures on how much the existing 1-800 number was costing the firm in terms of base service cost plus overrun charges for the amount of time people spent on hold waiting for service. He pointed out that the overrun charges were wasteful and could be put to better use in improved service. When asked why he did not provide this information earlier, he said he did not realize they did not have access to it.

Given this more specific guideline, the team was able to use a variety of problem-solving tools to determine the root cause of the problem. Analysis revealed that several key departments were not on the 1-800 number. Since callers on the 1-800 number could not be transferred to another number, the customer service representatives tried to solve a caller's problem by using internal lines to contact the appropriate departments, find answers to the caller's questions, and then relay the information back to the caller. This use of the middleman wasted time for everyone. The team used quality tools, including check sheets and Pareto diagrams, to determine which departments were most frequently called and thus should be on the 1-800 number. With this information they reorganized the existing 1-800 service to include the appropriate departments.

After making the appropriate changes, the team found that the 1-800 number served 50% more callers at 25% lower cost. The average wait time had been reduced from 25 minutes to fewer than 3 minutes during peak loads, such as immediately after a catalog is sent, and less than 1 minute otherwise. Callers, employees, and the vice president were all much happier with the revised 1-800 number.

HOW DO INDIVIDUAL PERSONALITIES AFFECT TEAM PERFORMANCE?

Teams are composed of individuals. Team leaders and members need to recognize each individual's knowledge and skill base so that the individuals can be assigned tasks that utilize these skills effectively. Less likely to be understood is that the individuals on any team possess a wide variety of personality traits. One key to helping teams function more effectively is to acknowledge those personalities and use them to the team's advantage.

As early as the 1870s, doctors recognized the concept of left-brain functions versus right-brain functions. As shown in Figure 7.8, left-brain-dominant individuals typically are more analytical and logical. Right-brain-dominant individuals, on the other hand, are more emotionally and visually oriented. In a team meeting left-brain-dominant activity may be recognized when a person focuses on listing or categorizing activities. These individuals like to arrange and coordinate schedules, which makes them good people to have on hand when analyzing processes

Left-Brain Functions	Right-Brain Functions
Logic	Feelings
Reasoning	Visualizing
Judgment	Insight
Analytical	Intuition
Mathematical Ability	Recognizing Similarities
Lists/Categories	Recognizing Patterns
Time Management Ability	Insight
Reading	Spatial Perception
Writing	Seeing Whole Things at Once
Speaking	Synthesizing
Verbal Memory	Visual Memory

Figure 7.8   Comparison of Left Brain versus Right Brain

and determining which activities add value and which do not. They are concerned about time and money estimates. Staying on track and on target is important to these individuals.

Right-brain-dominant activities may be evidenced by individuals who want to focus on discussing new ideas. Since they possess strong caregiver traits, they are concerned about other team members' attitudes and feelings. Because they like to talk things through, they are very good at evaluating customer needs and determining why the customer may want a certain function or service. They are good at visualizing the entire system in operation, and they can recognize patterns that may indicate the most promising steps to complete a task or process.

When asked to list which elements of left-brain or right-brain functions they possess, most individuals will come up with a combination selected from both the left- and right-brain functions. Those who study human personalities have proposed a variety of different theories on the makeup of people. One popular theory is the four-quadrant brain. This theory, first proposed by Ned Herrmann, describes four basic thinking preferences related to personalities (Figure 7.9).

Individuals displaying analytical, type A, personalities enjoy working with data, facts, and money. Their approach to problems is very logical. They are interested in answering the question "What?" as in "What are the facts here? What are the expectations? What needs to be done?" On a team they are the individuals who encourage the use of data and factual information. They enjoy digging deep to get the facts related to a situation. They also like to analyze these facts and the data surrounding them before making a decision. They are able to present the facts in a direct, clear, brief style which makes them easy to understand and follow during a presentation. When they are working on a project, their attitude will be realistic when it comes to meeting deadlines and budgets. Because they are good with numbers, they can serve as excellent individuals to monitor and audit project money issues.

Type B individuals are the planners and organizers. They are focused on getting things done. These individuals are interested in structure, details, plans, and schedules. They are very disciplined, so they like nothing better than developing and following clear plans and schedules. The step-by-step activities needed to complete a project are clearly visible to them. These are the people who are interested in creating the plan and then working the

A Analytical	Leader D
Wants to know "What?"	*Wants to know "Why?"*
Logical	Visual
Quantitative	Conceptual
Analytical	Holistic
Critical	Imaginative
Technical	Intuitive
Factual	Innovative
Data Focused	Future Oriented
Precise	Curious
Accurate	Risk Taker
Direct/To the Point	Impetuous
Structured	Communicative
Detailed	Interpersonal
Planning	Spiritual
Disciplined	Feeling
Organized	Sensory
Sequential	Emotional
Neat	Supportive
On Time	Talkative
Reliable	Sensitive
B Planner	**Communicator C**
Wants to know "How?"	*Wants to know "Who?"*

Figure 7.9   Herrmann's Four-Quadrant Personalities

plan. They want to answer the question "How?" as in "How are we going to get the work done?" They are thorough in their approach to projects and problem solving. They can be counted on to be on time and reliable. Type B people are great to have on a team because they keep projects on track and on schedule.

All teams need type C individuals. These individuals possess great interpersonal skills, so they provide two significant benefits to a team. They are the peacemakers who can talk to anyone on the team and keep the team working together. Their focus is on answering the question "Who?" as in "Who are the people affected by this project, decision, or problem?" Since they are sensitive to the feelings of others, they like to talk through issues. They are very expressive and are excellent speakers. They make good teachers. Because of their strong communication and interpersonal skills, they are also the people who can help a team determine what the customer needs, wants, and requires.

Type D individuals are natural leaders. Since they have far-reaching vision and see things in context, they have an eye for the future and are very interested in innovations. They possess lots of imagination and are quite spontaneous. They like to approach situations with the freedom to explore ideas. They like to have fun during problem-solving and project efforts. These attributes combine to make them excellent team members because they are the "idea people." When difficulties arise, they are the ones with the creative and innovative approaches to solving them.

Trustworthy
Show respect to other team members
Positive attitude
Self-motivated
Contributes positively
Listens well
Participates
Encourages the participation of others
Openly shares knowledge, concerns, expectations
Accepts the results of group decisions
On time to meetings or team activities
Attends meetings and team activities
Organized
Provides creative/intelligent approaches to solutions
High quality of work
Accomplishes tasks on time and on schedule
Communicates well
Can be trusted to get the work done
A team player
Shows initiative
Prepares well for activities and meetings
Stays on track

Figure 7.10   Positive Team Behaviors

It is easy to see how a team could benefit from having all four personality types present on the team. A type D individual can set the direction for the team and provide innovative and imaginative solutions to problems. A type B person can create and execute plans supporting the vision and goals set by the type D person. The role of information gatherer and analyzer falls to the type A person. Type A people are the ones who can tell you what can actually be done given the available time, money, and talents. A type C person is critical to the team because type C people can help the other members get along and they can communicate effectively with customers.

Not every individual fits neatly into one category. Each person possesses some traits from each of the categories. It is also possible, with time, experience, and training, to develop and enhance traits from different categories. For instance, a strongly type A person can learn to be a better listener, just as a strongly type B person can learn to think outside the box and be innovative. Regardless of a person's strengths or weaknesses, team members should practice positive team behaviors (Figure 7.10).

WHAT NEEDS TO HAPPEN IN ORDER TO HAVE EFFECTIVE MEETINGS?

Sharing information with employees is critical to organizational success. The right information, delivered at the right time, can show people how to become less wasteful and more efficient. Information is often shared in meetings. Unfortunately, meetings are often seen as long, boring, and unproductive time wasters, and they can be if the meeting lacks sound leadership.

Without good leadership, a meeting can become a series of digressions on why an issue exists rather than an emphasis on how the issue will be dealt with. Unfortunately, one bad meeting usually leads to another because in a poorly run meeting, decisions are rarely made and responsibilities rarely assigned. Poorly run meetings nearly always exhibit the same problems: no specific or clearly defined objective(s) for the meeting, leaders, or participants; no meeting agenda; unprepared leaders and participants; and the wrong choice of participants.

In order to have effective meetings, care should be taken to incorporate several principles:

1. Determine the objective of the meeting.
   - Why is the meeting going to be held?
2. Determine who should participate.
   - Who can influence the fulfillment of the meeting objective?
3. Set an agenda.
   - What is the plan of action for the meeting?
4. Prepare for the meeting.
   - What is needed in order to provide answers and save time? Experts? Information? Key participants?
5. Run the meeting.
   - How will the agenda be used to keep the meeting on track?
6. Make decisions.
   - How will the issues be dealt with?
7. Assign responsibilities.
   - Who will be responsible for achieving results?
8. Following the meeting, verify that individuals have dealt with their assigned issues.

These principles enhance meeting effectiveness, whether an individual is a meeting participant or a meeting leader. Preparation for a meeting makes the meeting work for, rather than against, a participant or leader. If all participants know why a meeting is being held, they can come prepared to participate in the meeting. If the leader knows what they want to achieve during as well as after the meeting, he or she can set the agenda to guide this effort. Meetings are more successful if the participants are limited to the individuals who are directly connected to the issue. They will be the ones who will take an active interest in finding a way to fulfill the objective of the meeting. Remember, the number of people in a meeting is often directly proportional to the length of the meeting.

Having an agenda and following it are key aspects of having a successful meeting. A good agenda will state:

- The objective of the meeting
- The issues that will be discussed
- The date and beginning and ending times of the meeting
- The location
- The participants
- The preparation expected of the participant

With an agenda, it is easier to keep a meeting on track and to reach a conclusion about the issues under discussion. Meetings are successful only when they bring about results.

Following a meeting, minutes should be prepared and distributed. These minutes should include:

The date
The attendees
The objective of the meeting
The major discussion points
The decisions reached
The actions to be completed including when and by whom
The time, date, and location of the next meeting if there is to be one

HOW CAN THE MALCOLM BALDRIGE NATIONAL QUALITY AWARD CRITERIA HELP AN ORGANIZATION'S HUMAN RESOURCES BECOME MORE EFFECTIVE?

Section 5 of the MBNQA criteria, Human Resource Development and Management, examines how an effective organization encourages its workforce to develop to its fullest potential. This section of the award concentrates on how the efforts of the workforce must be aligned with the organization's overall objectives. Applicants are asked to describe how their organization achieves a high level of performance through job and compensation structure and workforce practices. The criteria recognize that effective organizations have effective employees. They ask questions about employee education and training programs that build employee knowledge, skills, and capabilities. The criteria study how the work environment within an organization contributes to employee well-being and motivation.

Employee activities throughout an organization must be aligned with the goals and objectives of the organization. Ask the following questions based on the MBNQA criteria to help determine whether an organization has integrated the concepts of employee motivation and effectiveness throughout its organization:

Leadership

- Is leadership knowledgeable about employee attitudes?
- Are leaders converting policies into actions related to employees?
- Does leadership encourage communication at all levels?
- Does leadership share data openly at all levels where appropriate?
- Does leadership utilize consensus decision making?

Strategic Planning

- Has the strategic plan identified employee education and training supportive of the business objectives?
- Has the strategic plan mapped out a strategy for improving employee performance?

Customer and Market Focus

- Is the organization communicating useful customer and market information to the employees in a manner that improves employee job performance and enhances customer service?

Measurement, Analysis, and Knowledge Management

- How are employees gathering, analyzing, disseminating, and using information related to customers and key processes?
- Does the organization have communication systems in place that enhance its employees' abilities to obtain and use information?
- Do employees understand how their work efforts affect key performance measures?

Workforce Focus

- Do employee policies and reward systems support improvement activities related to key processes?
- Do employee policies and reward systems support education, training, and self-improvement?
- How does the organization engage the workforce to achieve organizational success?
- How does the organization engage the workforce to achieve personal success?
- How does the organization create a culture conducive to high performance?
- How does the organization encourage ethical business practices?
- How does the organization ensure and improve workplace health, safety, and security?
- How does the organization recruit, hire, place, train, and retain employees?
- How does the organization align day-to-day activities with core competencies and customer focus?
- Do employees receive feedback on their work? How?
- Do processes help employees learn from failure, or do they punish employees for failure?
- Does the organization place a high priority on enhancing employee capabilities?
- Does the organization vary the rewards to match rewards with contribution levels?
- Does the organization treat everyone with respect, acknowledging everyone as being equally important?

Process Management

- Have the key processes supporting a customer focus been identified and improved by enhancing employee involvement, skills, and capabilities?
- What process does the organization use to improve the work environment?
- Do the organization's policies concentrate on why problems occur rather than on who caused them?

Results

- Does the organization's skill at managing key processes include utilizing employees' skills, knowledge, and capabilities?
- Have the changes that have been made to employee policies improved business results?
- Has the organization successfully integrated employee policies, strategic planning, and customer and market information? Do the business results reflect this?

LESSONS LEARNED

Effective organizations follow good human resource practices. The involvement of leadership, through supportive actions, provides employees with the knowledge and skills they need to do their jobs and provide value for the customer. Effective organizations engage in a continuous cycle of learning and training. To support this cycle of continuous improvement, effective organizations design recognition and reward systems that motivate their employees. These rewards are aligned with customer needs, wants, and expectations as identified in the strategic plan and supported leadership actions. Effective organizations utilize teams to solve problems and enhance processes. Participants on these teams can expect their talents, skills, and personality types to be appreciated and used to their best advantage.

1. Employees play a significant role in creating value for customers.
2. Effective leaders communicate clearly and effectively.
3. Effective leaders understand how to motivate people.
4. Effective leaders know how to lead the change process.
5. Reward systems can be used to support change and organizational culture.
6. Effective organizations avoid organizational myopia by finding ways to teach their employees new skills.
7. Effective organizations use high-performance teams.
8. Effective leaders know how to run effective meetings.

Are You a Quality Management Person?

It's the little day-to-day ways that you treat people that identifies you as a quality management person. Check out the quality management person questions and find out whether or not you are a quality management person. How do you answer them on a scale of 1 to 10?

Do you treat other people with respect?

| 1 | 2 | 3 | 4 | 5 | 6 | 7 | 8 | 9 | 10 |

Do you treat other people fairly?

| 1 | 2 | 3 | 4 | 5 | 6 | 7 | 8 | 9 | 10 |

Do you actively listen to what other people are trying to communicate?

| 1 | 2 | 3 | 4 | 5 | 6 | 7 | 8 | 9 | 10 |

Do you admit your mistakes, determine the root cause, and learn from the experience?

| 1 | 2 | 3 | 4 | 5 | 6 | 7 | 8 | 9 | 10 |

Do you celebrate successes and improvements?

```
1  2  3  4  5  6  7  8  9  10
```

Do you provide positive feedback to the people you interact with on a day-to-day basis?

```
1  2  3  4  5  6  7  8  9  10
```

Are you careful about how you provide feedback to people?

```
1  2  3  4  5  6  7  8  9  10
```

Do you make changes and improvements in areas that are important to you?

```
1  2  3  4  5  6  7  8  9  10
```

Do you think about making a change but never do?

```
1  2  3  4  5  6  7  8  9  10
```

If you have had trouble making an important change, have you used a more formal change process to help?

```
1  2  3  4  5  6  7  8  9  10
```

Do you help others make changes and improvements in areas that are important to them?

```
1  2  3  4  5  6  7  8  9  10
```

Do you try to be a team player?

```
1  2  3  4  5  6  7  8  9  10
```

Do you strive to communicate clearly and effectively?

```
1  2  3  4  5  6  7  8  9  10
```

Do you recognize your personality type and use this information to help you get along better with others?

```
1  2  3  4  5  6  7  8  9  10
```

Chapter Questions

1. Find and read the article "On the Factory Floors, Top Workers Hide Secrets to Success" by Timothy Aeppel, *The Wall Street Journal*, July 3, 2003. What are your feelings about employees sharing their knowledge? How can leadership encourage this sharing?

2. Describe a situation where a leader motivated you. What did he or she do? How did you react?

3. Describe a change you were required to make. How did it go? Was the change accomplished? How? What motivated you?

4. Change one thing about yourself. Use the model provided in Example 7.3 as a guide for your change process. Recognize that making the change may take time, and set a timetable for your change.

5. Having changed something about yourself, how would you encourage someone else to make a change?

6. Describe the phases of team development.

7. For each of the phases of team development, provide an example from your own experience describing how your team got through the phase.

8. Study the left-brain/right-brain information presented in Figure 7.8. How would you describe your category? Provide examples showing how you know you fit in this category. Knowing this, how will you change your attitudes and behaviors when working on projects?

9. Study the quadrant types presented in Figure 7.9. How would you describe your personality? What type are you? Provide examples showing how you know that you fit in this category. Knowing this, how will you change your attitudes and behaviors when working on projects?

10. What is the difference between education and training? Why is it important to have both?

11. Describe a situation where you received (or did not receive) job skill training. Was it adequate? Why? Why not? What would you have done differently?

12. Describe a situation where you received (or did not receive) an educational experience. Was it adequate? Why? Why not? What would you have done differently?

13. What key items must be in place in order to have an effective meeting?

14. What is an agenda? How does it help create an effective meeting?

Section 5.0: Human Resource Focus

The Human Resource Focus section of the MBNQA criteria addresses how an organization motivates and develops employees to their fullest potential. It also examines how the organization builds and maintains a work environment that enables employees to do their jobs in a way that provides value to the customer while enabling the organization to grow.

The key goals of this section are to:

1. Examine the performance and improvement of Remodeling Designs, Inc. and Case Handyman Services in key human resources areas:
   a. Motivation
   b. Training
   c. Communication
   d Work systems
   e Work environment
   f. Employee well-being

This section of the review was created by Ryan McDonald, Brian Tobin, Joan Cordonnier, and Donna Summers.

 Human Resource Focus Questions

5.1 WORK SYSTEMS

1. Does your company have any incentive plans that might make your employees want to work harder?
2. Do your incentive plans/reward plans encourage customer focus? How is customer focus measured?
3. How can you get in touch with everyone in the company during work hours?
4. Do new employees have a mentor to teach them the ropes? If yes, how is the mentor selected?
5. What types of motivation techniques does your company have?
6. What is done to make managers encourage "above-and-beyond" work?
7. Do customers rate employees on their work?
8. What type of training do managers need?
9. What characteristics do you look for in employees?
10. What is the process for hiring new employees?
11. How does your company take in new and diverse ideas, even when going though hard times?

5.2 EMPLOYEE LEARNING AND MOTIVATION

12. What types of education and training do your employees get?
13. How do you make sure that your employees are trained correctly?
14. How do you make sure that employees' education and training continue to develop over time? Do you have continuous training?
15. How is information used to make training better for employees?
16. How do you make sure that your employees' skills keep up with more modern technology—training or inspection?
17. Do you teach employees the proper way to do their work to reduce their risk of injury on the job? If yes, how is this training done?
18. Do you have special initial training for someone who joins the organization?
19. Do you have a program whereby someone can learn how to become a better manager or leader?
20. How do you educate your employees? Do they learn from a book, or do they learn by actually doing the work?
21. Do you obtain any information on how education and training could be better? If yes, how is this done?
22. How do you evaluate the education and training to make sure that is it effective? Do you do this by individual employee performance or by overall organizational performance?
23. Do you have a mentoring program? If yes, how does this program affect the work of the employees?
24. What types of results demonstrate that your company is benefiting from training?
25. How do you measure training with regard to customer relations? What types of jobs are directly related to dealing with customers? What type of training do employees get on customer training?
26. How do you reinforce the use of knowledge and skills on the job?

5.3 EMPLOYEE WELL-BEING AND SATISFACTION

27. How do you know that you have all the necessary tools for the job that you are going to do?
28. What types of safety precautions does your company take? How is human error taken into consideration with your safety precautions?
29. Are there target times for completing certain jobs? Are employees rewarded for being on target?
30. Do employees fill out surveys to determine whether they are motivated and satisfied at their job?
31. What key factors affect employee motivation, satisfaction, and well being? How do you measure the effectiveness of the key factors? Are they the correct factors?
32. What types of measurements do you use to assess employee satisfaction and well-being?

33. How many times are your workers late or absent?
34. Do your workers know their job role and how it affects the direction of the organization?
35. Have you encountered any problems with employees and working conditions?
36. How are employee surveys used to help your business? How are these results communicated to others?

5.0 Human Resource Focus

5.1 WORK SYSTEMS

This section examines how your organization creates work systems and employee opportunities that enable employees to reach their fullest potential.

5.1a Organization and Management of Work

Because they are actually two companies, Remodeling Designs and Case Handyman are able to stay current and react flexibly to changes in the business world. Although they are two separate entities, their work systems and employee relations are similar. The work systems at Remodeling Designs and Case Handyman are designed to enable employees and the organization to achieve high levels of performance. Remodeling Designs and Case Handyman promote cooperation, initiative, and innovation by providing their employees with the tools and training they need.

Communication is a critical part of the job. It is easy to communicate with any employee during work hours because each one has a cell phone provided by the company. This connection to others enables employees to ask questions as they arise, order parts from a store, or ask advice from peers.

5.1b Employee Performance Management System

Employees are motivated to work to their full potential by being recognized for their workmanship. One way of making sure that customers are satisfied and employees are motivated is leadership's efforts to match personalities to the job. Erich and Mike assign jobs according to the customer's personality, the size of the job, and the employees available.

Employees' efforts on the job are measured by several techniques, including the use of client feedback evaluations. These forms let the managers know what the customers liked and disliked about the work that was done. The customer is also asked to answer a few questions relating to the attitude and performance of the worker. These forms and the information they contain are relayed back to the employee during regular employee meetings.

5.1c Hiring and Career Progression

Case Handyman advertises for new employees whereas Remodeling Designs utilizes a referral system. New employees for both companies are required to take a service skills self-evaluation as well as a carpentry knowledge survey. When these assessments reveal that the employee needs more training in a certain area, efforts are made to ensure that on-the-job training takes place. This training involves placing the new employee with an employee who is certified in that specific area. When employees become certified, they are able to do jobs on their own. This type of on-the-job training develops leadership skills among older employees.

A new employee "pitch" book is used during the hiring process. This pitch book explains the direction in which the company wants to go and the performance measures they use to evaluate their progress. It also provides succinct information about what Remodeling Designs and Case Handyman expect from their employees. Current employees participate in the hiring process, and after interviews they are asked to evaluate the potential employee for both technical and social skills. Since the company is not large, everyone must get along. Each new employee has a two-month probation period during which his skills and ability to fit into the organization are judged. The hiring process is shown in Figure 1.

Career progression is encouraged. Employees gain insight into their abilities through the customer surveys. Another measure that is used to judge performance is the Job Cost Profit sheets which show how each employee is performing on the jobs to which he or she has assigned. This evaluation takes place weekly. To improve their skills, employees are encouraged to take classes to attain Certified Lead Carpenter status. Remodeling Designs also encourages competition for awards offered by national remodeling associations.

5.2 EMPLOYEE LEARNING AND MOTIVATION

Training is crucial to Remodeling Designs and Case Handyman. Each Tuesday, training sessions, known as "Tool Box Talks," are held. During these talks, employees learn, for example, what chemicals are harmful, how to avoid accidents, and how to improve their productivity. Employees also share their knowledge, skills, and experience with each other. Employees are able to discuss issues that have arisen, including different ideas of the way to do a particular job. These meetings are excellent sources of information for all employees.

Remodeling Designs and Case Handyman encourage education and training. Much of the training that they sponsor consists of peer training and in-the-field training. Employees are encouraged to take classes related to their respective fields. The companies will pay for the classes as long as the employee receives a passing grade. For instance, a secretary took a grammar class to help with writing letters. Another employee took a class on how to use Microsoft® Project to help improve job scheduling and tracking. Each year, employees attend conventions and remodeling shows hosted by the National Association of Remodeling Industries. These education and training opportunities support the organization's overall objectives and contribute to high performance.

5.3 EMPLOYEE WELL-BEING AND SATISFACTION

Employees are motivated to reach their fullest potential by the way that they are recognized for their workmanship. Incentive plans are not part of the companies, but they do offer such perks as trips for five years of service, jackets for one year of service, and

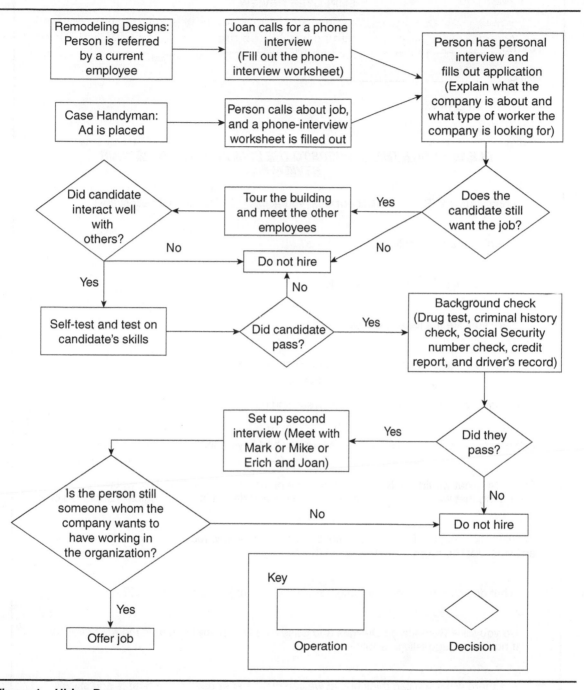

Figure 1   Hiring Process

EMPLOYEE REVIEW

NAME:_____

DATE:_____

REVIEW DATE/TIME:_____

PLEASE WRITE THE ANSWERS TO THE FOLLOWING BEFORE YOUR REVIEW.

Describe your job. (Really think about what you *actually* do. Use back if necessary.)

What are the most difficult parts of your job?

What do you like the most about your job?

What do you like the least about your job?

What is the greatest challenge you face in completing your job?

In the past year, what is the greatest success you had in your job?

In the past year, what is the greatest failure you had in your job?

For 2004, what would you like to most improve on related to your job? This would be something that we would look at a year from now and determine if you actually improved on it.

We will talk about the following questions during your review. You do not need to write down answers. Just be prepared to talk about them.

What do you need from us in order for you to do your job more effectively?

Do you think Remodeling Designs and Case Handyman are headed in the right direction? If not, what suggestions would you make?

Are there ways you can think of that we can make better use of your talents and skills?

Figure 2   Employee Review Form

company-sponsored awards. The company also sponsors summer fun events and a Christmas party.

Remodeling Designs and Case Handyman employee benefit policies help maintain a supportive work environment. Vacation days, personal leave time, education/training funding, and family health insurance are all part of the benefits offered.

This is a family-run company that is aware of family responsibilities. Employees who need to leave to deal with family issues or obligations are not penalized. When a situation arises, as long as the employee calls in and explains why he or she is unable to work on a given day, management tries to work with the employee.

When jobs are completed, the employees review those jobs. The review contains questions such as:

> How would you describe the job?
> What aspect of the job did you like the best?
> What did you like the least?
> What would you change if you could?
> What was your best success?
> What was your worst failure?
> What do you feel you should have been paid?

If the employees are unhappy or dissatisfied with some aspect of the job or future jobs, Mike and Erich work with them and attempt to make positive changes. Figure 2 shows the Employee Review form used in an employee's yearly review.

Turnover rate is very low. Most of the employees are there for life. Only one employee left; he did so because he wanted to further his career by starting his own remodeling company. Remodeling Designs and Case Handyman have been expanding and adding employees since their beginning.

8 Managing the Supply Chain

What is a supply chain?

What are the benefits of effective supply chain management?

What are the elements of effective supply chain management?

What is inventory management?

Why is information sharing so critical when creating an effective supply chain?

How does e-commerce relate to supply chain management?

What is logistics?

What role does purchasing play in supply chain management?

What challenges face the development of an effective supply chain?

How does supply chain management relate to the Malcolm Baldrige National Quality Award?

Lessons Learned

Learning Opportunities

1. To become familiar with the concept of supply chain management

2. To understand the benefits of effective supply chain management

3. To become familiar with the elements of supply chain management

4. To understand the value of strategic partnering

5. To understand the importance of information exchange across a supply chain

6. To know the steps and challenges to creating an effective supply chain

Companies don't compete against other companies; they compete against each others' supply chain.

WHAT IS A SUPPLY CHAIN?

Moving raw material through conversion to a product to the consumer takes effort. Consider what it takes to get a loaf of bread to a consumer. The wheat must be planted, cultivated, harvested, stored, and shipped to a mill. The mill refines the raw material, bags, and transports the flour to a bakery. The bakery combines a variety of other raw materials that have also come through a supply chain and bakes the bread. It is then wrapped, loaded onto a truck, transported to a store, unloaded, sorted, and shelved. Finally, it is available to the consumer for purchase. How many days does that take, from raw material to finished product? In the big scheme of things, the raw material cost is inconsequential to the total cost of traveling through the supply chain. Supply chains create value through two aspects: the physical movement of materials and exchange of information about those materials. As goods and services flow through the supply chain, various organizations add value to them. In today's global economy, a well-coordinated supply chain is a competitive necessity. Companies don't compete against other companies; they compete against each others' supply chains (Figure 8.1). Effective supply chains provide organizations with a strategic advantage.

A *supply chain is the network of organizations involved in the movement of materials, information, and money as raw materials flow from their source through production until they are delivered as a finished product or service to the final customer.* A supply chain may involve any number of organizations including multiple suppliers, wholesalers, manufacturers, warehouses, distribution centers, retailers, and customers (Figure 8.2). The links between these units must be strong in order for an effective supply chain to exist. Supply chains are driven by three factors: materials, money, and information. Throughout the chain, each organization is both a supplier and a customer (Figure 8.3). A complete supply chain starts with raw materials, flows through various organizations, and ends with the final customer (Figure 8.4). Raw material or parts or information flows in from the supplier side, value is added through internal processes, and then the item is distributed to the customer. Supply and demand issues occur throughout the chain. The supply chain links all the components in order to meet customer demand.

Effective organizations focus on supply chain integrity. This means they carefully manage the information, materials, and services that flow up and down the supply chain. A well-designed supply chain ensures that the right product or service is in the right place at the right time at an affordable price. When designing a supply chain, they ask questions like:

What products will travel through the supply chain?
What services will travel through the supply chain?
What information will travel through the supply chain?
What is the structure of the supply chain?
How will each segment of the supply chain be managed and optimized?

WHAT ARE THE BENEFITS OF EFFECTIVE SUPPLY CHAIN MANAGEMENT?

As discussed in previous chapters, today's effective organizations focus on creating customer value. For this reason, they optimize the value creation process by improving processes, implementing lean principles, optimizing asset utilization, and monitoring quality assurance. An

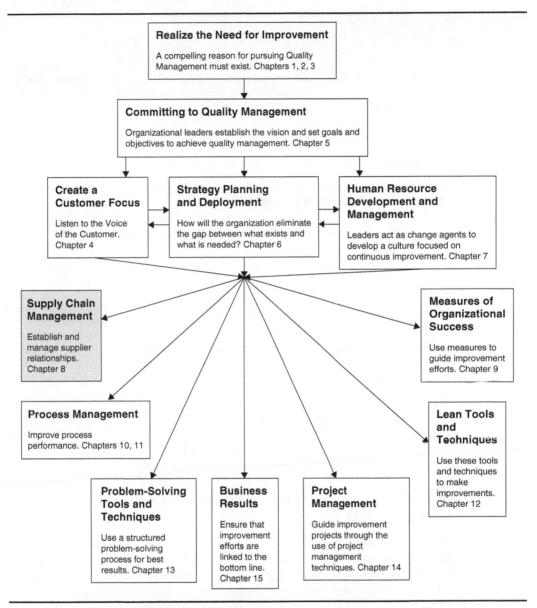

Figure 8.1   Achieving Organizational Success

effective supply chain supports the value creation process. Everything from the kind of weather to market demand can affect a supply chain. To perform well, an effective supply chain must be designed with quality, cost, flexibility, speed, and customer service in mind. A well-designed supply chain ensures reliability, adaptability, reduced costs, and appropriate asset utilization. Effective supply chains are able to provide order fulfillment, on-time delivery,

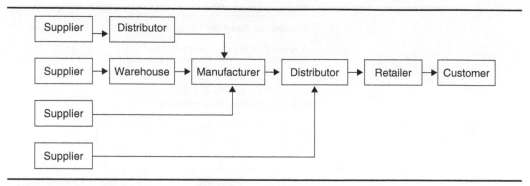

Figure 8.2 Participants in a Supply Chain

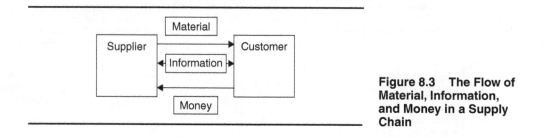

Figure 8.3 The Flow of Material, Information, and Money in a Supply Chain

short response time, high value-added input per employee, high inventory turns, and short cycle times at a reasonable cost. Well-managed supply chains benefit effective organizations by enabling them to meet the challenges facing them. They are able to achieve their objectives through improved supplier quality, delivery, and price.

WHAT ARE THE ELEMENTS OF EFFECTIVE SUPPLY CHAIN MANAGEMENT?

Effective organizations manage their supply chains. They do so in order to improve business operations, manage inventories, and reduce costs. Global economic conditions have combined to force companies to carefully consider their supply chains. As transportation costs increase and outsourcing opportunities arise, companies are faced with decisions affecting their ability to move their product to market. The rise of e-commerce has changed the way many companies do business from both a buying and a selling point of view. All of these changes have resulted in more complex supply chains. Typical issues that face any organization include determining what products or services are needed, how many, and when; in-sourcing or outsourcing the manufacture or processing of goods or providing of services; managing the trade-offs between market demand and the cost of holding inventory; evaluating, selecting, communicating, and working with suppliers; monitoring supplier quality and quality assurance; and deciding how to best package, transport, and store materials. *The critical elements of supply chain management are inventory management, information sharing, e-commerce, logistics, and purchasing.*

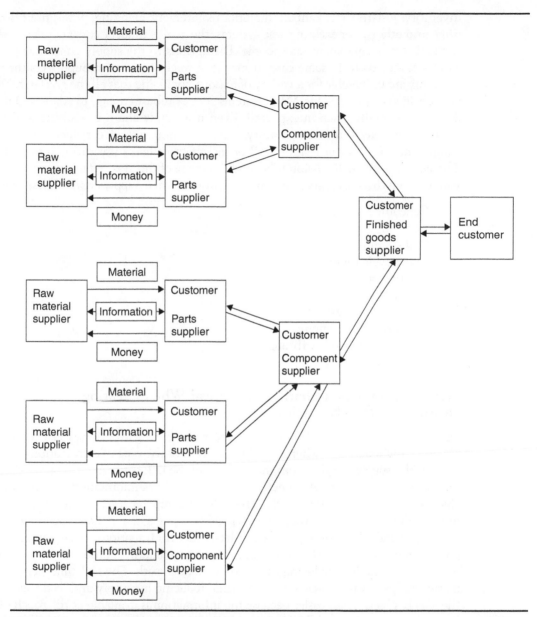

Figure 8.4 A Complex Supply Chain

What Is Inventory Management?

How much inventory should be held? Where should this inventory be stored? Is the inventory perishable? How much inventory needs to be held in order to satisfy the market and provide a good level of service? Organizations use a variety of systems in order to manage inventory: just-in-time, kanban, material requirement planning, enterprise resource planning, and others. For each of these systems, the goal is to meet customer demands through optimal use of equipment and people. Disruptions to the supply chain due to inventory shortages are costly. In some cases it may take weeks to recover. Supply chains in a global economy are vulnerable for a variety of reasons, including that all the risks involved cannot be identified or planned for. Many systems are not robust enough to adapt to different conditions, especially when unexpected. Even if a supply chain is working well, efforts are always under way to reduce inventory, improve customer service, reduce costs, and increase equipment and labor utilization. All of these factors make inventory difficult to manage. Though a complete discussion is beyond the scope of this text, effective organizations use a variety of methods to ensure against disruptions to their supply chain. These may include:

Schedule leveling
Schedule fixing
Pull systems
Short cycle times
Safety stock
Safety lead times
Small batch sizes
Flexible job sequencing
Workforce flexibility through cross-training
Capacity planning

Why Is Information Sharing So Critical When Creating an Effective Supply Chain?

Information is a key component of a supply chain. Organizations need to know what is needed, how much, and when. Careful selection of the type of information system used to manage the supply chain ensures the integrity of the link between upstream and downstream suppliers and customers. At effective organizations, this information is transferred through electronic data interchange (EDI). These direct computer-to-computer interactions can be interorganizational or external to individual customers or suppliers. Common information exchanged includes quotes, purchase orders, order confirmations, pricing information, shipping notices, and more. Electronic data interchange improves productivity and quality. Less paperwork is required, reducing the need for clerical work. The real-time exchange of information helps run just-in-time systems while reducing inventory and lead times. Managing the supply chain is also easier because the information is available at the touch of a button. Scanners, bar codes, and RFIDs (radio-frequency identification) simplify data acquisition, inventory control, and supply chain management. Technology allows real-time data gathering for the identification and tracking of products. Information about a products whereabouts, contents, and physical status are quickly available. Software applications like enterprise resource planning (ERP) systems also aid in managing the supply chain.

Information systems and networks provide increased data accuracy and reduced data transfer time. Unfortunately, these systems are vulnerable to equipment failure, power outages, and security threats. Effective organizations are careful to maintain information reliability. They recognize that the ability to rapidly disseminate information and capitalize on the knowledge it generates is a driving force in organizational success. Effective organizations design and manage their information management systems, make technological improvements, and measure system performance in order to ensure data accuracy, integrity, reliability, timeliness, security, and confidentiality.

How Does E-Commerce Support Supply Chain Management?

E-commerce is the use of electronic technology to aid business transactions. E-commerce allows an organization to reach global markets with little effort. Their products and services can be accessed through the web, allowing consumers to make their choices. E-commerce often provides the following benefits:

Faster purchasing cycle times
Reduced inventory
Fewer order processing people
Faster product or service search and ordering
Greater information availability
Automated validation and preapproved spending
Less printed material like order forms or invoices
Increased flexibility
Online communication

E-commerce affects the way an organization does business. The changes it requires changes the way people work. It takes time and training to bring people up to speed on the new technology. All of this new technology requires significant investment, too. Manual processes need to be studied and improved; otherwise, the organization runs the risk of computerizing an already poor manual process. Organizations need to decide just how pervasive the new technology will be. Many choose to integrate the entire organization, from the shop floor to accounting, payroll, and other systems. With all the systems linked within the plant and often to outside customers and suppliers, system integrity and security are an issue. Effective supply chains utilize technology to improve communication. This reduces the risk of production shortages, lost materials, or stock-outs.

What Is Logistics?

Logistics is the process of determining the best methods of procuring, maintaining, packaging, transporting, and storing of materials and personnel in order to satisfy customer demand. This physical side of supply chain management is sometimes called distribution. Logistics or distribution includes the material handling, packaging, warehousing, staging, and transportation of equipment, parts, subassemblies, tools, fuels, lubricants, office supplies, information, and anything else needed to keep the organization functioning. This means logistics manages the flow of materials and information within a facility and incoming and outgoing from that facility. When preparing a logistical strategy, frequently asked questions include:

What is being shipped?
How will it be packaged?

What type of handling care will be needed?
When will it be needed?
When will it be shipped?
What is the best way to ship it? Rail? Road? Air? Water? Pipeline? Satellite?
What are the optimal shipping practices?
What will it cost?

In effective organizations this movement should be smooth and expedient. In some organizations, time and effort is lost because of poor internal logistics. In Chapter 12, the JQOS feature discusses their internal logistical improvements. These problems can also be evidenced between organizations. Effective supply chains safeguard products while they are in transit. Though it adds time and cost to the process, security is vital as product cross multiple borders and great distances.

EXAMPLE 8.1   Logistics and a Tsunami

On December 26, 2004, an earthquake of magnitude 9.3 centered in the Indian Ocean near the northwest coast of the Indonesian island of Sumatra triggered a tsunami that killed nearly 230,000 people and destroyed homes throughout that part of the world (Figure 8.5). Caused by a slip of the Indian Plate under the Sumatra Plate, the earthquake deformed the ocean

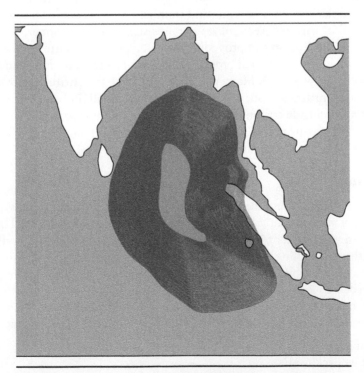

Figure 8.5   Earthquake Impact

floor, creating the tsunami (Figure 8.6). Forming a wave thought to be 80 feet high, it is estimated that nearly 200 billion gallons of water were dropped on the land in a single minute. In comparison, seismic activity for the 1906 San Francisco earthquake was 7.8 and that of Northridge, California, in 1994, 6.7. Hardest hit were Indonesia (167,739 dead or missing), Sri Lanka (35,322), India (18,045), and Thailand (8,212). The biggest challenge facing the humanitarian relief agencies responding to the situation was logistics. How can water, food, medicine, relief workers, and other critical supplies be brought to an area where much of the infrastructure had been damaged or destroyed? Normal supply chains simply did not exist.

Facing this logistical nightmare, relief organizations relied on those who are best in the business. Supply chain experts from around the world gathered together to help. The involvement of a number of firms enabled relief workers to set up temporary supply chains and use sea, air, and land assets of these organizations to move needed supplies from mainland warehouses to temporary warehouses built in the disaster areas. Within days, trucks, helicopters, planes, ships, landing craft, and ferry boats were mobilized to help cope with the disaster. As people and organizations from around the world contributed water, food, consumer products, tents, medicine, equipment, time, and money, this influx also needed to be managed. Logistics

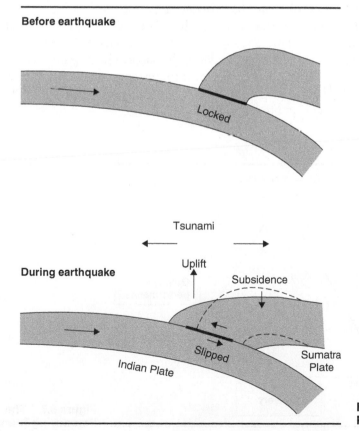

Figure 8.6   Slipping Plates Cause Tsunami

became a strategic weapon in the fight against further death and destruction. Logistics firms and their employees handled tons of supplies. Working 14- to 16-hour days, they coordinated the handling, stocking, warehousing, distributing, and shipping of lifesaving materials. Disaster response efforts lasted six months. Quick response to the crisis was possible because the chief relief organization, the United Nations World Food Programme has long partnered with logistics organizations like TNT Logistics. The strategy to respond to such a natural disaster was in place two years before disaster struck. One phone call activated a well-coordinated logistical effort to aid people in need.

What Role Does Purchasing Play in Supply Chain Management?

Materials, parts, supplies, and services are all purchases necessary for operating a business. In most companies, these costs comprise 40% to 60% of the final cost of the finished goods. *Purchases are driven by several factors, including quality of the goods or services, timing of the deliveries of the goods or services, and the costs associated with purchasing the goods or services* (Figure 8.7). The optimization of all three rarely occurs; effective purchasing happens when these three are balanced in a way that creates value for the customer. As business has become more global, purchasing has had to evolve. Originally, they focused on finding suppliers of goods and services. This developed to include managing deliveries using logistics. Now they are buying a more complete package, the supply chain. And supply chains must be improved, optimized, and managed. Business-to-business relationships are crucial when managing the supply chain. Purchasing has had to adapt to Web sources, global suppliers, improvement efforts, and lean production.

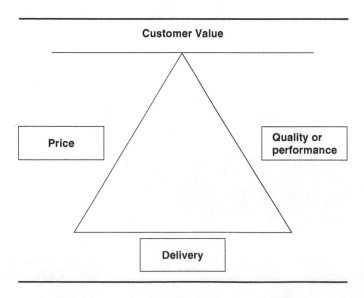

Figure 8.7 The Purchasing Triangle

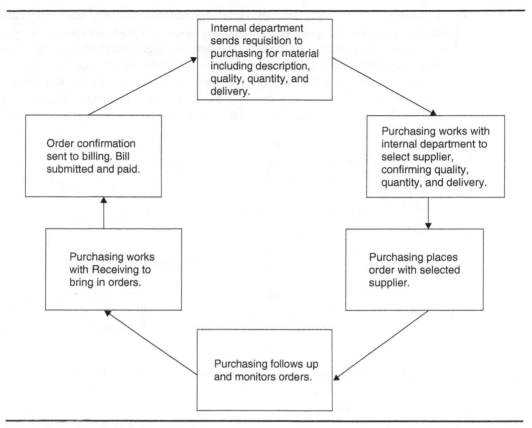

Figure 8.8   The Purchasing Cycle

The purchasing department is responsible for finding out what is out there, what the options are, how much it costs, and when it can be delivered. This means that they not only order goods and services but also select and manage suppliers. The purchasing cycle is summarized in Figure 8.8. Purchasing agents negotiate contracts and act as the liaison between the organization's internal departments and the external suppliers. Understanding the entire supply chain enables them to make effective purchases. In order to ensure that what they order is correct, they work closely with the internal departments making the requests for goods or services. These departments focus on the quality of the item purchased and the timeliness of the delivery. They provide purchasing with specifications and required delivery dates. In many cases they may also have an idea about where the items may be obtained. Purchasing adds value to the process by working to balance the costs associated with buying the goods and services. Efforts to reduce the material costs associated with providing a product or service often weigh heavily on the purchasing department.

Effective organizations seek to optimize the sourcing of their needed material, equipment, and services. To maintain an effective supply chain, these organizations develop a complete and accurate profile of their supplier base. As they judge supplier performance, they consolidate and grade suppliers, granting preferred status to those suppliers who perform best on meeting quality, quantity, performance, cost, and delivery requirements. Effective organizations develop and manage suppliers using several approaches. They evaluate suppliers using quality assurance systems like ISO 9000, the Malcolm Baldrige National Quality Award (MBNQA) criteria, or their own criteria. They run supplier certification or qualification programs to aid supplier development. These are followed by supplier audits to ensure compliance. Effective organizations ask questions like:

> How are suppliers selected?
> What will it take to become a preferred supplier?
> Who are the preferred suppliers?
> Why are these suppliers preferred?
> How does supplier interaction occur?
> How does communication with suppliers occur?
> Where is the supplier located?
> How will the supplier's location affect delivery performance?

Suppliers work closely with the purchasing department to determine material or service specifications, including quality, quantity, delivery date, and cost. Choosing reliable and trustworthy suppliers is crucial to maintaining the supply chain. Key performance factors to consider when choosing a supplier are:

> Type of product (new technology versus mature product)
> Quality
> Quality assurance
> Flexibility
> Reliability
> Location
> Price
> Delivery capability
> Lead times
> Financial stability
> Length of time in business
> Relationship strength
> Other factors, depending on the business

Supplier performance can be monitored through audits (Chapter 15) and certifications like ISO 9000 (Chapter 3). Effective organizations typically audit and review the following:

> Quality assurance activities
> Quality planning process
> Process flow diagrams for key processes
> Key product characteristic capability
> Key process capability
> Process planning

Process performance measures
Continuous improvement based on performance measures
In-process inspection and testing
Operator instructions
Visual factory
First-pass quality performance
Supplier assessment
Customer communication
Workforce stability or employee turnover
Value added per employee
On-time delivery performance
Quality certifications (like ISO 9000)

Supplier partnerships are pursued by many companies. The partnerships go beyond a price contract and focus on cooperation. The benefits associated with a partnership relate to improved operations. These long-term relationships help in maintaining the supply chain through improved delivery and quality, lower inventory, and lower costs. Computerized interaction on set schedules helps reduce purchasing, inventory, and transportation costs. These same suppliers also provide information about item improvements, new products, or other market data. Strategic partnerships may also provide the confidence an organization needs in order to invest resources in a particular product or service line. When deciding to partner, both organizations study the cost benefit trade-off. They ask questions like:

How critical to quality is what we are buying?
What are we buying through this partnership? Performance? Delivery? Savings?
How much is being bought?
What will the time investment be?
What will the dollar investment be?
What benefits will we receive through this partnership?

As with any partnership, there can be drawbacks. Though the relationship can be mutually beneficial, often investments in computer systems or dedicated equipment are necessary and costly. If the supply chain is complex, not all costs associated with a firm contract can be controlled. Sometimes, even the culture of the two businesses can be conflicting. Effective supply chain managers understand that people cannot be standardized and that there may be cultural differences. They standardize processes to reduce this difficulty. When trying to understand organizational and cultural needs, expectations, and differences, clear communication is essential. Product requirements start the supply chain process, but communication enables it.

There are advantages to having a single supplier and advantages for multiple sources. Single-supplier relationships are stronger, each side showing a greater commitment to the other. There are economies of scale present with a single supplier. Naturally, having only one other organization to communicate with simplifies information exchange and makes it easier to keep that information confidential. There are sound reasons for picking multiple suppliers too. Having several suppliers minimizes the risks associated with having only a single source. If something affects a particular supplier's ability to deliver, like a strike or a flood, then the supply chain is frozen until another source can be found. Multiple suppliers prevent this from happening. Multiple suppliers also encourage competition, which can bring about lower prices or better delivery schedules or other benefits. Multiple suppliers also provide more flexibility when demand increases unexpectedly. They also provide access to a wider variety of goods and services.

Selecting a supplier requires making an informed decision. Careful purchasing agents find out as much as they can about potential suppliers before making their selection. If the item being purchased is critical to quality or involves significant funds, it is not unusual for the purchasing agent to make a site visit. They come prepared to audit the supplier in order to determine whether the supplier is a good match for their organization. Suppliers that do not offer a strategic advantage are not chosen. Desperate manufacturers or service providers are also avoided. Their willingness to quote any part or service at any price when they may not have the expertise, capacity, or systems in place to support the demand can signal significant trouble later. An effective organization seeks suppliers who have the same mind-set that they do. They choose companies that want and value their business. Audits and site visits reveal whether a supplier has a focus on continuous improvement, value creation, and quality assurance. Effective organizations take time to ensure that the supplier they choose will be a good fit.

How Do We Know It's Working?

JQOS would like to achieve a two-week lead time from order placement to order receipt. They can't do this without managing their complete supply chain. Raw material must be available just in time for new orders. The shop must be prepared to handle the incoming orders. Shipping must deliver the parts quickly and safely. At JQOS, the chief material needed is steel bars made to specification and cut to length. They use a variety of steel, including 1045, 1144, 12L14, 300 series stainless, and exotics like Monell Hastway. JQOS makes small batches of parts to order and is too small to impact the purchase price of their steel raw material. For this reason, they partner with suppliers who can create value for them in other ways. Raw material suppliers regularly visit JQOS. Since they are in touch with lots of other shops, they provide information about where the industry is going. They have data on the industry that JQOS uses to improve and adapt to the market.

Since JQOS does not make large purchases, they shop every order among their six suppliers. They do not have specific contracts because of fluctuations in steel prices. There are six suppliers because JQOS recognizes that some suppliers are better at producing certain steel than others and chooses accordingly. Once the supplier meets quality assurance requirements, they are chosen on the basis of delivery and price. JQOS's inventory system is a pull system; raw material orders are triggered by customer order. For many customers, their orders repeat monthly. For these customers, a blanket raw material purchase order is in place with a single vendor.

Consistent mistake-free communication is crucial to ordering supplies and shipping parts. Improvements focused on modernizing JQOS computer capabilities. The new system standardizes order processing and links all computers in the plant. Order processing efficiency increased dramatically since automated spreadsheets allows quotes to be adjusted quickly to respond to customer requests. When accepted by the customer, quotes are turned into orders, complete with shop traveler, invoice, packing list, and material purchase order. When the material purchase order is sent via computer to the steel vendor, an e-mail reply notifying receipt of the order is requested. Confirmation of the order to the customer is sent via e-mail. Job progress through the plant can be tracked as the parts move from operation to operation. Shipments are also tracked. Taken together, the new order processing system supports the entire supply chain by tracking and confirming the order process from customer quote to final part delivery.

Working with suppliers requires a lot of negotiating. Negotiation is the process of making joint decisions when the people involved have different preferences. Simply taken, negotiating involves presenting the facts, discussing the facts, discussing the viewpoints on the facts, and reaching a rational agreement. Anyone who has participated in a negotiation will tell you that, in reality, negotiating is tough. There are at least two sides to the situation. This means that there are at least two sets of desires, interests, objectives, and agendas. The type of product or service involved can make negotiating more complex. If the product or service is highly technical, then the negotiation will take on more of a consulting nature. In this situation, the supplier provides information showing their in-depth knowledge of the product or service and their willingness to share it. This type of negotiating is different than one for a commodity product.

The outcome of any negotiation is significantly influenced by the amount of advance preparation on both sides. To conduct successful negotiations, prepare thoroughly in advance. Good preparation, including determining the other side's expectations, means fewer surprises. Outside the negotiation, when the pressure is off, information can be more clearly evaluated. While preparing in advance, good negotiators generate a variety of scenarios and alternatives to see how each one relates to their desired result. They come to the negotiation knowing which aspects of the negotiation must absolutely go your way and which ones are less essential.

Well-prepared negotiators have their thoughts organized. Advance preparation means that during the negotiation they can be factual rather than emotional. They have learned to focus on the situation and not on the attitudes of the people involved. They also can recognize where the two sides have matters of common interest and agreement, smoothing the negotiating process. Good negotiators are good listeners, too. They take the time to listen carefully to proposals and counterproposals. This means that they don't hurry the negotiations along, and they remain as flexible as possible. They allow people to save face and don't ignore the needs of others. They conduct their negotiations in an atmosphere of trust and confidence. This gives them a reputation for being fair but firm. Once the negotiation is complete, good negotiators take the time to set everything in writing. They avoid problems later by being sure that the signed agreement conveys the intent of both parties.

There are false perceptions about the negotiating process. Good negotiators realize that negotiation is never a win-lose confrontation. Each side must be able to feel good about the end results. Too big an imbalance, and future transactions may suffer. There is an old adage that what goes around comes around. Care should be taken to strike a mutually beneficial bargain because each negotiation is rarely an isolated transaction. Another false assumption about negotiating is that it is all about obtaining the lowest possible price. As this next JQOS feature shows, supplier negotiations usually have three aspects: quality, delivery, and price.

How Do We Know It's Working?

JQOS decided to shop around and compare their shipping company with others in the market. Their current shipping company has performed well on price and delivery. However, one of the performance measures they track, product damage during shipment, has been creeping up over the past several months (Figure 8.9). While recognizing that their heavy product is relatively easy to damage, they would like to keep the number of shipments damaged to one shipment damaged in any three-month time period, the level of the previous year's performance. Even better would be to reduce this damage. They are also concerned with the responsiveness of the shipper to their requests for meetings concerning damaged product. The company has been evasive and has not dealt with insurance claims quickly. Damaged parts are a significant burden on the JQOS manufacturing facility. Since they and most of their customers operate on a just-in-time basis, damaged parts seriously disrupt the supply chain. Customers have been sympathetic, knowing that JQOS did not release damaged parts to be shipped; however, they have production to get out and need good parts. It is JQOS's responsibility to provide good parts ready for use when needed. If JQOS cannot manage their part of the supply chain, then the customers will look around for someone who will.

Since they already have a clear idea of what they would like—maintain current price and delivery times while reducing shipment damage—they know what they need to discuss when they meet potential new suppliers. They have narrowed their candidates by preferring suppliers with the following characteristics:

> Quality (low damage rate)
> Quality assurance (ability to monitor and improve shipping process)
> Flexibility (able to deliver to a wide variety of customers)
> Reliability (on-time deliveries regardless of weather or other problems)
> Price (competitive)
> Delivery capability (able to manage shipments of JQOS size and weight)
> Lead times (able to make regular deliveries in reasonable times for distances)
> Financial stability (large enough to have and maintain effective shipping equipment and employees)
> Length of time in business (not new to business or new to products of this type)

Several companies were selected for on-site visits. There, leadership from JQOS studied the companies' activities and processes in order to verify the following:

> Quality assurance activities (how does the organization capture the voice of the customer or JQOS's shipping needs?)

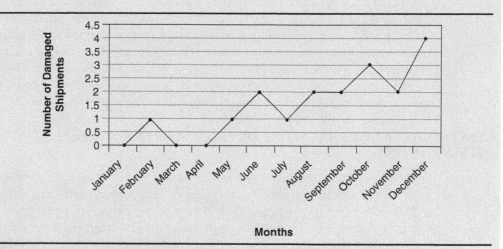

Figure 8.9   Damaged Shipments

Quality planning process (a plan for future support of customers, including improvements)

Process flow diagrams for key processes (how are shipments handled?)

Key process capability (what is their performance level with other similar shipments?)

Process performance measures (how do they measure their performance?)

Continuous improvement based on performance measures (what improvements are they planning in the near future?)

Operator instructions (what are their drivers told to do, and how are they trained?)

Visual factory (what does their equipment, trucks, look like?)

Customer communication (how will information be shared?)

Workforce stability or employee turnover

On-time delivery performance

Careful study was made of the results of their supplier visits. In the final analysis, two companies stood out. JQOS leadership invited each to visit their plant to discuss their requirements. The negotiation moved quickly past the pricing question because both companies were very competitive. On the critical performance measures—on-time delivery rate and damage during shipment—both had similar delivery performance too. When negotiations focused on performance, the two companies differed widely. One of the companies negotiated their involvement in the packing process. They had clearly prepared for the negotiation in advance. Their in-depth understanding of JQOS's real expectations became obvious. They clearly evaluated the situation and generated several alternatives focused on JQOS's desired result: no damage during shipment, on-time delivery, and competitive pricing. They understood what was essential, and this enabled them to prepare a package detailing packaging changes that could better protect the parts. The package they presented even ended up saving JQOS money on packaging materials.

Purchasing plays a key role in organizational effectiveness. It should not be taken lightly. Purchasing agents must do their homework when selecting suppliers, outsourcing activities, or developing partnerships. These relationships must be assessed and audited for both new and existing suppliers. Site visits ensure that they are able to meet their commitments. Measures of performance apply to external as well as internal activities. Negotiations with chosen suppliers ensure value creation by balancing delivery, price, and quality. Effective organizations choose their suppliers carefully, knowing that they are not all created equal.

WHAT CHALLENGES FACE THE DEVELOPMENT OF AN EFFECTIVE SUPPLY CHAIN?

Effective organizations treat their supply chains as processes that must be managed and improved. Process improvement techniques support a quality management focus that ensures reliability and high quality delivered to the customer. Since suppliers play a large role in providing a quality product or service to the customer, effective organizations ensure that suppliers share their philosophy. This means that the suppliers they chose conform to standards such as ISO 9000. Quality assurance audits confirm or deny good suppliers. Effective organizations appraise the performance of the supply chain. Performance parameters are negotiated in advance and are measured and tracked. Supply chains are continuously improved and made lean.

When a supply chain lacks synchronization, problems arise. The longer the supply chain, the greater the effects of changes in demand. Disruptions to any one of the organizations on the supply chain seriously affect its performance capabilities. These problems can be reduced by communication and information sharing throughout the supply chain. Information availability also helps to reduce lead times and supports smaller batch sizes.

HOW DOES SUPPLY CHAIN MANAGEMENT RELATE TO THE MALCOLM BALDRIGE NATIONAL QUALITY AWARD?

As presented in Chapter 3, the MBNQA criteria support several core values, including agility and valuing partners. Specific portions of several of the seven criteria deal with information sharing, e-commerce, external partnerships, and supply chain management. The MBNQA criteria recognize the need for flexibility in a rapidly changing world. As designed, the criteria judge all activities within an organization's overall value chain. According to the criteria, supply chains should be designed to enable more rapid, flexible, and customized responses to both suppliers and customers. They should also be designed to significantly reduce costs and time to market. Supply chains should enable suppliers and customers to communicate on a continuous 24/7 basis, including automated information access and transfer.

The organizational profile section includes a focus on supplier-dependent organizations. The criteria ask the organization to discuss how they create and maintain a sustainable competitive advantage through fundamentally sound supply chain management. In section 4.1, Measurement, Analysis, and Improvement of Organizational Performance, under 4.1.b., Performance Analysis, Review, and Improvement, the criteria study the organization's ability to share information with suppliers, partners, and collaborators in order to ensure organizational alignment of the supply chain. In Section 4.2 Management of Information, Information Technology, and Knowledge, the criteria recognize that data and information availability is critical to the supply chain. Organizations are asked to provide insight into how they ensure valid and reliable data disseminated in a timely fashion. The ability of a firm to maintain data integrity is also questioned. The criteria revisit supply chain management in Section 6.1, Work Systems Design. This section recognizes that supply chain management is an important factor in achieving productivity improvements, profitability goals, and organizational success. The criteria ask that organizations confirm that they have processes in place that enable them to help improve supplier performance in order to help them contribute to the organization's improved work systems. Information about an organization's process for supplier selection is also sought. The Baldrige Award encourages the reduction of the total number of suppliers and the increase in the number of preferred supplier arrangements. Taken as a complete package, the MBNQA criteria encourage effective supply chain management.

LESSONS LEARNED

Supply chains play a strategic role in organizational success. Effective supply chains can be judged by a variety of metrics including its ability to provide order fulfillment, on-time delivery, lead time, value added per employee, inventory turns, stock-outs, quality, response time, and cycle times. A supply chain must be designed with quality, cost, flexibility speed, and customer service in mind. In this global economy, supply chains know no boundaries. To be effective, supply chains must be designed with the organization's strategy in mind. The supply chain should correspond to the organization's views on outsourcing and in-sourcing. Effective organizations in-source tasks and processes related to their core competencies and outsource tasks that are not critical. For instance, at JQOS, their core competencies revolve around their ability to machine parts. They outsource payroll operations because an organization whose primary function is payroll generation and tracking will perform it more efficiently and cost effectively.

Effective supply chains are really networks (Figure 8.4). The functions cannot be managed as separate activities but as one in a series of events. Decisions concerning the supply chain must be carefully considered for their effects on other aspects of the chain. For this reason, effective organizations chose their supply chain partners carefully, knowing that effective supply chains are built on relationships. Business partners in the supply chain maintain clear communication channels and have a distinct interest in each other's success. Whether these organizations are worldwide or local, they display traits that strengthen relationships even when difficulties occur. Suppliers have a reputation for providing quality products or services on a timely basis. They are reliable and consistent on an ongoing basis.

1. A supply chain procures material, performs value-added processing, and distributes products or services to customers.
2. A supply chain supports organizational success by ensuring reliability, adaptability, low costs, and appropriate asset utilization.

3. The elements of a supply chain are inventory management, information sharing, e-commerce, logistics, and purchasing.

4. Information management and maintaining data integrity and security is crucial to the supply chain.

5. E-commerce refers to the electronic integration of information into the supply chain.

6. Distribution or logistics management refers to material handling, warehousing, packaging, and transportation.

7. The role of purchasing is to procure materials needed by an organization.

8. Negotiating is the process of making a joint decision about quality, delivery, and pricing.

9. The MBNQA criteria support effective supply chain management.

Are You a Quality Management Person?

The things you use in your life come to you through supply chains. Do you manage those supply chains to improve your life? Check out the following questions and find out whether or not you are a quality management person. How do you answer them on a scale of 1 to 10?

Do you recognize the supply chains that exist in your life?

| 1 | 2 | 3 | 4 | 5 | 6 | 7 | 8 | 9 | 10 |

Do you base your purchases on the initial price or on the overall long-term value you may receive?

| 1 | 2 | 3 | 4 | 5 | 6 | 7 | 8 | 9 | 10 |

Do you manage key points of your supply chain?

| 1 | 2 | 3 | 4 | 5 | 6 | 7 | 8 | 9 | 10 |

Do you choose suppliers on the basis of needs that you have identified, like hours, or employee knowledge or a special service supplied?

| 1 | 2 | 3 | 4 | 5 | 6 | 7 | 8 | 9 | 10 |

Do you consistently frequent a particular set of suppliers for your daily needs?

| 1 | 2 | 3 | 4 | 5 | 6 | 7 | 8 | 9 | 10 |

Do you try to create a relationship with your suppliers?

| 1 | 2 | 3 | 4 | 5 | 6 | 7 | 8 | 9 | 10 |

Do you make an effort to get to know the employees at the places you shop?

| 1 | 2 | 3 | 4 | 5 | 6 | 7 | 8 | 9 | 10 |

Do you communicate your specific needs to your suppliers?

| 1 | 2 | 3 | 4 | 5 | 6 | 7 | 8 | 9 | 10 |

Do the suppliers you choose respond to your stated needs?

| 1 | 2 | 3 | 4 | 5 | 6 | 7 | 8 | 9 | 10 |

Do you manage your supply chains to improve your life?

| 1 | 2 | 3 | 4 | 5 | 6 | 7 | 8 | 9 | 10 |

Chapter Questions

1. What is a supply chain?
2. What are the benefits of a well-managed supply chain?
3. What are the objectives of effective supply chain management?
4. What organizations might be involved in a supply chain?
5. What do organizations involved in a supply chain transfer back and forth?
6. Why is information so critical to supply chain management?
7. What role does information play in a supply chain?
8. What role does e-commerce play in supply chain management?
9. What are the challenges to creating an effective supply chain?
10. Describe the role that purchasing plays in the supply chain.
11. Describe the purchasing cycle.
12. What is meant by logistics?
13. Why does logistics play a key role in supply chain management?
14. From a firm you have visited or worked with, describe the logistical activities there.
15. From a firm you have visited or worked with, describe supply chain activities, including who is involved and what they provide.

Wave of Relief

The power of public-private partnership in speeding critical supplies to tsunami-ravaged villages

BY LUKE DISNEY

Dec. 26, 2004 is a day that changed the world. An earthquake of monstrous proportions struck and triggered a tsunami that ravaged the coastlines along the Indian Ocean. Entire coastal villages and towns from Asia to Africa were swept away in a matter of minutes. This devastation left millions of survivors behind in its wake. In short order, they lost their families, friends, homes, and communities.

A number of relief agencies quickly rallied their resources to help the victims of this natural disaster. Among those agencies was the United Nations World Food Programme (WFP), which is the United Nations' frontline agency in the fight against global hunger. Within days of the disaster, WFP launched a massive emergency operation valued at over 300 million euros to speed the flow of critical supplies to those areas hit hardest by the tsunami.

The world's largest humanitarian agency, WFP provides food aid to an average of 90 million people in more than 80 countries annually. According to those inside WFP, this disaster posed the humanitarian organization with one of the most logistically challenging operations in its 40-plus-year history. What made this relief effort particularly difficult was that WFP and its partners would have to work in locations where so much of the infrastructure had been damaged or destroyed, where many areas are remote and cut off from normal supply lines, and where existing capacities were already stretched thin.

WFP had two objectives going into the relief: Deliver life-saving food to tsunami survivors and establish a logistics support system that would enable workers to facilitate a rehabilitation operation that could be active for months, maybe even years.

Three months after the disaster, WFP had provided more than 60,000 metric tons of emergency food relief to nearly 2 million people in Indonesia, the Maldives, Myanmar, Somalia, Sri Lanka, and Thailand. This was accomplished through the extensive use of sea, air, and land assets, which were part of a supply chain that stretched from the ports and warehouses of origin to temporary warehouses erected in the disaster area. The transport assets used ranged from aircraft, helicopters, ships, landing craft, ferry boats, and trucks.

In its relief operations, WFP received crucial support from partners to get the job done. One of those partners is TNT Logistics, which helped with the logistics part of the project that encompassed the main modes of transport. This is where humanitarian experience intermeshed with supply chain expertise to help those in need.

EMERGENCY RELIEF OPERATIONS RELY ON LOGISTICS

TNT is a global mail, express, and logistics company that became a leading corporate partner with WFP about three years ago. In 2002, TNT made a long-term commitment to WFP aimed at fighting world hunger. By teaming up with WFP, TNT had already committed itself to sharing its resources and know-how to make a lifesaving difference. Logistics is an essential weapon in the fight against hunger and a key part of the supply chain that lies at the heart of WFP's mission.

Using emergency operational procedures that had already been put in place, TNT moved quickly to get emergency response staff, food and other goods to assist those countries most affected by the tsunami. These logistics efforts were also supported by financial support and personal sacrifice.

Despite the Christmas holiday, TNT powered an immediate and widespread response to the tsunami. For example, in the Dutch town of Heerlen, postman Rick Fijnaut appealed to his colleagues to convert some of their holiday pay into donations. Staff from across the Benelux responded, resulting in a donation of 800,000 euros, which was matched by the employer. In the Asia Pacific region, TNT employees made a donation of 225,000 euros. Elsewhere in the world, fundraising activities had added a further 70,000 euros to the company's contributions within weeks.

TNT reserved 2.3 million euros for in-kind support to WFP's relief efforts in the region. The funds fall under the emergency response initiative of the TNT/WFP Moving the World Programme. Within weeks, TNT had committed about 1.56 million euros in response to requests from WFP and its implementing partners. The remaining funds were to be spent on additional efforts aimed at providing the best possible support to WFP and to those in need.

A LOGISTICS NIGHTMARE

Destruction on the coastlines around the Indian Ocean resulted from approximately 200 billion gallons of water being dropped on land within a single minute, according to scientists. Within days of the tsunami disaster, supply chain experts in Dubai, Indonesia, Malaysia, Sri Lanka, and Thailand swung into action to help WFP mount relief efforts. In Indonesia, TNT ran the first food convoys to Banda Aceh, the hardest hit area in the region.

An average of 100 trucks a day parked at WFP's disposal to move food to those who needed it in Aceh. Daily road convoys traveled between Medan, Banda Aceh, Meulaboh, and Singkil carrying food and equipment. By the end of January 2005, there were 35 TNT teams as far away as Finland, Belgium, Germany, Holland, and Australia on the ground in Indonesia to work with WFP on the relief operation. Early in the emergency, TNT made a Mi-8 helicopter available to WFP in Medan to move food to hard-to-reach areas of Sumatra's west coast. In addition, the company made more than, 4,000 square meters of warehousing facilities available to WFP and its implementation partners in Medan, Jakarta, and Sinkil. Airport handling and clearance facilities were provided in Medan, while ramp handling services functioned in Banda Aceh.

After Indonesia, with a death toll of many thousands, Sri Lanka suffered the worst.

When Mohan Perera heard that a seismic wave had hit Sri Lanka, he immediately contacted WFP to offer assistance. Mohan is employed by TNT partner Cargo Private Ltd. as the company's country manager in the area. "Our contact revolved around the distribution of food to the affected areas in the south and east of Sri Lanka," Mohan says. "We worked from our office in Colombo and from the WFP warehouse in the suburbs. The Dutch head office also contacted us very quickly to see what they could do."

Mohan recalls that January 9, 2005 was an important day. "That's when a TNT aircraft landed in Colombo with 33 tons of temporary food storage tents. We handled the cargo and handed it over to WFP and then we used both their and our trucks for the distribution."

TNT also contributed coordination and support for the airport emergency team in Colombo and dedicated space for the clearance of the tsunami relief goods that poured in. A colleague flew in form Dubai to join the team, which was supplied with stationery, communication links, and bicycles, among other necessities. Teams were made up of various logistics and transport industry players. Planning and coordination skills were essential; only dedicated specialists could give these areas the attention they deserved.

The stories survivors of the tsunami told Mohan when he visited the affected areas have left a profound impression. "The people were still in a state of shock. We managed to get the food out to them, but the psychological trauma will take a long time to heal," Mohan notes. "At TNT, we made plans to start a community center in one of the hardest hit areas to help the victims overcome their trauma."

David Stenberg, TNT's country operations director in Indonesia, looks back on a month he and his team will remember for the rest of their lives.

On Dec. 27, 2004, Stenberg was looking forward to a round of golf when he received a call on his cell phone. A few hours later, the UN was briefing all its partners on a coordinated response to the tsunami that had hit. "Initially, the media were reporting casualties of 3,000 to 4,000," Stenberg recalled. "We know from experience that was grossly underestimated."

The amount of work Sternberg and his team performed in the next few days challenges the imagination. "We were authorized by the Indonesian military to coordinate operational handling at the airport at Banda Aceh," he explained. "We'd fly in two teams from Europe in order to turn the planes around faster and ease congestion."

Since that fatefull telephone call, Stenberg's life revolved around the disaster. A month after the waves struck, 14-hour working days was still the rule rather than the exception for the entire team. During trips to Medan and Banda Aceh, he saw the worst hit sites from the air. "It was a sad and eerie experience to see the devastation and talk to some of the survivors," Stenberg said. "But there was so much to do . . . there was simply no time to sit back and let it sink in.

"In Aceh, we weren't restricted to working with WFP. We also worked with many other agencies, from UN bodies to Oxfam and the Red Cross. By the 29th, we started to equip the emergency response center in Jakarta, and within two or three days it was operational."

Although the job was far from over at the time, Stenberg feels proud of what the team had already achieved. The office in Medan had taken over the day-to-day running of operations, for example. "The emergency phase lasted for about six months," he explained, "so we'd need an infrastructure even when daily convoys are no longer necessary."

A Mighty Chain of Management

Today's supply chains know no geographic boundaries. Customers and suppliers are situated throughout the world. To take advantage of this ever-expanding network, consider the following ways to manage a global supply chain more effectively.

Prepare a strategy that reflects both outsourcing and in-sourcing. Success is often determined by how well you define your core competencies and the extent to which you let others handle activities not central to your principal business. Consider in-sourcing the areas that are critical to making and selling your products and outsourcing the rest to functional experts. Examples of things to outsource in your global supply chain are product packaging, inbound and outbound transportation, warehousing, and the technology needed to connect all these together. Providing the internal resources to get this done is expensive and redundant with solutions that are openly available in the marketplace. Focus on what you know how to do well and let others do the rest.

Pick suppliers by reputation and quality, not by location and cost. The best measurement for quality is customer satisfaction, and customer satisfaction is usually a determinant of how happy customers are with a product at a particular price point. Those who buy from only lowest-cost vendors find themselves changing suppliers all the time, which means origin points for materials are constantly in a state of flux. This complicates supply chain logistics and can produce extra freight costs that exceed the savings from buying from the cheapest vendor. Solid reputations are based on quality, which is perceived on a consistent, ongoing basis, not transaction by transaction.

Build relationships for endurance. Work through inevitable glitches in relationships. Choose business partners who are in it for the duration and work together to improve the collective performance of the supply chain. While periodic mistakes will occur, these shortcomings can be overlooked if there is steady and consistent communication and the partners demonstrate a distinct interest in the success of one another's business. Identifying these traits is often difficult going into a relationship, so references from other business partners are the best way to determine the potential.

Choose logistics providers that possess a global footprint. Third-party logistics has evolved greatly. Acquisitions and mergers have created several capable 3PLs whose service portfolios now boast a global footprint. Take advantage of their services, such as materials management, to add value to your global supply chain. Significant value is added by selecting a provider that can manage local pickup and delivery, line-haul ground transportation and air freight port-to-port coordination, ocean freight, and customs clearance on either end of the shipment. The alternative is to manage these activities separately, which may be more costly and without timely information.

Manage networks, not silos. Treat warehousing and transportation as a single function and don't manage them as independent activities. Minimizing the cost of one will likely increase the cost of the other. Make decisions that lower total cost and improve overall levels of service. These decisions start at the top by defining the requirements for the complete supply chain and executing each of the pieces with all the other pieces in mind. For example, when selecting locations for distribution centers, consider the service and cost impact on transportation to and from these operations.

Measure outputs and share results. Global supply chains require a greater degree of measurement because additional links have been added between origin and destination. Identify those measures, implement them, communicate the results to suppliers and customers, and then take corrective action as needed. The most important measure is how long it takes to get a finished product to the point of consumption. Don't underestimate the role of technology. Knowing inventory status ensures that product is available when and where customers demand it. Use technology to meter the flow of materials and product through the supply chain to avoid production shortages or stockouts.

Have an effective security strategy in place. Cargo security became a key part of the supply chain more than five years ago as a result of the events of Sept. 11, 2001. Safeguarding products while they're in transit adds cost and time to the process. Programs such as C-TPAT have evolved over the past three years to help better secure cross-border freight. It's essential to understand how these measures affect your supply chain.

Product life cycles affect your supply chain. Functional products don't change much over time and have stable, predictable demand and long life cycles. They usually have low profit margins due to heavy competition. Innovative products have higher profit margins, but short life cycles and unpredictable demand. Each requires a different kind of supply chain. Functional product managers should concentrate on minimizing physical costs, while innovative product managers should concentrate on market availability.

Recognize cultural differences. You can standardize processes, but you can't standardize people. There will always be cultural differences, whether you live in North America, Europe, Asia, or elsewhere. Understand those differences because others will have a different viewpoint of life in the global supply chain.

—*Mark Morrison is senior vice president and general manager at CEVA Logistics, formerly TNT Logistics.*

Communications of Rescue

The success of this initiative—and of the relationship between WFP and TNT—can be traced to the same core elements of business: communications, preparation, and training.

Disaster recovery efforts were aided by the fact that relationship was already well established, starting two years before the tsunami hit. When key executives from both sides met in the beginning, they established relationships that would make it easy to get processes rolling in the event of a disaster.

Another major factor in this success was the preparation that enabled swift action. Contact lists were already in place, the lines of communications were pre-defined, and everyone knew who to contact about what. In addition, WFP already knew what TNT could offer WFP in the event of a crisis, such as detailed information about warehouses and space availability, transportation capabilities, and the existing skill sets available in the respective countries where TNT operates.

In the two years since these relief efforts began, the millions of survivors of the 2004 tsunami have come to depend on the relief efforts of humanitarian organizations such as WFP. But these efforts couldn't go filled without the aid of key partners to direct the tasks. The logistics partnership that the WFP enjoys with TNT epitomizes a powerful blend of humanitarian heart and supply chain expertise.

Luke Disney is global director of communications for TNT's Moving the World, the foundation that runs TNT's partnership with the United Nations World Food Programme. Disney is also director of fundraising and external relations for the North Star Foundation, which aims to bring the transport industry together to address its role in the spread of HIV and other illnesses in low- and middle-income countries.

Case Questions

1. What happened?
2. Who responded to the disaster?
3. What role did logistics play?
4. What challenges and difficulties did the logistics firm have to deal with?
5. The article provides information on how to manage the supply chain more effectively. Combine their information with that of Chapter 8 and discuss one of their suggestions.
6. Research the tsunami on the Web. What other logistics firms were involved? What role did they play? What challenges did they face? How did they use logistics to overcome the challenges?

9 Measures of Organizational Success

Why measure?

What are good measures of performance?

How are measures of performance utilized in an effective organization?

What are the goals of a measurement system?

What role does the cost of quality information play in an effective organization?

How are quality costs defined?

What types of quality costs exist?

What does a formal quality cost measurement system look like?

How are quality costs used for decision making?

How do the Malcolm Baldrige National Quality Award criteria support measurement, analysis, and knowledge management for effective organizations?

Lessons Learned

Learning Opportunities

1. To become familiar with measures and measurement systems
2. To understand how measures serve as guides for effective organizations
3. To understand costs of quality
4. To understand the role costs of quality play in decision making

During the 1990s the U.S. Health Care Financing Administration (HCFA) and the University of Wisconsin–Madison's Center for Health Systems Research and Analysis (CHSRA) developed 24 indicators covering 12 areas of care to be used to assess the quality of nursing home care. The 12 areas of care are number of accidents, behavioral and emotional patterns, clinical management, cognitive patterns, elimination and incontinence, infection control, nutrition and eating, physical functioning, psychotropic pharmaceutical use, quality of life, sensory functioning, and skin care. Software developed by CHSRA allows participating nursing homes to review their own improvement progress and compare their efforts with the performance of other facilities within their state.

Paraphrased from:
"Quality Initiative Deemed a Success,"
IIE Solutions, *February 2002, page 18*

WHY MEASURE?

> If you don't drive your business, you will be driven out of business.
>
> *B. C. Forbes*

Very simply, what gets measured gets done. Effective organizations recognize that if they cannot measure, they cannot manage. They know that if they do not have sufficient information about a process, product, or service, they cannot control it. If a process cannot be controlled, the organization is at the mercy of chance. Dr. Deming's favorite question was, "How do you know?" Without measures of performance, this question can't be answered.

Measures are indicators of performance. Leaders of effective organizations ask themselves, "How are we doing?" every day. They understand that without measures, they don't know what they don't know. They can't act on what they don't know. They won't search for answers unless they question, and they won't question what they don't measure. Properly designed measures compare past results with current performance, enabling leaders to answer the question, "How do we know how well we are doing?" Leaders receive a wide variety of information each day. Having enough information usually isn't the problem. Having *useful* information is. Mark Twain once said, "Data is like garbage. You had better know what you are going to do with it before you collect it." As discussed in Chapter 6, effective strategic plans contain performance measures chosen for their ability to quantify information about the critical success factors. Information concerning these critical success factors enables leaders to make better, more informed decisions. Leaders use measures of performance to ensure alignment among their organization's mission, strategy, values, and behavior. Measures of performance enable effective organizations to define the meaning of success numerically. As Lord Kelvin, the British physicist, stated in 1891, "When you can measure what you are speaking about, and express it in numbers, you know something about it; but when you cannot measure it, when you cannot express it in numbers, your knowledge is of a meager and unsatisfactory kind: it may be the beginning of knowledge, but you have scarcely in your thoughts, advanced to the stage of science."

To remain competitive, effective organizations must manage their employees, processes, scheduling, production cycle times, supplier partnerships, shipping, and service contracts far better than their competition. Effective performance measurement systems are used to understand, align, and improve performance at all levels and in all parts of the organization (Figure 9.1). As the nursing homes did in the introduction to this chapter, in order to know how they are doing in key areas that affect their customers' value perceptions, organizations must select and then track indicators of their performance efforts.

WHAT ARE GOOD MEASURES OF PERFORMANCE?

In any organization, employees recognize the importance of working on activities valued by leadership. Leaders use measures of performance to communicate what activities are important. Good measures of performance are designed based on what is valued by the organization and its customers. Well-designed measures encompass the priorities and

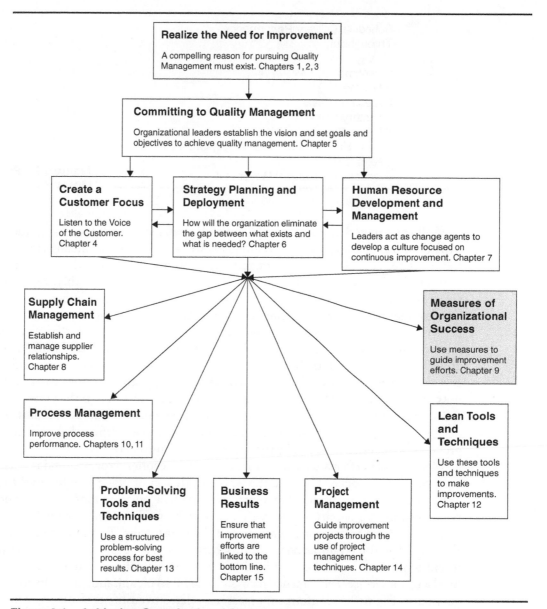

Figure 9.1   Achieving Organizational Success

values of both. Essentially, performance measures enable an organization to answer the following questions:

- How well is something performing its intended purpose?
- Is the organization able to measure the impact of the changes being made?
- How does the organization know that it has allocated its assets correctly?

Schedule/Delivery Performance
Throughput
Quality
Downtime
Idle Time
Expediting Costs
Inventory Levels
Work-in-Process Levels
Safety, Environment, Cleanliness, Order
Use of Space
Frequency of Material Movement

Figure 9.2   Process Measures

Well-constructed measures are aligned with strategic goals of the organization as well as with their customers' priorities. Willingness to use performance measures increases when the measures are relevant to the organization operationally and, where applicable, to the individual personally. Usability is also a function of understandability. Measures that are clearly written and focused are more powerful than ones that are oblique or lengthy.

Customer needs, wants, requirements, and expectations must translate into measures of performance. Effective organizations ask themselves the question, "Have we identified, selected, and measured factors that reflect what the customers want?" Critical to the success of any organization is the ability to determine what their customers want and to find the best way to get it to them. Well-designed measures align strategic objectives with the priorities of the customer. These indicators measure what is of value to the customer. Measures should also provide information about undesired customer outcomes that must be avoided, or eliminated, or at least minimized. Measures should also define the product, service, or process characteristics valued by the organization's customers.

Measures of performance can be found in two categories: processes and results. Processes exist to get work done. They are the activities that must take place in order to produce a product or provide a service. Since processes are how organizations do their work, process measures monitor operational activities, or how the work is done (Figure 9.2).

Results relate to both organizations and their customers. To an organization, results are the objectives the organization wants to achieve. From a customer point of view, results represent what they hope to obtain by doing business with the organization, whether it be a product or a service. Performance measures related to organizational results focus on strategic intent (Figure 9.3). Performance measures related to customer results concentrate on the

Market Share
Repeat/Retained Customers
Product Line Growth
Name Recognition
Customers/Employees Ratio
Pretax Profit

Figure 9.3   Organizational Results Performance Measures Focused on Strategic Intent

Scrap
Rework
Downtime
Repair Costs
Warranty Claims
Complaints
Liability Costs

Figure 9.4 Customer Results Performance Measures Related to Product or Service Attributes

attributes of products and services (Figure 9.4). Products and services are created by the organization and purchased by customers. Products can be tangible manufactured items or information products such as reports, invoices, designs, or courses. Services received by customers are even more varied. Organizations may provide anything from information to dental work to entertainment products such as movies, games, or amusement park rides.

Measures need to be integrated and utilized throughout the entire organization. Traditionally, organizations have focused their attention on measures related to the financial aspects of doing business, such as revenues, profits, and earnings. Effective organizations realize that the outcomes, processes, products, and services of their business must be measured. Figure 9.5 provides some traditional as well as broader measures.

Traditional Measures
Revenues
Profits
Growth
Earnings
Return on Investment (ROI)
Sales Revenue
Total Expenses
Number of Customers
Number of Repeat Buyers
Payroll as a Percent of Sales
Number of Customers per Employee
Number of Customer Complaints
Complaint Resolution Rating
Schedule Achievement

Overall Organization Measures
Employee Satisfaction, Growth, and Development
Customer Survey Results
Number of Completed Improvement Projects
Cost of Poor Quality (COPQ) Reduction
Vendor Quality Rating
Return on Process Improvement Investment
Safety, Environment, Cleanliness, Order
Condition and Maintenance of Equipment/Tools

Figure 9.5 Traditional and Broader Performance Measures

The Balanced Scorecard method, introduced by Robert Kaplan and David Norton, goes beyond financial measures and integrates measures from four areas. These measures focus on key business processes and are aligned into a few manageable indicators of performance so that management is able to quickly assess the short- and long-term health of the organization. The Balanced Scorecard combines and categorizes process and results measures into four areas: Customer Focus, Internal Processes, Learning and Growth, and Financial Analysis. Taken as a whole, the Balanced Scorecard lets leaders see how changes and improvements affect all areas of their business. This decreases the chance of suboptimization of results.

When designing measures related to Customer Focus, an organization emphasizes connecting with the customer, determining what the customer is interested in achieving, and using that information to translate the mission and strategy statements into specific market- and customer-based objectives. In other words, these measures ask, "How do customers see us? What must the organization deliver to its customers to achieve high degrees of satisfaction, retention, acquisition, and, eventually, market share? What is the organization going to do for its customers that will make them want to buy from the organization time and time again?" They identify and monitor the level of value propositions the organization delivers to targeted customers and market segments. *Value propositions are the product or service attributes that meet customer needs, wants, and expectations.* They are indicators that customers are satisfied. These measures are in three categories:

1. Product or service attributes related to functionality, price, and quality
2. Customer relationship attributes such as delivery, response time, convenient access, responsiveness, and long-term commitment
3. Image and reputation attributes which are more intangible factors that attract a customer

Internal Processes measures study the effectiveness and efficiency of the processes performed by the organization in order to fulfill customer requirements. They answer the question, "What must our organization excel at?" When designing these measures, effective organizations identify the internal processes that are most critical for achieving customer and shareholder objectives. Once these key processes have been identified, measures are developed that concentrate on monitoring the improvement efforts in the areas of quality, response time, and cost. This information is used to determine whether the current processes serve the customer effectively, whether these processes are the best, and whether they are operating at their best.

Learning and Growth measures monitor the individual and group innovation and learning within an organization. These measures focus on the ability of the organization to enhance the capability of their people, systems, and organizational processes. They answer the question, "How can our organization continue to improve and enhance customer value?" Recognizing that this is a long-term effort, organizations emphasize the development of employee capabilities; the development of information system capabilities; and employee motivation, retention, productivity, and satisfaction. These measures judge whether the organization's employees have sufficient information and the right equipment to do their jobs well. They are also interested in whether employees are involved in the decisions affecting them and how much recognition and support employees receive. They can be used to evaluate employee skills and competencies and compare this information with what employees will need in the future. In addition, these measures can be used to monitor the morale and climate within the organization.

Financial Analysis measures are perhaps the measures with which people are the most familiar. These measures track an organization's performance in the financial arena. They are focused on monitoring financial performance. They answer the question, "How do we appear to our stakeholders?" Examples of financial measures include revenue levels, cost levels, productivity, asset utilization, and investment risk. Examples of each of the four types of measures are provided in Figure 9.6. In the following "How Do We Know It's Working?" feature, examples of how measures relate to goals are shown in Figure 9.7.

Financial Measures
Sales Revenues
Total Expenses
Pretax Profit
Return on Investment

Customer Measures
Number of Customers
Number of Repeat Buyers
Customer Survey Results
Number of Customer Complaints
Complaint Resolution Rating
Name Recognition
Price Differentials
On-Time Delivery
Response Time

Internal Measures
Payroll as a Percent of Sales
Number of Customers per Employee
Cost of Poor Quality
Employee Survey Results
Throughput Yield
Quality Level
Product/Service Cost
Productivity
Employee Morale

Learning and Growth Measures
Number of Teams
Number of Completed Projects
Number and Percentage of Employees Involved
Number and Percentage of Employees Involved in Educational Opportunities

---

Figure 9.6   Measures of Performance

Customer Perspective

Goal: Reduced Lead Time
Measure: Lead time

Goal: Customer Partnership
Measure: Number of orders from same customer
Measure: Number of repeat orders

Goal: Increased Sales
Measure: Number of new customers
Measure: Number of orders/week

Goal: Improve Quality
Measure: First-pass quality

Learning and Growth Perspective

Goal: Improved Employee Flexibility and Agility
Measure: Number of employees cross-trained and proficient at at least one other position

Goal: Improved Use of Technology
Measure: Number of employees trained and proficient at using Job Boss software

Goal: Increased Safety
Measure: Number of employees attending safety training
Measure: Number of incidents and accidents

Internal Processes

Goal: Improved Use of Technology
Measure: Number of orders completed using Job Boss software
Measure: Number of employees using Job Boss software
Measure: Number of machines linked to Job Boss software

Goal: Reduced Order Processing Time
Measure: Order processing time

Financial Perspective

Goal: Prosper
Measure: Quarterly financial performance
Measure: Quarterly sales performance
Measure: Profit
Measure: Accounts receivable: Number of days outstanding
Measure: Sales per employee

Figure 9.7   JQOS Balanced Scorecard

How Do We Know It's Working?

JQOS leadership develops measures of performance to support the goals and objectives that they laid out in their strategic plan. Their measures relate to all aspects of their business: customers, employees, and finances. Figure 9.7 provides a brief summary of several of their goals and measures. Using these measures, they are able to respond to questions like:

> How do customers view JQOS?
> What must JQOS excel at as an organization?
> How can JQOS's people continue to improve and enhance customer value, both internally and externally?
> How does JQOS appear to our investors?

Improvements and changes cost money. JQOS's investors have raised concerns over the money being spent, but management response is to show how the cost of not running parts, not making quality parts, and not meeting customer expectations far outweighs wise improvements to the business. As effective leaders, they watch their cash flow. Positive cash flow means that the organization has money to grow on. They won't need to borrow for growth, and they have enough to meet their debt obligations. Several of their measures support this, including cash flow and inventory. Cash flow comes from what an organization sells. It represents money coming in. Once the product is complete, they ensure that customers are paying on a timely basis. A high value in the accounts receivable column isn't always good. Effective organizations can collect on these accounts. They don't have long-past-due accounts.

Since inventory that sits around unused ties up money, effective leaders find improvements to increase inventory turnover. This means increasing the flow of product through the plant. It also means that material and component purchasing is timed carefully with customer orders. In service-related activities, careful management of human resources and the supplies or information they need is important. Having who you need and what you need in order to service customers means that customers are served more quickly. Serving more customers quickly means that more customers can be served.

Their measures of performance support investment in the organization too. Sales have risen from $4 million to $8 million a year. The number of new, long-term customers has risen from 30 to 36. The number of orders handled weekly by the plant has risen from 30 to 60 per week. Lead time has fallen from four to six weeks to nearly two weeks. First pass quality has increased from 80% to 87%. Profit has risen from 2% to 8%, a figure that includes money spent on improvements. The number of employees has risen from 26 to 38, primarily due to adding a second shift to handle the increase in customer orders. Even with the additional employees, the ratio between the JQOS's sales and the number of people they employ is strong, $4 million/26 to $8 million/38. By making improvements to their billing and account receivable processes, despite the significant increase in sales, the dollar amount of outstanding receivables has not increased.

Measures are not without their pitfalls. The number of measures used by an organization must be reasonable to be effective. Measures must focus on what is important to the organization, not simply on what is easy to measure. It is critical to determine what needs to be measured and why it needs to be measured before designing a measure of performance. Measures for measurement's sake must be avoided. Too many measures or unfocused measures lead to hesitation in the users. Often, it is difficult to establish meaningful measures to act upon. For instance, if the measure is stated as "Improve contact with customers," there is little room for defining what is actually being measured or for determining the effectiveness of activities designed to enhance contact with the customers. Measures need to be specific and quantitative to be effective. For example, "Visit five customer accounts per month" or "Decrease scrap on Line 2 by 5% in six months" is significantly more specific, more easily acted upon, and more measurable compared to the generic statements "Improve contact" or "Reduce waste." It is also critical to ensure that the correct things are being measured. Sometimes the choice is made to measure the easy actions rather than the important actions, leading to measures that are merely counts of actions rather than indicators of business opportunities. In other cases, measures may be too abstract, and although they may sound good on paper, they cannot be achieved because few people understand what the measures mean. One key question to ask is, "If only one item, parameter, type of result, and so on, could be measured in order to assess the performance of the organization at a given level, what would it be?"

EXAMPLE 9.1   Indicators or Measures of Performance

Leaders of PM Printing and Design created their strategic plan in Example 6.4. Using the plan, they hope to communicate to their employees the importance of creating and maintaining a customer-focused process orientation as they improve the way they do business. Their indicators or measures of performance have been taken from their strategic plan and organized into the four categories discussed in this chapter.

Customer Measures

Results measures:	Overall customer satisfaction
	Market share
	Number of customers
	Number of repeat customers
	Number of new customers
Process measures:	Changes in customer requirements versus changes in processes to serve customers
	Improvements to processes critical to serving customers

Financial Measures

Results measures:	Cost per printed unit (impression)
	Profitability
	Return on investment (ROI)
Process measures:	Cost avoidance (performing work in-house versus contracting work)

Internal Measures

Results measures:	Improvements in hours paid versus hours billed
Process measures:	Improvements in job turnaround time (cycle time reduction/removal of non–value-added activities)
	Improvements in billing lag time (cycle time reduction/removal of non–value-added activities)
	Improvements in first-time-through quality (reduction in rework/scrap)

Learning and Growth Measures

Results measures: Improvements in employee retention

Process measures: Progress toward cross-training goals for critical processes as identified by customers

---

When developing measures, consider the following:

What does the organization need to know?

What is currently being measured?

How does what the organization needs to know compare with what it is currently measuring?

How is this information being captured?

Is the information currently being captured useful?

Is the information currently being captured being used?

What old measures can be deleted?

What new measures are appropriate?

Do the identified, selected, and measured factors reflect what the customers need, require, and expect?

Are these measurements being captured over time?

Can these selected factors be acted upon within the organization?

Can the impact of the changes made be measured?

Have the organization's assets been allocated correctly?

Measures must be defined in objective terms. The best measures are easy to express as a numerical value. Key measures focus on the organization's desired outcomes, undesired outcomes that the organization wants to avoid or eliminate, and the desired product, service, or process attributes or characteristics. Valuable measures show progress toward the milestones of strategic goals and objectives. These types of measures will receive more attention than those that do not. Decision makers must find a performance measure valuable. Measures are useful and effective when those using them can identify how the measures enable them to make better decisions in relation to their day-to-day activities. All measures must be evaluated to determine their usefulness.

EXAMPLE 9.2   Measures of Patient Care

Consider the process of admitting a patient to the hospital. Patients are focused, rightfully so, on their health issues. Upon admission, they are faced with the need to complete insurance paperwork, get preliminary testing, provide medical history, receive a room assignment, and schedule necessary procedures. Consider the total cycle time of this process. It can seem overwhelming.

Hospitals use many measures to help them determine where process improvement is needed. These measures may include the length of time for a nurse to respond to a patient call button, the accuracy of diagnosis, billing errors, and patient pain and comfort. Sometimes people focus on only the tough measures, such as pain and comfort, and use that as an excuse to ignore other measures.

In response to customer complaints about substandard treatment, hospitals are taking steps to measure patient satisfaction and make improvements based on the measures. To create care that is more patient centered, hospitals need to understand patient experiences by looking at processes from the patients' point of view. Measuring patient-centered care is more complex than the traditional measures of performance that hospitals have used in the past, requiring a new set of performance measures. The new measures of performance focus on more subjective patient issues, such as how patients feel about the way they are treated and how the kindness and courtesy of the staff affected their experience. Previously, most of the measures focused on safety and quality standards, such as how many infections were contracted during patient hospital stays, or the number of heart-bypass surgeries or C-sections a hospital performs in a year. Although there is a

Financial Measures
Total Expenses
Return on Investment (ROI)

Customer Measures
Number of Patients Treated
Patient Survey Results
Number of Patient/Customer Complaints
Complaint Resolution Rating
Kindness Rating
Staff Courtesy Rating
Patient Information Availability
Patient Comfort Rating
Patient Pain Relief
Patient Readmittance Rate

Internal Measures
Medicine-Dispensing Error Rate
Number of Patients per Employee
Length of Stay
Quality Level
Safety Rating
Number of Surgeries
Infection Rates

Learning and Growth Measures
Number of Teams
Number of Completed Projects
Number and Percentage of Employees Involved
Number and Percentage of Employees Involved in Educational Opportunities

Figure 9.8   Measures of Performance for a Hospital

tendency to view the "soft stuff" as unimportant, more and more hospitals are developing measures of performance that focus on the patient. These measures of performance that focus on patient care may include how well informed of their condition and treatment patients felt they were, patients' physical comfort and pain relief, the emotional support patients received, the respect they received for their preferences and expressed needs, and how well hospitals prepared them for caring for themselves at home. Hospitals use this information to improve the level of care they offer their patients. Some examples of measures of performance are given in Figure 9.8.

HOW ARE MEASURES OF PERFORMANCE UTILIZED IN AN EFFECTIVE ORGANIZATION?

Information is instrumental to guiding an effective organization. Leaders of effective organizations analyze information provided by performance measures to assess and understand their organization's overall performance. This information enables leaders to make appropriate decisions about the actions they must take to ensure organizational success. Performance measures can be used to align day-to-day activities with the strategic plan and improve organizational performance at all levels and in all parts of the organization (Figure 9.9).

The quality management system and Six Sigma programs used by effective organizations emphasize good management of information and knowledge. Through the use of performance measures, leaders use the systems and programs to select projects ensuring that all decisions are tied to business results. The measures used by these systems and programs seek to optimize overall business results by balancing cost, quality, features, and availability of products and their production. The measurement phase of any problem-solving process should require a complete evaluation of key process variables through the use of measures. This enables those working on projects to understand how the measures work together within a complex system to produce good products in a timely manner, at the best cost, and in a way that meets the needs of the customer and the company.

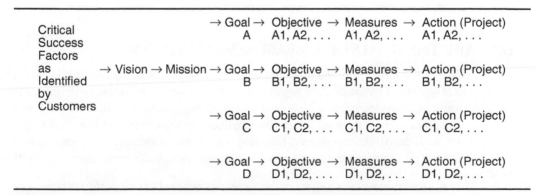

Figure 9.9   Measures and Their Relationship to Strategic Planning

EXAMPLE 9.3   Hospitals Reduce Emergency Room Wait Time

When a medical emergency occurs, the last thing a person wants to do is wait in an emergency room to be seen. On average it takes 49 minutes for a doctor to see a patient in a U.S. emergency room, according to studies by VHA Inc., an alliance of community hospitals who track such information. Hospitals are responding to patients' desire to be seen quickly by listening to the customer, acting on the information they gather, and making significant changes to the way they do business. Because of these efforts, short waits in the emergency room may become the norm rather than the exception. As reported in *The Wall Street Journal*, July 3, 2002, one hospital, Oakwood Hospital and Medical Center in Dearborn, Michigan, went so far as to promise that anybody taken to the emergency department would be seen by a doctor within 30 minutes—or he would get a written apology and two free movie passes.

To reduce emergency room wait times, hospitals may seek to improve patient check-in procedures, billing, record keeping, and laboratory processes; they may also upgrade their technical staff. Measures of performance guide improvement efforts by enabling hospitals to track indicators such as patient wait time before being seen; causes for delay, such as incomplete paperwork or doctor unavailability; and the amount of time a physician spends with patients. These measures also serve as a guideline for the hospital, pointing out where improvements need to be made and how effective their changes have been. These measures will also tell a hospital whether it can actually offer the guarantee. Some hospitals are redesigning their emergency departments along a dual-track system to speed up their handling of lesser problems. Acute-care beds are set aside for the real emergencies, and an adjacent "fast track" section is designed to treat lesser emergencies and to get patients treated and released quickly. This type of arrangement has been shown to reduce the length of stay in the emergency unit by as much as half. Since Oakwood implemented the 30-minute guarantee, at all four of Oakwood Healthcare System's hospitals, patient satisfaction levels have soared. The performance measures revealed that only 0.9% of the 191,000 emergency room patients (about 1,700 people) were eligible for free tickets. By studying their performance measures, Oakwood was able to determine that the average wait is now 17 minutes between patient arrival and a physician's examination. Who knows—with further process improvements, maybe they will start offering a 15-minute guarantee.

WHAT ARE THE GOALS OF A MEASUREMENT SYSTEM?

One of the chief goals of a measurement system is to provide leaders with a multidimensional and qualitative view of their organization. A measurement system is a critical element in the strategic planning process because it allows an organization to measure progress toward goals and objectives. Performance measures are decision tools that enable leaders to link their strategy with day-to-day operations. Effective organizations measure the performance of areas that the organization values the most. Measurement systems allow effective organizations to:

- Determine that a gap exists between desired and actual performance
- Determine the root cause of the gap

■ Determine the necessary corrective action to eliminate the root cause of the gap
■ Determine whether the corrective actions eliminated the root cause and closed the gap between the actual and desired performance

Gap analysis—the study of the difference between actual and planned progress—allows effective organizations to learn where their efforts should be focused concerning their strategic plans. Gap analysis is a critical component of a measurement system because it drives organizational change. For instance, suppose an organization's strategic plan focused on five activities: health and safety, quality, cost reduction, throughput, and environmental issues. Then in order to know whether progress was being made toward these goals, measures such those in Figure 9.10 must be in place. Effective organizations utilize measures to change the way they do business. Measures of performance that reinforce the strategic plan enable these organizations to guide the direction their business is taking.

Health and Safety

Goal: Reduce reportable and lost-time incident rate by 33%
      Measure: Lost-time incident rate
Goal: 100% employee compliance in audit of ear and eye protection
      Measure: Count of employees without ear and eye protection during random monthly inspection

Quality

Goal: Reduce PPM by 50% from last year's level
      Measure: PPM defect rate
Goal: Complete PDSA problem-solving training for all members of engineering department in four weeks
      Measure: Number of people who have completed training per week

Cost

Goal: Reduce waste from last year's level by 33%
      Measure: Scrap rate
      Measure: Rework rate

Throughput Objectives

Goal: Up-time objective for Line 1—improve 10% from actual performance level of 70% to 80% by year-end
      Measure: Up-time performance level for Line 1

Environment

Goal: Reduce PPM exhaust rate by 33% by year-end
      Measure: PPM exhaust rate

Figure 9.10   Measures of Performance

EXAMPLE 9.4   Measures and Six Sigma: Justifying a Project

Queensville Manufacturing Corporation creates specialty packaging for automotive industry suppliers. The project team has been working to improve a particularly tough packaging problem involving transporting finished transmissions to the original equipment manufacturer (OEM). Company management has told the team that several key projects, including theirs, are competing for funding. In order to improve the chances of their project being selected, the team developed strong metrics to show how investment in their project will result in significant cost savings and improved customer satisfaction through increased quality. After brainstorming about this project, the team developed the following list of objectives and metrics for their project:

By retrofitting Packaging Machine A with a computer guidance system the following will improve:

Capacity:

KPOV: Downtime reduction from 23% to 9% daily
(downtime cannot be completely eliminated due to product changeovers)

KPIV: Resource consumption reduction
20% less usage of raw materials such as cardboard and shrink-wrap due to an improved packaging arrangement allowing five transmissions per package instead of four previously achieved

Customer Satisfaction:

KPOV: Improved delivery performance
Improved packaging arrangement integrates better with customer production lines saving 35% of customer production time

KPIV: Reduced space required for in-process inventory
Improved packaging arrangement allowing five transmissions per package instead of four previously achieved saving 10% of original factory floor space usage

KPOV: Reduced defect levels due to damage from shipping
Improved packaging arrangement provides better protection during shipment saving 80% of damage costs

Revenue:

KPIV: Reduced costs
$600,000 by the third quarter of the following fiscal year given a project completion date of the fourth quarter of this fiscal year

KPOV: Reduced lost opportunity cost
Increased customer satisfaction will result in an increased number of future orders

This project was selected as a focus for Six Sigma improvement because it could be justified by the impact it would have an overall organization performance.

WHAT ROLE DOES THE COST OF QUALITY INFORMATION PLAY IN AN EFFECTIVE ORGANIZATION?

What drives a company to improve? Increased customer satisfaction? Greater market share? Enhanced profitability? Companies seek to improve the way that they do business for many reasons, and one of the most important is the cost of quality. Quality costs are the costs that would disappear if every activity were performed without defects every time.

Products featuring what the customer wants at a competitive price will result in increased market share and therefore higher revenue for a company. But this is only part of the equation. Companies whose products and processes are free of defects enjoy the benefits of faster cycle times, lower warranty costs, and reduced scrap and rework costs. These lead to lower total cost for the product or service, which leads to more competitive pricing, which results in higher revenue for the company. Companies that manage their processes improve the bottom line on their income statements.

EXAMPLE 9.5   Costs of Quality

Over an 11-year period, airlines have been trying to reduce the number of lost bags (Figure 9.11). As of the year 2001, the average number of lost bags per 1,000 passengers was 3.79. In other words, 99.62% of passengers and their luggage arrive at the same airport at the same time. Why would an airline company be interested in improving operations from 99.62%? Why isn't this good enough? What are the costs of quality in this situation?

In the case of lost luggage, the types of quality costs are numerous. For instance, if a passenger is reunited with her luggage after it arrives on the next flight, the costs of quality include passenger inconvenience, loss of customer goodwill, and perhaps the cost of a gesture of goodwill on the part of the airline in the form of vouchers for meals or flight ticket upgrades.

As the scenario becomes more complex, a wider array of quality costs exists. For example, if an airline must reunite 100 bags per day with their owners at a hub airport such as Atlanta or New York, the costs can be enormous. The expenses associated with having airline employees track, locate, and reroute lost bags might be as high as 40 person-hours per day at $20/hour for a cost of $800/day. Once located, the luggage must be delivered to the owners, necessitating baggage handling and delivery charges. If a delivery service were to charge $1 per mile for the round-trip, in larger cities the average cost of delivering that bag to a customer might be $50. One hundred deliveries per day at $50 per delivery equals $5,000. Just these two costs total $5,800/day, or $2,117,000 per year! This doesn't include the cost of reimbursing customers for incidental expenses while they wait for their bags, the costs of handling extra paperwork to settle lost baggage claims, loss of customer goodwill, and negative publicity. Add to this figure the costs associated with bags that are truly lost ($1,250 per bag for domestic flights and $9.07 per pound of checked luggage for international travelers), and the total climbs even higher. In addition, these figures are merely the estimated costs at a single airport for a single airline!

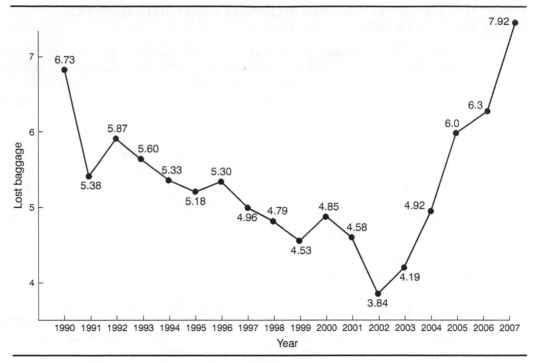

Figure 9.11   Mishandled Luggage (per 1,000 Passengers)
Source: K. Choquetter, "Claim Increase for Lost Baggage Still Up in the Air," *USA Today*, March 17, 1998, and
http://airconsumer.ost.dot.gov/.

Consider the more intangible effects of being a victim of lost luggage. If customers view airline baggage-handling systems as a black hole where bags get sucked in only to resurface sometime later, somewhere else, if at all, then consumers will be hesitant to check bags. If the ever-increasing amount of carry-on luggage people board the plane with is any indication, fears of lost luggage do exist. Carry-on baggage presents its own problems, from dramatically slowing the process of loading and unloading of the plane, to shifting in-flight and falling on an unsuspecting passenger when the overhead bins are opened during (or after) the flight, to blocking exits from the airplane in an emergency. These costs, such as fees for delayed departure from the gate or liability costs, can be quantified and calculated into a company's performance and profit picture. Is it any wonder that airlines have been working throughout the past decade to reduce the number of lost bags?

Two factions often find themselves at odds with one another when discussing quality. There are those who believe that no "economics of quality" exists, that it is never economical to ignore quality. There are others who feel that it is uneconomical to have 100% perfect quality all the time. Cries of "Good enough is not good enough" will be heard from some, while others will say that achieving perfect quality will bankrupt a company. Should decisions about the level of quality of a product or service be weighed against other factors such as meeting schedules and cost? To answer this question, an informed manager needs to understand the concepts surrounding the costs of quality. Investigating the costs associated with quality provides managers with an effective method to judge the economics and viability of a quality improvement system. Quality costs serve as a baseline and a benchmark for selecting improvement projects.

HOW ARE QUALITY COSTS DEFINED?

Cost of quality, poor-quality cost, and *cost of poor quality* are all terms used to describe the costs associated with providing a quality product or service. A quality cost *is considered to be any cost that a company incurs to ensure that the quality of the product or service is perfect. Quality costs are the portion of the operating costs brought about by providing a product or service that does not conform to performance standards. Quality costs are also costs associated with the prevention of poor quality.*

The most commonly listed costs of quality include scrap, rework, and warranty costs. As Figure 9.12 shows, these easily identified quality costs are merely the tip of the iceberg. Quality costs can be measured and tracked and then used to guide improvement efforts.

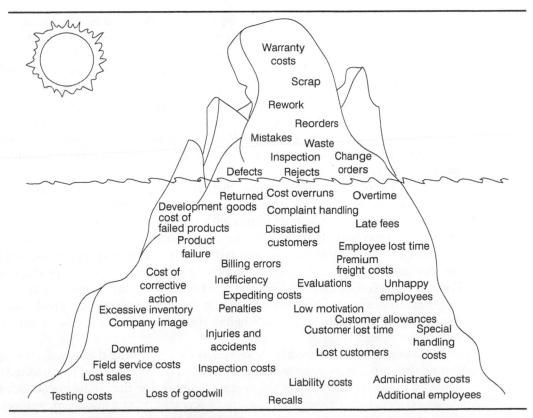

Figure 9.12   The Iceberg of Quality Costs

Rewriting an insurance policy to match customer expectations
Redesigning a faulty component that never worked right
Reworking a shock absorber after it was completely manufactured
Retesting a computer chip that was tested incorrectly
Rebuilding a tool that was not built to specifications
Repurchasing because of nonconforming materials
Responding to a customer's complaint
Refiguring a customer's bill when an error was found
Replacing a shirt the dry cleaner lost
Returning a meal to the kitchen because the meat was overcooked
Retrieving lost baggage
Replacing or repairing damaged or lost goods
Correcting billing errors
Expediting late shipments by purchasing a more expensive means of transportation
Providing on-site assistance to customers experiencing problems in the field
Providing credits and allowances

Figure 9.13   Examples of Quality Costs

Quality costs can originate anywhere within a company. No single department has exclusive rights to making mistakes that might affect the quality of a product or service. Even departments far removed from the day-to-day operations of a firm can affect the quality of a product or service. For example, a receptionist, often the first person with whom a customer has contact, can affect a customer's perceptions of the firm. The cleaning people provide an atmosphere conducive to work. Anytime a process is performed incorrectly, quality costs are incurred. Salespeople must clearly define the customer's needs as well as the capabilities of the company. Figure 9.13 provides a few more examples of quality costs. Every department within a company should identify, collect, and monitor quality costs that are within its control.

Quality, in both the service and the manufacturing industries, is a significant factor in enabling a company to maintain and increase its customer base. Measuring and tracking key customer success indicators begins with the prevention of poor quality. As a faulty product or service finds its way to the customer, the costs associated with the error increase. Preventing the nonconformity before it is manufactured or prepared to serve the customer is the least costly approach to providing a quality product or service. Potential problems should be identified and dealt with during the design and planning stage. Effort is required to solve an error at this phase of product development, but in most cases the changes can be made before costly investments in equipment or customer service are made.

If nonconformity of a product or service is located during the manufacturing cycle, or behind the scenes at a service industry, then the nonconformity can be corrected internally. The cost is greater here than at the design phase because the product or service is in some stage of completion. Product may need to be scrapped or, at a minimum, reworked to meet the customers' quality expectations.

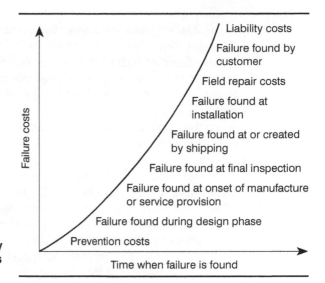

Figure 9.14 Increases in Quality Costs as Faulty Product Reaches Consumer

If a nonconforming product or service reaches the customer, the company that is providing it incurs the greatest costs. Customers who locate this nonconformity will experience a variety of feelings, the least of which will be dissatisfaction. Even if the customers are later satisfied by being provided with a new product, a repaired existing product, or a repetition of the service, the damage has already been done to the company's reputation. Disgruntled customers rarely fail to tell others about their experience. Mistakes that reach the customer result in the loss of goodwill between the present customer, future customers, and the provider. Failure cost increases as a function of the failure's detection point in the process (Figure 9.14). Mistakes, scrap, and rework are measures of performance that most organizations use to indicate where improvement is needed.

EXAMPLE 9.6   Human Error or Poor Design?

Repainting a room is a time-consuming if not costly task. Paints come in a wide variety of colors, types, and finishes. Consistently mixing paints to specific formulas allows for uniformity even when multiple individual gallons of paint are purchased. To ensure true colors, careful research by paint developers often results in different tint bases needed for different colors.

Recently, a home owner purchased four gallons of a custom-mixed, off-white paint in order to paint their front room and entry hall. During the painting process, the paint developed noticeable variations: large areas looked more gray than white. This was odd since the previous color on the wall was the same off-white and the four new gallons were all mixed and purchased at the same time. A very close inspection of the fine print on the label of one of the new cans of paint revealed that the tint base used for that can was different for the other three cans. Two types of tint base were available for this paint, one for whites and one for colors. The clerk at the paint store had inadvertently mixed an incorrect can.

The costs of quality associated with this situation are many: the customer's time to return to the paint store for another can of paint, the time to repaint the walls, the store's time to mix another can of paint, the cost of refunding the money on the first can of paint, record keeping for inventory and billing, and so on. This situation was further compounded when the original store discovered it did not have any cans of the correct tint base available. They had to send an employee to a store 30 minutes away to retrieve additional inventory.

At first glance, these costs of quality should be placed under "operator error." Doing so would result in follow-up training for the operator who inadvertently mixed the wrong tint base. Chastised, the operator would pledge not to make such a mistake in the future. However, follow-up of this sort may fall short of preventing other errors of the same kind at other stores by other employees. Why?

Assigning operator error to these costs of quality fails to uncover the root cause of the problem. Remember, it took "a very close inspection of the fine print on the label" to discover the paint tint-base type. At first, second, and even third glance, the labels of the two tint-base types appear identical. Both labels are the same color, red with gold lettering. Boldly displayed on both labels, the company logo, the name of the paint, and the luster level of the paint (flat, semi, or glossy) are identical in color, font, font size, and word choice. Only a single word in white in 10-point font size in a phrase at the bottom of the label is different. It took careful inspection, following the painting error, to determine the difference. On a busy day, it's easy to see that the single tiny word might be overlooked by an overworked clerk. The design of the label abounds in human factors or ergonomics errors. The key to eliminating the costs of quality in this situation is to redesign the label to make reading it error proof. This would be the only way to avoid these costs of quality in the future, regardless of the store, the employee, or the customer request.

WHAT TYPES OF QUALITY COSTS EXIST?

The costs associated with prevention of nonconformities, with appraisal of products or services as they are being produced or provided, and with product failure can be defined. Less well defined but equally important are intangible costs, those that involve a company's image. With an understanding of these four types of costs, a manager can make decisions concerning the implementation of improvement projects. Using quality costs, a manager can determine the usefulness of investing in a process, changing a standard operating procedure, or revising a product or service design.

Prevention Costs

Prevention costs are costs that occur when a company is performing activities designed to prevent poor quality in products or services. Prevention costs are often seen as front-end costs designed to ensure that the product or service is created to meet customer requirements. Examples of such costs are design reviews, education and training, supplier selection and capability reviews, and process improvement projects. Prevention activities must be reviewed to determine whether they truly bring about improvement in the most cost-effective manner.

Prevention efforts try to determine the root causes of problems and eliminate them at the source to avoid recurrences. Preventing poor quality keeps companies from incurring the cost of doing it over again. Essentially, if they had done it right the first time, they would not have to repeat their efforts. The initial investment in improving processes is more than compensated by the resulting cost savings.

Appraisal Costs

Appraisal costs are costs associated with measuring, evaluating, or auditing products or services to ensure that they conform to specifications or requirements. Appraisal costs are the costs of evaluating the product or service during the production of the product or the providing of the service to determine if, in its unfinished or finished state, it is capable of meeting the requirements set by the customer. Appraisal activities are necessary in an environment where product, process, or service problems are found. Appraisal costs can be associated with raw materials inspection, work-in-process (activities-in-process for the service industries) evaluation, or finished product reviews. Examples of appraisal costs include incoming inspection, work-in-process inspection, final inspection or testing, material reviews, and calibration of measuring or testing equipment. When the quality of the product or service reaches high levels, then appraisal costs can be reduced.

Failure Costs

Failure costs occur when a completed product or service does not conform to customer requirements. Two types exist: internal and external. *Internal failure costs are costs associated with product nonconformities or service failures found before the product is shipped or the service is provided to the customer.* Internal failure costs are the costs of correcting the situation. The failure costs may take the form of scrap, rework, remaking, reinspection, or retesting. *External failure costs are costs that occur when a nonconforming product or service reaches the customer.* External failure costs include the costs associated with customer returns and complaints, warranty claims, product recalls, or product liability claims. Since external failure costs have the greatest impact on the corporate pocketbook, they must be reduced to zero. Because they are highly visible, external costs often receive the most attention. Unfortunately, internal failure costs may be seen as necessary evils in the process of providing good-quality products or services to the consumer. Nothing could be more false. Doing the work twice, through rework or scrap, is not a successful strategy for operating in today's economic environment.

Intangible Costs

How a customer views a company and its performance will have a definite impact on the company's long-term profitability. *Intangible costs—hidden costs associated with providing a nonconforming product or service to a customer—involve the company's image.* Because they are difficult to identify and quantify, intangible costs of poor quality are often omitted from quality cost determinations. However, these costs must be neither overlooked nor disregarded. Is it possible to quantify the cost of missing an important deadline? What will be the impact of quality problems or schedule delays on the company's image? Intangible costs of quality can be three or four times as great as the tangible costs of quality. Even if these costs can only be named, with no quantifiable value placed on them, it is important that decision makers be aware of their existence.

The four types of quality costs are interrelated. In summary, **total quality costs** *are considered to be the sum of prevention costs, appraisal costs, failure costs, and intangible costs.* Figure 9.15 shows some of the quality costs from Figure 9.12 in their respective categories. The costs associated with poor quality are tangible and measurable. Poor quality costs an organization and its customers time and money. Poor quality affects the success of any business. Costs of quality come from a wide variety of sources. Some are the result of basic

Prevention Costs

Quality planning
Quality program administration
Supplier-rating program administration
Customer requirements/expectations
   market research
Product design/development reviews
Quality education programs
Equipment and preventive maintenance

Appraisal Costs

In-process inspection
Incoming inspection
Testing/inspection equipment
Audits
Product evaluations

Failure Costs

Internal:
Rework
Scrap
Repair
Material-failure reviews
Design changes to meet customer
   expectations
Corrective actions
Making up lost time
Rewriting a proposal
Stocking extra parts
Engineering change notices

External:
Returned goods
Corrective actions
Warranty costs
Customer complaints
Liability costs
Penalties
Replacement parts
Investigating complaints

Intangible Costs

Customer dissatisfaction
Company image
Lost sales
Loss of customer goodwill
Customer time loss
Offsetting customer dissatisfaction

Figure 9.15   Categories of Quality Costs

organizational capabilities or the lack of them. These can be remedied only by improving the skills and tools of the people involved so that they can improve business processes. Variation is another source of quality costs. Poor industrial processes that are not capable of producing products or providing services to customers show high levels of scrap, rework, and field failures. Sometimes this is due to a lack of engineering design, and sometimes customer requirements have not been clearly transmitted throughout the organization. In the service part of an organization, business process variations lead to poor financial performance. Process improvements are necessary to streamline service processes and eliminate errors. Poor product engineering or process design results in quality costs. Within these systems there are non–value-added activities, poorly documented processes, and inadequate procedures. Process management and improvement (Chapter 11) is crucial to eliminating these sources of quality costs. Supplier capabilities play a role in quality costs. The lack of quality suppliers can lead to late deliveries, poor-quality components and parts, and poor service. Effective organizations identify and address the quality costs that plague their organization.

Investments made to prevent poor quality will reduce internal and external failure costs. Consistently high quality reduces the need for many appraisal activities. Suppliers with strong quality systems in place can reduce incoming inspection costs. High appraisal costs combined with high internal failure costs signal that poor-quality products or services are being provided. Efforts made to reduce external failure costs will involve changes to efforts being made to prevent poor quality. Internal failure costs are a portion of the total production costs, just as external failure costs reduce overall profitability. When dealing with quality costs, companies need to ensure that appraisal costs are well spent. Companies with a strong appraisal system need to balance two points of view and consider the trade-off: Is the company spending too much on appraisal for its given level of quality performance, or is the company risking excessive failure costs by underfunding an appraisal program? In all three areas—prevention, appraisal, and failure costs—the activities undertaken must be evaluated to ensure that the efforts are gaining improvement in a cost-effective manner. Figure 9.16 reveals that as quality

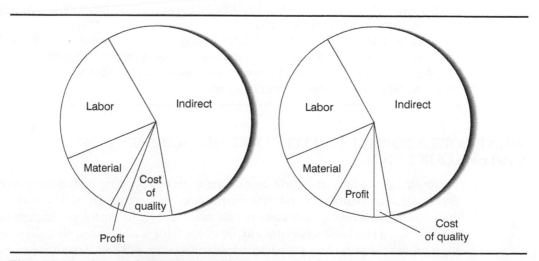

Figure 9.16   Cost of Quality versus Profit

costs are reduced or, in the case of prevention quality costs, invested wisely, overall company profits will increase. The savings associated with doing it right the first time show up in the company's bottom line.

EXAMPLE 9.7 Uncovering Quality Costs

A sheet metal stamping firm recently bid $75,000 as the price of providing an order of aluminum trays. After stamping a large number of these trays, the company investigated their costs and determined that if production continued, the cost of providing the order would be $130,000! This was nearly double what they had bid.

They immediately set out to figure out what was causing such a dramatic increase in their costs. They investigated their product design and found it to be satisfactory. Prevention costs associated with product design and development were negligible. The customer had not returned any of the completed trays. There were no customer complaints and therefore no corrective-action costs. Their investigation showed that external failure costs were zero. Appraisal costs were normal for a product of this type. Incoming material and in-process inspection costs were similar to the amount projected. The product had not required any testing upon completion.

It was only when they investigated the internal failure costs that the problem became apparent. At first, they felt that internal failure costs were also negligible since there was almost no scrap. It wasn't until a bright individual observed the operators feeding the trays through the stamping operation twice that they were able to locate the problem. When questioned, operators told the investigators that they knew that due to inappropriate die force, the product was not stamped to the correct depth on the first pass. Since the operators knew that the trays would not pass inspection, they decided among themselves to rework the product immediately, before sending it to inspection. The near doubling of the manufacturing costs turned out to be the internal failure costs associated with reworking the product.

Essentially, the company incurred a cost of quality equal to the standard labor and processing costs of the operation. To meet specifications, a significant number of trays were being handled and processed twice. Only by preventing the need for double processing can the costs of quality be reduced.

This type of internal failure cost actually began as a prevention cost. If the manufacturer had investigated the capability of the stamping operation more closely, they would have determined that maintenance work was necessary to improve the machine's performance and therefore improve product quality.

WHAT DOES A FORMAL QUALITY COST MEASUREMENT SYSTEM LOOK LIKE?

In the struggle to meet the three conflicting goals of achieving high quality, low cost, and timely delivery, identifying and then quantifying quality costs helps to ensure that quality does not suffer. When quality costs are discussed as a vague entity, the importance of their relation to cost and delivery is unclear. By quantifying quality costs, all individuals producing a product or providing a service understand what it will cost if quality suffers. True cost reduction occurs when the root causes of nonconformities are recognized and eliminated.

A quality cost measurement system should be designed to keep track of the different types of quality costs. The ability to define and quantify quality costs enables organizations to manage quality more effectively. Once quantified, these costs can be used to determine which projects will allow for the greatest return on investment and which projects have been most effective at improving processes, lowering failure rates, and reducing appraisal costs. A quality cost measurement system should use quality costs as a tool to help justify improvement projects.

Quantifying quality costs will help organizations zero in on which problems, if solved, will provide the greatest return on investment. A quality cost measurement system should try to capture and then reduce significant quality costs. Effective cost reduction occurs when the processes providing products or services and the related support processes are managed correctly. Measures are key to ensuring that the processes are performing to the best of their capability. After all, what is measured can be managed. Once the cost of quality data has been measured and tabulated, the data can be used both to select quality improvement projects and to identify the most costly aspects of a specific problem. Failure costs are prevalent in all aspects of providing goods and services, from design and development to production and distribution. Support processes, such as marketing and human resources, can also be the source of quality costs. When justifying improvement projects, look for rework, waste, returns, scrap, complaints, repairs, expediting, adjustments, refunds, penalties, wait times, and excess inventory. Cost improvement efforts should focus on locating the largest cost contributors. Try not to be tempted to spend time chasing insignificant, incremental measures of quality costs. Use a Pareto analysis to identify the projects with the greatest return on investment, and tackle those first.

Those establishing a quality cost measurement system need to keep in mind that the success of such a system is based on using quality cost information to guide improvement. Each cost has a root cause that can be first identified and then prevented in the future. Remember that preventing problems in the first place is always cheaper. Those implementing a quality cost measurement system need to establish a basic method of identifying correctable problems, correcting the problems, and achieving a new level of performance.

HOW ARE QUALITY COSTS UTILIZED FOR DECISION MAKING?

All initiatives undertaken by an organization must contribute to its financial success. Quality costs can be used as justification for actions taken to improve the product or service. Typically, investments in new equipment, materials, or facilities require the project sponsor to determine which projects will provide the greatest return on investment. These calculations traditionally include information on labor savings, on production time savings, and on the ability to produce a greater variety of products with better quality. The "better quality" aspect of these calculations can be quantified by investigating the costs of quality, particularly the failure costs. It is important to determine the costs of in-process and incoming material inspection, sorting, repair, and scrap, as well as the intangible costs associated with having a nonconforming product or service reach the customer. Making a decision with more complete quality information, such as product appraisal costs, can help determine the true profitability of a product or service.

Costs of Quality

Prevention	**Appraisal**
Process planning	Inspection
Process control	Auditing
Training	
Internal Failures	**External Failures**
Scrap	Cost to customer
Rework	Complaint adjustments
Overtime	Returned material
	Costs to do it over again
Intangible Costs	
Reputation	

Figure 9.17   JQOS Costs of Quality

Once quality costs have been identified, those reviewing the project can determine whether the money for the project is being well spent to contribute to the growth of the company. Identifying and then quantifying quality costs has a twofold benefit: Cost savings are identified, and quality is improved. By improving the quality performance of a company, the company also improves its quality costs.

With any quality measurement or improvement system, the information and improvements realized need to be applied to any future activities. Dr. Deming's Plan-Do-Study-Act cycle plays a significant role. Once an organization has planned and implemented improvements, it must act upon the improvements to ensure that future activities maintain the new level of performance and the decreased level of quality costs. Identifying quality costs allows management to judge improvement investments and profit contributions. A quality program is only as valuable as its ability to contribute to customer satisfaction and, ultimately, to company profits.

How Do We Know It's Working?

The leaders at JQOS recognize the importance of tracking costs of quality. They have measures of performance related to the five types of quality costs: prevention, appraisal, internal failure, external failure, and intangible costs (Figure 9.17). They use this quality cost information for decision making. As with their other measures, as they study process performance, they make changes and improvements that align with their strategic plans.

For instance, they have made significant changes in how shafts are packaged and shipped, significantly reducing the damage caused by poor handling after the shafts leave the plant. This decision came about after the owner visited a very good customer. During a courtesy tour of the plant, they visited shipping. There on the table, to his horror, lay a broken box of very expensive shafts. The box could hardly be called a box any more, consisting of mangled cardboard held together by bands of steel. The shafts, their protective wrapping sleeves missing or torn, were covered in nicks, scratches, and dents caused by the bands of steel rubbing against their unprotected surfaces.

With his heart in his throat, the owner profusely apologized to the plant manager. The manager responded that they had never had trouble before. The total cost of the lack of quality in this one incident: $15,000. This included money for the scrapped parts, original shipping and reshipping costs, money spent to process the parts through the plant the first time and then to do it over again, money spent to rearrange production in the plant and work overtime to create replacement shafts, and money spent to make repeat customer visits to show support and check on future shipments.

A root cause investigation into the situation revealed several problems. A new and incompletely trained shipping clerk had been left on his own despite specific instructions that he was to ship no parts on his own. This new clerk had not wrapped the protective sleeves around the parts before putting them in the box and banding it. The second critical error occurred when the shipping company contracted to ship the parts subcontracted the job to a less qualified shipper. This shipper significantly damaged the box in shipment and then placed the metal bands around the parts to hold them together until they got to the customer's plant.

The changes made to the process included enhanced training procedures and a sign-off form from the supervisor for shipments prepared by any new clerk until that clerk completed their training. Also, a new type of protective sleeve is now being used. Shipments of shafts over a certain size now are packaged completely differently, allowing more support for the heavy weight of the shafts. A different shipper has been contracted and provided with specific shipping instructions. In the year since this event occurred, no shafts have been damaged in shipping. Rather than react to this incident as a single occurrence, cost-of-quality information was used to guide a permanent improvement.

HOW DO THE MALCOLM BALDRIGE NATIONAL QUALITY AWARD CRITERIA SUPPORT MEASUREMENT, ANALYSIS, AND KNOWLEDGE MANAGEMENT FOR EFFECTIVE ORGANIZATIONS?

The Malcolm Baldrige National Quality Award (MBNQA) criteria encourage organizations to take a careful look at their knowledge management systems, particularly the measures of performance used as indicators of process improvement. In Section 4.0, Measurement, Analysis, and Knowledge Management, organizations are asked to describe how their performance measurement system operates. They are asked to provide details about how their organization gathers, analyzes, and uses performance data and information. Organizations must provide information about their performance measurement system, including how it is used to effectively understand, align, and improve performance at all levels and in all parts of the organization. They are asked to describe how their organization analyzes performance data and information to assess and understand overall organizational performance.

Knowledge management and information analysis are crucial for creating and maintaining an effective organization with an unwavering focus on the customer. Gathering and analyzing information are not limited to one area of the organization. When evaluating how effective an organization's knowledge management and information analysis system is, ask these questions based on the MBNQA criteria:

Leadership

How does leadership analyze and use information?
What information are leaders analyzing and using?
Are leaders converting information into appropriate policies and actions?

Strategic Planning

How does the strategic plan analyze and use information?
How does the strategic plan react to new information?

Customer and Market Focus

Is the organization gathering useful customer and market information?
How is this information gathered and disseminated?
Is useful customer and market information communicated to employees in a manner that improves their job performance and enhances customer service?

Measurement, Analysis, and Knowledge Management

How does the organization measure, analyze, and improve performance?
How does the organization select, collect, analyze, align, integrate, and use information in day-to-day operations?
What comparative data is used? Why was it chosen?
When and how are organizational performance and capabilities reviewed?
How are review findings integrated into operations?
How do employees access needed information?
How is information kept reliable, secure, and accessible?
How are information accuracy, integrity, reliability, timeliness, security, and confidentiality kept?
How are emergencies dealt with?
How are employees gathering, analyzing, disseminating, and using information related to key processes?
Does the organization have communication systems in place that enhance the employees' ability to obtain, analyze, and use information?

Workforce Focus

Do employee policies and reward systems support the use and analysis of information related to key processes during day-to-day activities?

Do employee policies and reward systems support the use and analysis of information during continuous improvement activities?

Process Management

Have key processes supporting the analysis and use of information been identified and/or improved?

Is the organization gathering appropriate information concerning key processes?

Is the information gathered being used to support appropriate improvement efforts?

Results

Do the business results associated with key processes reflect the judicious analysis and use of information?

Is the successful integration of customer and market information into the strategic plan shown by the business results?

LESSONS LEARNED

Metrics are powerful tools. Metrics help effective organizations see the difference between desires, perceptions, intuitive guessing, and reality. Fact-based decision making is sound decision making. Dr. Shewhart showed that metrics can be used to determine whether a process is stable and predictable. He proved that metrics show how much variation is present in a process. Because they help people understand processes, metrics help identify problems that might have been overlooked. Metrics validate what is important and what is not. They also validate whether our processes are performing to expectations. Metrics are the link between process inputs and process outputs. By establishing a baseline, they help us evaluate the current situation and communicate the changes necessary. Metrics show whether the gains that were made have been retained. Metrics don't leave room for excuses. They show process performance at its best and at its worst.

Effective organizations have learned that it is essential to measure performance in all areas of the company. Performance measures allow for appropriate accountability and provide important information about organizational performance in all areas of the organization, including finance, operations, and support services. Measures enable effective organizations to complete their goals and objectives as established in their strategic plan. Cost-of-quality studies can reveal areas where organizational improvement is needed.

Measures are indicators of performance.

1. Good measures of performance tell how well something is performing its intended purpose.
2. Good measures enable organizations to determine the impact of changes.

3. Good measures tell leaders whether assets have been correctly allocated.

4. The Balanced Scorecard approach emphasizes the importance of having financial, customer, internal processes, and learning and growth.

5. Gap analysis is the study of the difference between what was planned and what actually occurred.

6. Costs of quality are the costs associated with providing a product or service.

7. Five types of quality costs exist: prevention, appraisal, internal failure, external failure, and intangible costs.

8. Quality costs should be utilized for decision making.

Are You a Quality Management Person?

Measures of performance and costs of quality can play a role in your day-to-day activities. Check out these questions and find out whether you are a quality management person. How do you answer them on a scale of 1 to 10?

Can you identify measures of performance for your key day-to-day processes?

1 2 3 4 5 6 7 8 9 10

Do you regularly measure whether you are meeting your personal goals and objectives?

1 2 3 4 5 6 7 8 9 10

Do you regularly measure whether you are meeting your work-related goals and objectives?

1 2 3 4 5 6 7 8 9 10

Do you measure the impact of changes that you have made in your life?

1 2 3 4 5 6 7 8 9 10

Do you think about balancing your scorecard? (What measures would be on your scorecard?)

1 2 3 4 5 6 7 8 9 10

Do you perform a gap analysis to determine the gap between what you had planned and what you actually accomplished?

1 2 3 4 5 6 7 8 9 10

Have you identified any prevention costs associated with the activities you perform?

1 2 3 4 5 6 7 8 9 10

Have you identified any appraisal costs associated with your activities?

1 2 3 4 5 6 7 8 9 10

Have you identified any internal failure costs associated with your activities?

1 2 3 4 5 6 7 8 9 10

Have you identified any external failure costs associated with your activities?

1 2 3 4 5 6 7 8 9 10

Have you identified any intangible costs associated with your activities?

1 2 3 4 5 6 7 8 9 10

Do you consider your identified costs of quality when making decisions or changes?

1 2 3 4 5 6 7 8 9 10

Chapter Questions

1. What did B. C. Forbes mean when he said, "If you don't drive your business, you will be driven out of business"?
2. How do effective organizations use performance measures?
3. What is the difference between process and results measures?
4. Why is an effective performance measurement system necessary?
5. What performance measures does the organization you work for use?
6. How does your organization use performance measures?
7. Create a set of measures, based on the Balanced Scorecard and Example 9.1, for a fast-food restaurant.
8. Create a set of measures, based on the Balanced Scorecard and Example 9.1, for a movie theater.
9. What is a prevention cost? How can it be recognized? Describe where prevention costs can be found.
10. What is an appraisal cost? How can it be recognized? Describe where appraisal costs can be found.

11. What are the two types of failure costs? How can they be recognized? Describe where these costs can be found.

12. Describe the relationship among prevention costs, appraisal costs, and failure costs. Where should a company's efforts be focused? Why?

13. How are quality costs used for decision making?

14. What should a quality cost system emphasize?

15. What are the benefits of having, finding, or determining quality costs?

SECTION 4.0: MEASUREMENT, ANALYSIS, AND KNOWLEDGE MANAGEMENT

The Measurement, Analysis, Knowledge section of the MBNQA criteria addresses an organization's information management and performance measurement systems. It also examines how an organization analyzes and utilizes performance data and information.

The key goals of this section are to:

1. Examine the key aspects of the information management and performance measurement systems of Remodeling Designs, Inc. and Case Handyman Services.
2. Examine how Remodeling Designs/Case Handyman analyze performance data and information.

This section was created by Jen Meyer, Molly Schuetz, Erich Eggers, Joan Cordonnier, Mike Cordonnier, and Donna Summers.

QM   Information and Analysis Questions

1. What methods do you use to gather performance data and information? Are the methods the same for both Case Handyman and Remodeling Designs?
2. How large a role does the actual data gathered play in your overall organization? Where exactly in your organization is the data used?
3. Who in your company collect/compile data? What information is important to them? What do they need to know about your company? What do they need to know about your customers? What do they need to know about your competitors?
4. Give examples in which your companies use the data collected to make management (leadership) decisions.
5. How do your companies ensure that the data gathered is helping you work toward your mission statement?
6. How does the analysis of your information help to ensure customer satisfaction?
7. How do your companies use the results found to show where you are now and to project where you will be in the future?
8. How do your companies gather data for monitoring daily operations?
9. How do your companies communicate important data and information between and among:
   a. employees?
   b. suppliers?
   c. customers?

10. How do your companies use your data to track trends of problems within the company?
11. How do your companies ensure that the data/information gathered is accurate?
12. How do you make sure that the data collected is kept up to date and responded to in a timely manner?
13. Do your companies use benchmarking to compare themselves against other companies' data?
14. How do your companies gather, analyze, disseminate, and use information related to your key processes?
15. Has the information that you gathered, analyzed, disseminated, and encouraged enabled your companies to improve their performance as exhibited by the bottom line?

4.0 Measurement, Analysis, and Knowledge Management

The Measurement, Analysis, Knowledge Management category examines the information management and performance measurement systems of Remodeling Designs/Case Handyman, including how these organizations analyze performance data and information.

4.1 MEASUREMENT AND ANALYSIS OF ORGANIZATIONAL PERFORMANCE

This section describes how Remodeling Designs/Case Handyman provide effective performance management systems for measuring, analyzing, aligning, and improving performance at all levels and in all parts of the companies.

4.1a Performance Measurement

Remodeling Designs' and Case Handyman's first essential employee performance measurement tool is the Employee Review worksheet (see Figure 2 in the Chapter 7 Case Study). Annually, each employee is given this review sheet to complete. The employees use this review sheet to constructively rate themselves and identify improvements needed for the following year. They then discuss the review with either Mike or Erich in a performance review meeting. This worksheet changes yearly based upon what information the managers need that year to maintain happy employees and to complete jobs successfully.

The second performance measurement tool is the Valued Client Feedback survey. This survey provides Remodeling Designs feedback from the customer on completed jobs. The original format, based on another company's template, simply had the customer circle Yes/No and did not allow customers to truly rate the company's performance. After reviewing their company's needs, Remodeling Designs developed a new survey in order to track data that was important to them. The new survey allows customers to rate the company's performance on a scale from 1 = Strongly Disagree to 5 = Strongly Agree. The new format targets and supports the mission statement.

The Daily Log records the daily progression of a remodeling job. This log contains details such as weather conditions, temperature, inspections, labor on-site (subcontractors and company employees), daily safety inspections, calls made, communications with customers, and a detailed listing of jobs performed. The log is essential to the company's alignment and communication processes. It allows Remodeling Designs and Case Handyman to track the amount of time needed for different phases of a project and helps them project costs for future projects.

Both companies use benchmarking to make improvements. Attending the multicompany national meeting Remodeler's Executive Roundtable (RER) allows Remodeling Designs to benchmark itself against other remodeling companies. This meeting has brought many successful changes to Remodeling Designs, for example, revision of the customer contract, increase in salaries, increase in the price of services, and the idea of incentive trips for employees. Another key item from RER is the "Percent-of-Completion" accounting system, a standard program used in the construction industry. This program shows the profit/loss ratio and allows Remodeling Designs to see a clearer picture of where they stand as an organization. Attendance at the RER meeting allows them to improve their performance as exhibited by the bottom line. Case Handyman is part of a 30-company franchise in which they benchmark against one another.

4.1b Performance Analysis

Remodeling Designs and Case Handyman use data to make management decisions in order to move their companies forward and be competitive in the market. Examples of how management uses this data are reallocation of manpower, preparation of financial data/budget, and advertisement/marketing. In addition, when Remodeling Designs and Case Handyman receive a lead, they conduct an extensive interview of what the customer's needs are, and they use this information to make managerial decisions on whether to accept or reject a job.

Remodeling Designs and Case Handyman are lacking in the area of data tracking. The items they do track are gross profit and referral/repeat business leads. The items they would like to track are the approval rating with customers (discussed in Section 4.1a) and the warranty costs. Case Handyman tracks performance data and collects data through a database. Remodeling Designs uses verbal communication and finds it difficult to track numbers. Thus, both companies collect pertinent information which they could track but do not. Time and manpower are also an issue. By tracking information, the companies could make valuable changes to increase the bottom line.

The second area for improvement is tracking trends of problems within Remodeling Designs. The Case Handyman office has a white board on which employees write problems. Remodeling Designs does not have an internal system for tracking problems. Externally, Remodeling Designs uses warranty to track problems after the fact. The warranty form is completed and submitted to the section in the binder.

4.2 INFORMATION AND KNOWLEDGE MANAGEMENT

This section describes how Remodeling Designs and Case Handyman ensure the quality and availability of needed data and information for employees, suppliers, and customers.

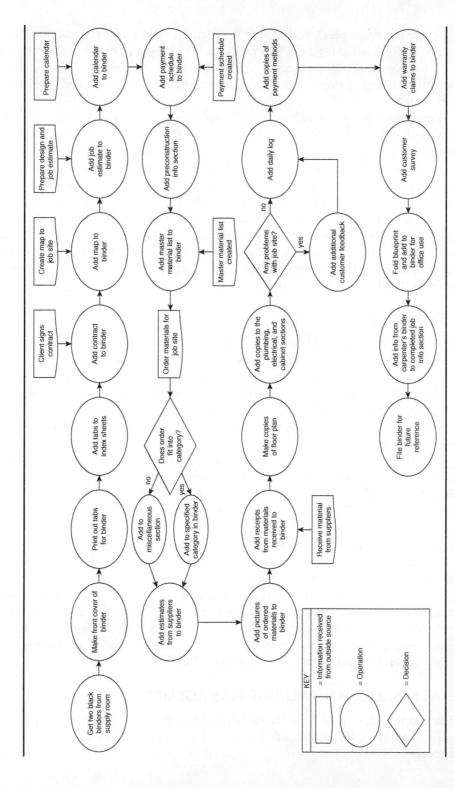

Figure 1 Process Map for Remodeling Designs Binder

286

4.2a Data Availability

Remodeling Designs and Case Handyman retain information about a job and its organization in a binder. The process of compiling the binder is an ongoing task throughout a job (Figure 1). From the moment that Remodeling Designs accept the job until the job is completed, this binder stores all information that the company might need. After the job has been completed, the binder is stored and is readily available for later referral. The company begins the process map by starting a binder for every job accepted and making the appropriate cover and tabs. Then they use the map to explore how all of the information is added to the binder, including the contract map to the job site, job estimate, calendar of events, payment schedule and copies of payment methods, estimates from suppliers, master list of materials, copies of the design, daily log, warranty claims, and customer survey. As mentioned above, the map ends with the binder being put into storage for future reference in case the customer needs an item repaired or wants to have another remodeling job completed.

The weekly meetings help employees exchange positive and negative aspects of jobs. Employees discuss issues such as job progression and problems at the site, and they forecast task completion for the following week. This exchange of information helps employees to communicate and to be in alignment with one another.

4.2b Hardware and Software Quality

In general, Remodeling Designs and Case Handyman use word processing, database, and presentation software to complete daily tasks. They also use Microsoft Project® to show a timeline of individual jobs. These jobs are tracked in areas such as what job has been completed, what needs to be completed, and which employee performs which key tasks. Remodeling Designs and Case Handyman also use the Percent-of-Completion accounting system to show their profit/loss ratio and to provide a clearer picture of where the company stands financially.

10 Benchmarking

How does benchmarking help an effective organization measure its success?

What is the purpose of benchmarking?

What types of benchmarking can be performed?

What are the benefits of benchmarking?

What are the different standards for comparison?

How is benchmarking done?

How do the Malcolm Baldrige National Quality Award criteria view benchmarking?

Lessons Learned

Learning Opportunities

1. To understand why benchmarking can be used to increase organizational effectiveness

2. To become familiar with how benchmarking is performed

Are we keeping up with the Joneses?

HOW DOES BENCHMARKING HELP AN EFFECTIVE ORGANIZATION MEASURE ITS SUCCESS?

Many firms choose to compare their performance against that of another firm in order to learn how they are performing in the marketplace. This act of comparing is called *benchmarking*. As illustrated in the Chapter 9 introduction, nursing homes can readily access their performance measures information and compare that information with the performance of other nursing homes in their state. Process improvement ideas can come from all difference sources. Effective organizations seek out ideas through benchmarking. When a process has been chosen for study, it's wise to seek out and investigate the companies who do it best. Benchmarking answers the question "How do they do it?" This information can then be used to fill the gap between "How do we do it now?" and "What do we need to do differently?"

During a benchmarking process, a company compares its performance against a set of standards or against the performance of best-in-field companies. With the information provided by the comparison, a company can determine how and where to improve its own performance. Benchmarks serve as reference points. The information gathered and the measurements taken are used to make conclusions about an organization's current performance and any necessary improvements (Figure 10.1). Although companies may choose different aspects of their operations to benchmark, typical areas include procedures, operations, processes, quality improvement efforts, and marketing and operational strategies.

How Do We Know It's Working?

Leaders at JQOS understand the importance of keeping in touch with best practices both within their organization and beyond. For this reason, the president has joined several local small business alliances similar to the AMIBA (American Independent Business Alliance) or BALLE (Business Alliance for Local Living Economies). At meetings, members of his groups share best practices and valuable business advice. They have visited each other's facilities and freely shared information about their better business practices. They are also able to pool resources and create joint purchasing groups to receive volume discounts once available only to larger corporations.

He has even gone so far as to start up his own industrial roundtable. Members of this group share similarities, like business size and product lines, but none of them compete for the same customers. This enables them to freely share information without worrying about competitive intelligence or information related to pricing or strategies being jeopardized. At a recent meeting, their discussions focused on finding ways for job shops to enhance their scheduling abilities. Follow-up meetings included visits by several industry representatives who described computer-based scheduling packages available on the market.

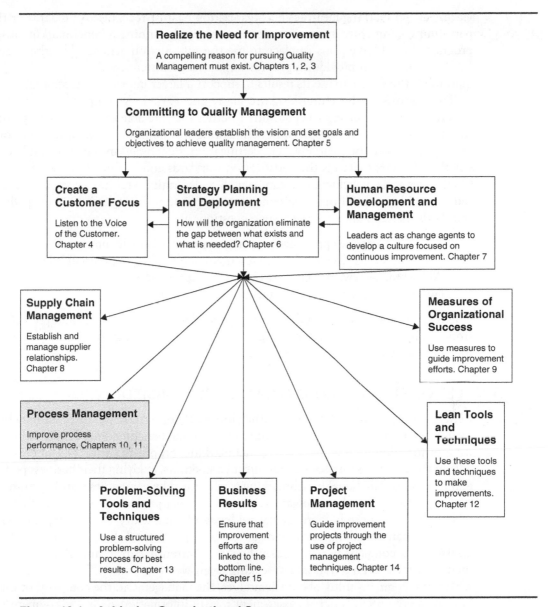

Realize the Need for Improvement

A compelling reason for pursuing Quality Management must exist. Chapters 1, 2, 3

Committing to Quality Management

Organizational leaders establish the vision and set goals and objectives to achieve quality management. Chapter 5

Create a Customer Focus

Listen to the Voice of the Customer. Chapter 4

Strategy Planning and Deployment

How will the organization eliminate the gap between what exists and what is needed? Chapter 6

Human Resource Development and Management

Leaders act as change agents to develop a culture focused on continuous improvement. Chapter 7

Supply Chain Management

Establish and manage supplier relationships. Chapter 8

Measures of Organizational Success

Use measures to guide improvement efforts. Chapter 9

Process Management

Improve process performance, Chapters 10, 11

Lean Tools and Techniques

Use these tools and techniques to make improvements. Chapter 12

Problem-Solving Tools and Techniques

Use a structured problem-solving process for best results. Chapter 13

Business Results

Ensure that improvement efforts are linked to the bottom line. Chapter 15

Project Management

Guide improvement projects through the use of project management techniques. Chapter 14

Figure 10.1   Achieving Organizational Success

WHAT IS THE PURPOSE OF BENCHMARKING?

Effective organizations use benchmarking to compare their key measures of performance with those of others in order to determine where improvement opportunities exist. Companies planning a benchmarking assessment should carefully consider the motivating factors—specifically, "Why is the company planning to do this, and what do leaders hope to learn?" Benchmarking measures an organization against recognized standards or the best-performing companies in the industry. Companies beginning a benchmarking assessment program should have plans in place to use the information generated by the comparison. Benchmarking will provide targets for improved performance. A major pitfall of benchmarking is the failure to use its results to support a larger improvement strategy.

The reasons for benchmarking are many and varied. A company may embark on a benchmarking assessment to determine whether they are able to comply with performance standards set by their customers. Benchmarking will point out areas where improvements are needed before the company seeks certification. Benchmarking against standards verifies whether a company meets the certification standards and qualifications set by a customer. On a larger scale, benchmarking can be used to determine whether a company's quality systems are able to fulfill the requirements appropriate to meet ISO 9000 or quality award standards. Benchmarking answers such questions as:

> Are the company's processes properly constructed and documented?
> Are systems in place to allocate resources and funding appropriately?
> Which areas have the greatest improvement needs?
> What are our internal and external customer needs?

The information gathered in the assessment guides objectives, plans, and projects for continuous improvement. Benchmarking is never about collecting competitive intelligence or information related to market research.

WHAT TYPES OF BENCHMARKING CAN BE PERFORMED?

Benchmarking can be performed either internally or externally. Internal benchmarking focuses on activities within the organization. External benchmarking can be either competitive or functional. In competitive benchmarking, an organization focuses on companies within their own market, sometimes direct competitors, studying their business performance and processes. Functional benchmarking is performed by companies wanting to study a particular process. They choose organizations with similar processes regardless of their industry.

Benchmarking can be done at several levels of complexity. Some companies choose to conduct a benchmarking assessment at the perception level. From a *perception benchmark assessment*, a company hopes to learn how it is currently performing. A perception assessment can focus on internal issues, seeking to answer questions related to what the people within the company think about themselves, the management, the company, or the quality improvement process. Such an assessment reveals information about the company's current performance levels. This assessment may later serve as a baseline to compare with future benchmarking experiences.

Companies beginning the quest for ISO 9000 certification, qualified supplier certification, or a quality award may choose to perform a *compliance benchmark assessment*. This more in-depth benchmarking experience verifies a company's compliance with stated requirements and standards. The information gathered will answer questions about how a company is currently performing compared to the published standards. The assessment will also help a company locate where compliance to standards is weak.

A third type of benchmarking assessment investigates the effectiveness of a system that a company has designed and implemented. An *effectiveness benchmark assessment* verifies that a company both complies with the requirements and also has effective systems in place to ensure that the requirements are being fulfilled. For example, a quality manual may be published, but unless systems are in place to ensure the effective use of the manual, the requirement has not been fulfilled.

A fourth type of benchmarking assessment deals with continuous improvement. A *continuous improvement benchmark assessment* verifies that continuous improvement is an integral and permanent facet of an organization. It judges whether the company is merely providing "lip service" to process improvement issues or instead is putting into place systems that support continual improvement on a day-to-day basis.

Perceptions, compliance, effectiveness, and continuous improvement can be verified by conducting a thorough review of the existing business practices. This review should follow an organized format. Activities can be judged on the basis of visual observations by the reviewers, interviews with those directly involved, personal knowledge, and factual documentation.

WHAT ARE THE BENEFITS OF BENCHMARKING?

The primary benefit of benchmarking is the knowledge gained about where a company stands compared to standards set by its customers, by the company itself, or by national certification or award requirements. With this knowledge a company can develop strategies for meeting its own continuous improvement goals. The benchmarking experience will identify existing assets within the company as well as opportunities for improvement. Most quality assurance certifications involve discovering how a company is currently performing, strengthening the company's weaknesses, and then verifying the company's compliance with the certification standards. Because benchmarking assessment provides an understanding of how the company is performing, it is a valuable tool to use throughout the certification process.

WHAT ARE THE DIFFERENT STANDARDS FOR COMPARISON?

Typically, when a company chooses to perform a benchmarking assessment, one of the following standards is chosen: the Malcolm Baldrige National Quality Award, the International Organization for Standardization ISO 9000 standard, the Deming Prize, supplier certification requirements, or other companies who are the best in their field. Several of these benchmarking comparisons were covered in greater detail in Chapter 3; all are discussed briefly in the following paragraphs.

Malcolm Baldrige National Quality Award

The Malcolm Baldrige National Quality Award (MBNQA) criteria are used by companies pursuing the United States' highest award for quality management systems. The award criteria are also a popular and rigorous benchmarking tool for companies seeking a better understanding of their performance. Using the following specific categories, companies can discover their present capabilities and the areas for improvement:

Leadership
Strategic Planning
Customer and Market Focus
Measurement, Analysis, Knowledge Management
Workforce Focus
Process Management
Results

Because these criteria provide complete and thorough coverage, they are a valuable tool for benchmarking.

ISO 9000

Benchmarking assessments using ISO 9000 documentation are often used when a company is seeking ISO 9000 certification. The ISO 9000 standard was developed to help companies effectively document the quality systems that they need to create and implement in order to maintain an effective total quality system. The standard covers areas such as process management and quality assurance and serves as a guide during benchmarking. Comparisons can be made between the company's existing systems and those required by the ISO 9000 standard.

Deming Prize

The Deming Prize, including its guidelines and criteria, is overseen by the Deming Prize Committee of the Union of Japanese Scientists and Engineers. The award guidelines are rigorous and can be used to judge whether an organization has successfully achieved organization-wide quality.

Supplier Certification Requirements

Several larger manufacturers have supplier certification requirements (such as ISO 16949) that a supplier must meet in order to maintain certified or preferred supplier status. A company wishing to remain a supplier must conform to the expectations set by its customers. Companies often use these guidelines to assess where they currently stand compared to the requirements and to determine where they need to make improvements. Once the improvements are in place, the requirements serve as a benchmark against which company performance will be measured by the customer.

Best-in-Field Organizations

Comparing one's performance against companies judged best in their field can be a powerful tool for companies wishing to improve their position in the marketplace. By comparing their own performance with that of the market leader, companies can better understand their own assets and capabilities as well as the areas where they need improvement. It is important to realize that the companies against which to benchmark are those that perform well in the area under study; they may not necessarily be competitors. For instance, a corporation interested in benchmarking its packaging and shipping system may choose to gather information from an organization in an unrelated field that is known to possess an excellent system. In another example, Southwest Airlines based its improvements on airplane turnaround time by comparing its work with the efforts of Indianapolis 500® race car pit crews. The effectiveness of comparison-company benchmarking is limited by the ability to obtain performance information from the comparison company.

HOW IS BENCHMARKING DONE?

A variety of different plans and methods are available to aid interested companies in the benchmarking process. Certain steps and procedures are part of every benchmarking experience. The following steps are usually facets of benchmarking (Figure 10.2).

1. *Determine the focus.*   At the beginning of a benchmarking experience, those involved must determine what aspect of their company will be the focus of the study. The focus may be based on customer requirements, on standards, or on a general continuous improvement process. Information gathered during the benchmarking experience should support the organization's overall mission, goals, and objectives. Be aware that benchmarking and gathering information about processes are of greater value than merely focusing on metrics. A narrow focus on metrics can lead to stumbling over apples-and-oranges comparisons, whereas a focus on processes encourages improvement and adoption of new methods. To be of greater value, benchmarking should be a tool used to support a larger strategic objective.

2. *Understand your organization.*   Defining and understanding all aspects of a situation are critical to any process. Individuals involved in the process need to develop an

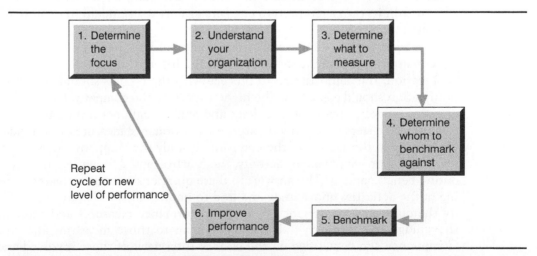

Figure 10.2   Flowchart of the Benchmarking Process

understanding of their company. To create a plan and conduct the process, companies must gather information concerning the external and internal customers and their major inputs and outputs. Often this step receives less attention than it should. Since people working for the company are performing the benchmarking assessment, the company is a known entity. Avoid the tendency to treat this step trivially. Using flowcharts to describe the processes involved will greatly enhance everyone's understanding of the system to be studied during benchmarking.

3. *Determine what to measure.*   Once an understanding of the systems present in a company has been gained, it is time to determine the measures of performance. These measures will allow those conducting the benchmarking assessment to judge the performance of the company. This is the time to define what is truly critical in order for the company to remain competitive. These critical success factors will be supported by standards for procedures, processes, and behaviors. Benchmarking will pinpoint questions to be answered and issues to be resolved, as well as processes and procedures to be improved. It is important to identify the questions pertinent to the company's particular operations. A well-developed list of items to be benchmarked will result in more consistent assessments and comparisons.

4. *Determine whom or what to benchmark against.*   Organizations should choose whom to benchmark against by considering the activities and operations under investigation, the size of the company, the number and types of customers, the types of transactions, and even the locations of facilities. Careful attention should be paid to selecting appropriate companies. Similarities in size and types of transactions or products may be more important in some instances than selecting a competitor. For instance, the bursar's office at a university may choose to benchmark against successful banking operations, not other universities' bursars' offices. A manufacturing company studying inventory control may be interested in the inventory control activities of a mail-order catalog operation. If the company is interested in beginning a quest for ISO 9000 certification, then the company might systematically select areas within its own operations to compare against the standard in order to check for compliance.

5. *Benchmark.*   The areas of the company that have been chosen for the benchmarking assessment should be notified before the process begins. To ensure cooperation, the authorization to proceed with the process should be obtained and notification should come from the highest levels of the company. During the benchmarking process, investigators collect and analyze data pertaining to the measures established in step 3. The investigators use performance measures and standards that are critical to the success of the company to study the company. As with any problem-solving or improvement activity, the 5 'whys' and 2 'hows' are questions asked during benchmarking. The answers to these questions provide a clearer understanding of the activities under study and lead to improvements (Figure 10.3). They verify the company's compliance with the performance measures and standards, and they judge the company's ability to perform to those measures and standards. Compliance can be verified on the basis of interviews of those involved and direct observation of the process.

	Questions to Be Asked
Who?	Who performs the process?
	Who is best to benchmark?
What?	What steps are being taken?
	What needs to be done?
	What is being done better?
	What could be done differently?
	What purpose does it serve?
Where?	Where does the activity take place?
	Where does it need to take place?
When?	When do the activities take place?
	When should the activities take place?
	When is the right time for the activities to take place?
Why?	Why are we doing this?
	Why is it important?
	Why must it be done?
How?	How is the work being done?
	How could it be done differently?
	How can it be changed to match or exceed the best?
How much?	How much does the old method cost?
	How much will the new method cost?

Figure 10.3 The Five Ws and Two Hs

6. *Improve performance.* Once the data and information have been gathered, a report summarizing the significant strengths and weaknesses of the area under study is created. In this report the gap between the existing and the desired levels of performance is documented. An effective report will focus on patterns of standards violations and missing elements. The report does not need to detail each of the observations made by the investigators. It should not be a list of all the observed infractions.

In a successful benchmarking experience, the final report becomes a working document to aid the continuous improvement process. The information gathered in this report is used to investigate and solve root causes, reduce process variation, and establish systems to prevent the occurrence of nonconformities. The benchmarking document is a power customer feedback tool and should be used accordingly. The case study at the end of this chapter leads the reader through a benchmarking experience.

HOW DO THE MALCOLM BALDRIGE NATIONAL QUALITY AWARD CRITERIA VIEW BENCHMARKING?

The MBNQA criteria support the use of benchmarking because organizations engaging in benchmarking activities are increasing their awareness of current levels of world-class performance. Through the use of comparison data, benchmarking enables organizations to achieve breakthrough improvements. No company exists in a vacuum; knowledge and insight into the performance of others direct organizational growth. Throughout the criteria, references are made to the need to know how one's own organization processes and results compare with those of other outstanding organizations. Benchmarking shows organizations whether they are making progress. Though applicable in any section, benchmarking is referenced or discussed in sections 4.1, 6.2, and 7 of the MBNQA criteria.

LESSONS LEARNED

Leaders of organizations that use benchmarking can be heard to remark that they search out and shamelessly borrow good ideas from other businesses and cultures. They are not caught up in the myopic "not invented here" syndrome. They are always on the lookout for improvement ideas that can be integrated into their own processes. They test these new ideas and keep the ones that will work in their own organizations. Benchmarking can also serve as a spur to drive continual improvement; effective organizations must invent new ideas faster than these ideas can be copied by competitors. Benchmarking is an essential tool for effective organizations because it helps these organizations determine the most important opportunities for improvement. Correctly performed, benchmarking identifies best practices. It is an effective tool for guiding improvements.

1. Benchmarking searches for best practices.
2. Benchmarking can be either functional or competitive.
3. During the benchmarking process, a company compares its performance against a set of standards or against the performance of best-in-field organizations.
4. Benchmarking provides valuable insight into how an organization is performing against either standards or best-in-field companies.
5. Benchmarking is never about collecting competitive intelligence or information related to market research.
6. Benchmarking is a six-step process: determine the focus, understand your organization, determine what to measure, determine whom to benchmark, benchmark, and improve performance.
7. Effective organizations must invent new ideas faster than these ideas can be copied by competitors.

Are You a Quality Management Person?

Do you compare yourself to others? Do you use that information to make changes? If so, then you benchmark. Check out the following questions and find out whether you are a quality management person. How do you answer them on a scale of 1 to 10?

Do you compare your activities to someone else performing a similar activity?

| 1 | 2 | 3 | 4 | 5 | 6 | 7 | 8 | 9 | 10 |

Do you improve how you do things based on what you learn by comparing with others?

| 1 | 2 | 3 | 4 | 5 | 6 | 7 | 8 | 9 | 10 |

Do you sometimes see someone do something and see similarities between their actions and your activities?

| 1 | 2 | 3 | 4 | 5 | 6 | 7 | 8 | 9 | 10 |

Do you read about others who are doing the same activities as you are and make comparisons?

| 1 | 2 | 3 | 4 | 5 | 6 | 7 | 8 | 9 | 10 |

Do you seek to improve upon how you do things?

| 1 | 2 | 3 | 4 | 5 | 6 | 7 | 8 | 9 | 10 |

Do you ever compare yourself against standards?

| 1 | 2 | 3 | 4 | 5 | 6 | 7 | 8 | 9 | 10 |

Do you seek out new methods in order to learn and grow?

| 1 | 2 | 3 | 4 | 5 | 6 | 7 | 8 | 9 | 10 |

Do you seek additional training or advice to help improve?

| 1 | 2 | 3 | 4 | 5 | 6 | 7 | 8 | 9 | 10 |

Chapter Questions

1. Why would a company be interested in benchmarking?
2. What does a company hope to gain by benchmarking?
3. Describe the difference between internal and external benchmarking.
4. Describe why an organization would want to benchmark against an industry leader.
5. Describe the steps involved in benchmarking.
6. How does asking the 5 'whys' and 2 'hows' facilitate the improvement process?

7. What types of questions do the whys and hows answer? Can you think of any others?

8. How would you choose a particular company or group of companies to benchmark against?

9. Study the MBNQA criteria (www.baldrige.nist.gov). Cite and discuss specific sections where benchmarking could be applied.

10. Select a company, either service or manufacturing oriented, and describe the steps and activities involved in a comparison benchmarking. Be sure to indicate specifically who they should benchmark against, what they should study, and how they should go about gathering the information.

Benchmarking at TST Performing Arts

PART 1

TST, a local performing arts association, has been experiencing several years of financial prosperity. Their board of directors feels that it is time to update the formal strategic long-range plan (SLRP) by including benchmarking. The purpose of the SLRP is to focus and guide the ideas and activities of the board of directors and the staff of the organization. The implementation of the SLRP is the responsibility of the board as well as of the staff.

Members of the board are very enlightened about total quality management and are determined to put their knowledge to use in creating the new SLRP. They begin their revision with the creation of mission and vision statements to guide the activities of the organization.

The Vision

Our performing arts association will advance a national reputation of excellence to ever-expanding local and touring audiences. In addition to performing, the performing arts association will embrace vigorous and respected educational and community outreach programs, which will include a school and associated preprofessional performance troupes.

The Mission

The purpose of the performing arts association is to educate, enlighten, and entertain the widest possible audience in the city and on tour by maintaining a professional company characterized by excellence. Its purpose is also to maintain a preprofessional training company and a school. The performing arts association will develop and maintain the artistic, administrative, technical, and financial resources necessary for those endeavors.

After developing the mission and vision statements, the board has decided to determine the viability of their organization compared with other, similar performing arts associations. For this reason, they have chosen to do benchmarking.

Questions

1. What type of benchmarking will TST be performing?
2. What benefits will TST see from benchmarking?
3. What steps should TST take in the benchmarking process?

PART 2

Following the steps suggested by texts on benchmarking, TST begins its benchmarking process with a brainstorming session to determine the *focus* of the benchmarking activities. The three general areas that the board feels are indications of viability are resource development, artistic excellence, and marketing effectiveness.

Resource development involves the financial aspects of the performing arts association. The association receives income from several sources, but all revenue can be classified as either unearned income—corporate and individual gifts, agency/government funding, and contributions—or earned income—ticket sales. Revenue is used to pay staff, purchase shows, employ performing artists, maintain the theater, market upcoming shows, and sponsor educational programs that are presented before a performance. Marketing efforts include advertising and promotional and discounted tickets. Staff members, as well as volunteers, devote a significant amount of time and effort to solicit contributions from agencies, corporations, and individuals. These efforts are vital to the association's ability to offer a wide variety of performances.

Artistic excellence involves such aspects as the program mix offered during the year, the performers themselves, the performers' age and training, the number of community outreach programs, and preperformance lectures. Marketing effectiveness is based on the relationship of attendance, ticket sales, and dollars spent on marketing.

The results of the discussions are helpful in *understanding the company*—how it is viewed both internally and externally. Using their focus areas (resource development, artistic excellence, and marketing effectiveness) as a beginning, the board must *determine what to measure*.

 Question

4. For each of the three areas—resource development, artistic excellence, and marketing effectiveness—brainstorm what you consider to be appropriate measures of performance (or indicators of viability).

PART 3

To identify the appropriate measures of performance, the board holds a brainstorming session. In the session the board recognizes the following indications of viability:

1. Efficiency of programs
2. Marketing effectiveness
3. Educating the audience
4. Other measures of performance

 Question

5. The board has listed a large number of potential measures of success or indicators of viability. To seek all of this information from the associations chosen as

benchmarks would require a significant amount of time. Which would you select as the key benchmarks? Why?

PART 4

The board holds an additional session to determine the key benchmarks indicating the viability of the association. The following areas are chosen to investigate as key benchmarks:

Resource Development

- Percentage of earned versus contributed income
- Percentage of contributed income from corporations, funding agencies, and individuals

Artistic Excellence

- Number of performances on main stage
- Number of outreach programs
- Preperformance lecture attendance

Marketing Effectiveness

- Attendance as percentage of house
- Percentage of total tickets that are complimentary or discounted tickets
- Percentage of earned income dedicated to marketing efforts

Now that the performance measures have been selected, before performing the actual benchmarking, the board needs to determine *whom to benchmark against*.

 Question

6. On what factors should TST base their selection of whom to benchmark against?

PART 5

The board members decide to base their selection of performing arts associations to benchmark on several factors. From this they are able to select five performing arts associations to compare themselves with.

The next step is to begin *benchmarking*. One of the committee members is chosen to contact each of the selected associations and gather the information. At the next board meeting, this individual will share the information with the other board members. The information from their benchmarking efforts will be used to *improve performance*.

 Question

7. Study Figures 1 through 8. How well is TST doing compared with other best-in-field associations?

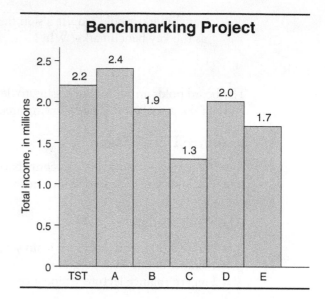

Figure 1   Marketing Expense as a
Percent of Ticket Revenue

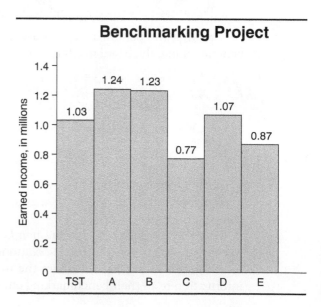

Figure 2   Marketing Expense as a
Percent of Total Budget

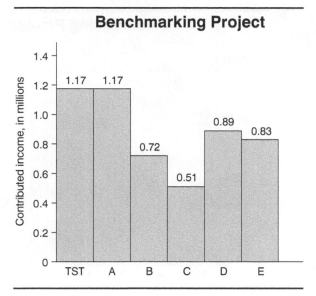

Figure 3   **Total Tickets Available per Season**

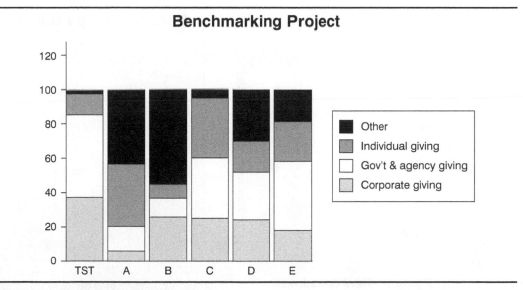

Figure 4   Performance Hall Size

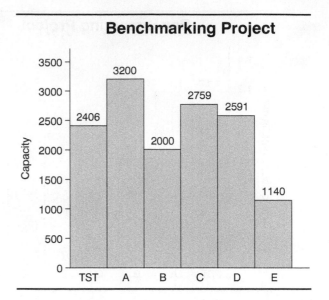

Figure 5 Composition of Contributed Income

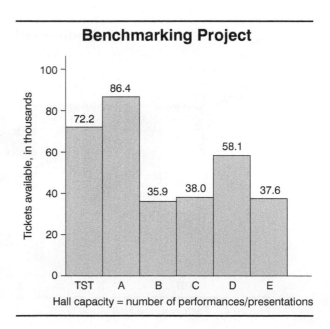

Hall capacity = number of performances/presentations

Figure 6 Total Contributed Income (in Millions)

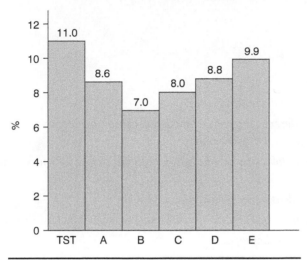

Benchmarking Project

Figure 7   Total Earned Income (in Millions)

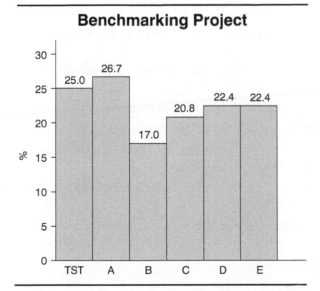

Benchmarking Project

Figure 8   Total Income (in Millions)

11 Process Management

What is a process?

How are key processes identified?

What is the difference between a functionally structured organization and a process-focused organization?

Why do functionally structured organizations have more difficulty focusing on key processes?

What benefits does a process orientation provide?

How do well-managed processes create value and generate customer satisfaction?

How are processes improved?

Why is it important to determine the objective of the process as it relates to the customer?

Why is it important to identify the boundaries of the process?

Why is it important to involve individuals from key activities in the process improvement effort?

What is meant by process ownership?

Why is it necessary to create a process map by identifying all of the activities in the process?

Why should an effort be made to separate the non–value-added activities from the value-added activities?

Why should non–value-added activities be eliminated?

What is variation, and why should it be eliminated?

Why is it critical to determine whether the remaining value-added activities are truly the "best practice"?

Why should the process be redesigned using the knowledge gained during the improvement effort?

How do the Malcolm Baldrige National Quality Award criteria support process management?

Lessons Learned

Learning Opportunities

1. To become familiar with processes

2. To become familiar with how processes are mapped

3. To become familiar with how processes are managed and improved

It's all about processes.

WHAT IS A PROCESS?

The emphasis throughout this text has been on delighting the customer. In order to delight its customers, a company must have in place processes and systems that perform as the customer expects, the first time, every time. A **process** *takes inputs and performs value-added activities on those inputs to create an output* (Figure 11.1). Every business, whether in manufacturing or the service industries, has key processes that it must absolutely perform well in order to attract and retain customers to whom to sell their products or services (Figure 11.2).

People perform processes on a day-to-day basis without realizing it. Even something as simple as going to the movies requires a process. The input required for this process includes information about what the show times and places are, who is going, and what criteria will be used for choosing a movie. The value-added activities are driving to the movie theater, buying a ticket, and watching the movie. The output is the result, the entertainment value of the movie. Organizations have innumerable processes that enable them to provide products or services for customers. Think about the number of processes necessary to provide a shirt by mail order over the Internet. The company must have a catalog website preparation process, a website distribution process, a process for obtaining the goods it plans to sell, an ordering process, a credit-check process, a packaging process, a mailing process, and a billing process, to name a few. Other processes typically found in any organization include financial management; customer service; equipment installation and maintenance; production and inventory control; employee hiring, training, reviewing, firing, and payroll; software development; and product or service design, creation, inspection, packaging, delivery, and improvement. The key or critical business processes work together within an organization to carry out the organization's mission and strategic objectives. If an organization's processes do not work together, or if they work ineffectively, then the organization's performance will be less than optimal. Effective organizations recognize that in order to supply what the customer needs, wants, and expects, they must focus on maintaining and improving the processes that enable them to meet these needs, wants, and expectations.

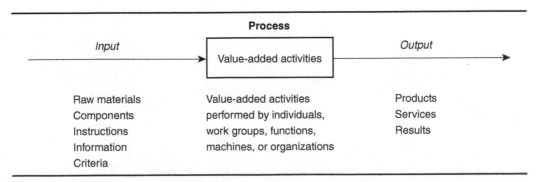

Figure 11.1   Processes

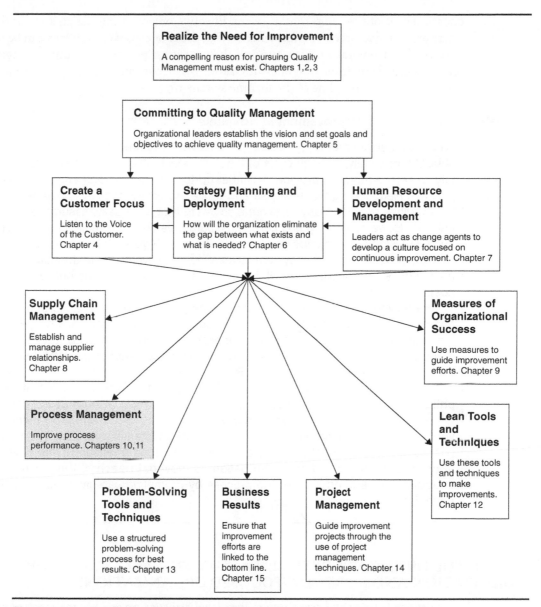

Figure 11.2   Achieving Organizational Success

HOW ARE KEY PROCESSES IDENTIFIED?

Key processes are the business processes that have the greatest impact on customers' value perceptions about the product or service and the greatest impact on customer retention. Effective organizations concentrate system and process improvement efforts on business processes that will increase their competitiveness. The challenge for today's organizations is to implement systems that reduce the frequency of human errors and to devise ways of limiting the consequences of the errors that do occur. Effective organizations design systems that investigate and analyze process performance so that the root causes of problems can be found and corrective actions taken. By managing their business processes effectively, organizations realize significant improvements in overall organizational performance that make an impact on the bottom line of the income statement.

EXAMPLE 11.1   Process Improvement at Hospitals

When a medical emergency strikes, patients want the newest health care options available. Unfortunately, the high costs of these high-tech offerings force hospitals to continually generate funds to provide the most up-to-date technology and services to their patients. Recognizing that mistakes and waste are costly, hospitals have been changing the way they operate. Many have done so by studying the quality assurance philosophies of Dr. W. Edwards Deming and Dr. Joseph Juran, as well as investigating the activities of such Six Sigma companies as Motorola Inc. and Toyota Motor Corporation. Echoing Dr. Deming's sentiment that system failures, not employees, are at the heart of problems, many hospital administrators feel that process management is the only way to enhance hospital performance.

Dispensing medications incorrectly has the potential to create a life-threatening situation. In order to improve the process of medication dispensing, many hospitals are taking action to identify where errors might occur, and they are finding ways to error-proof medication dispensing processes. One technique, reported in the *Wall Street Journal* article "ICU Checklist System Cuts Patients' Stay in Half," is to provide a checklist to accompany each patient's medication. This checklist is used to verify that the correct patient is receiving the correct medication in the correct dosage at the correct time. Other hospitals are using checklists developed during process improvement efforts to verify that it is appropriate to move a patient from an intensive care unit to a regular hospital room. This checklist provides a systematic method of ensuring that all the necessary actions are taken before the patient is moved. In each of these examples, the checklist serves to error-proof the process by prompting caregivers to remember vital patient needs.

WHAT IS THE DIFFERENCE BETWEEN A FUNCTIONALLY STRUCTURED ORGANIZATION AND A PROCESS-FOCUSED ORGANIZATION?

In a functionally structured organization, functional activities are grouped together and managed as separate entities (Figure 11.3). That is, people or machines that perform similar activities are grouped together and are overseen by a manager for their unit. As each person or machine completes its function (activity), the item is handed off to the next function. In a functionally structured organization, management boundaries are clear. Like activities are clearly grouped into an individual department, with each department containing its own manager, staff, supplies, budget, equipment, and specialized duties. Since the work is divided into distinct activities, people become specialists in their jobs and only their jobs. Some traits of a functional organization are shown in Figure 11.4.

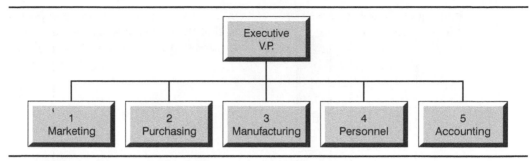

Figure 11.3   Functional Organization

A process organization is arranged according to processes or product lines (Figure 11.5). The organization focuses on the key business processes that it must absolutely do well in order to gain, satisfy, and retain customers. Traditional departmental boundaries are blurred or nonexistent in a process-oriented company. A process-oriented organization is, by its very nature, flexible. Individuals work toward completing an overall process rather than one particular activity. For this reason, people are cross-trained and are aware of all steps in the process of providing a product or service for the customer. Resources such as materials and information flow through the process to where they are needed. Process managers are responsible for the overall process and what it produces. Individuals in the process are judged on the basis of their contribution to the overall process and what it produces. Some traits of a process-oriented organization are shown in Figure 11.6.

WHY DO FUNCTIONALLY STRUCTURED ORGANIZATIONS HAVE MORE DIFFICULTY FOCUSING ON KEY PROCESSES?

Functional departments are relatively autonomous and have an internal rather than an external focus. This internal focus makes the department very good at doing their particular job, but that doesn't necessarily mean that members of the department are able to see how

Individual specialization
Departmental specialization
Improvement efforts focused internally to the department
Limited or no cross-functional training
Underutilization of personnel and/or equipment
Lack of understanding of overall organization mission and objectives
Limited communication with other departments
Narrow accountability and responsibility
Limited flexibility and agility to deal with changes
Department focus versus organization focus
Optimizing department performance
Limited gathering and use of feedback from customers
Barriers between departments

Figure 11.4   Traits of a Functional Organization

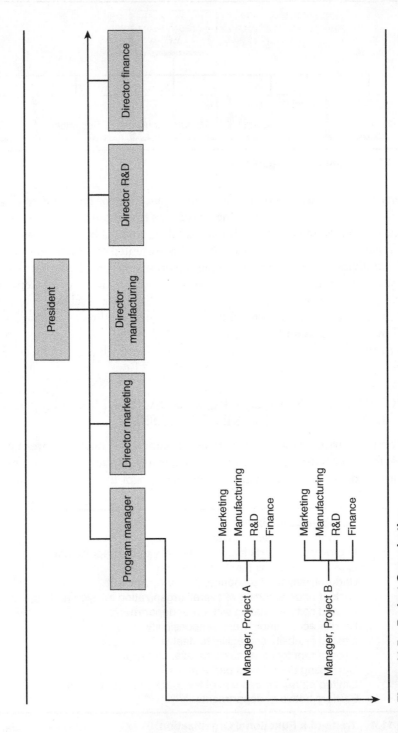

Figure 11.5   Project Organization

Broad competencies
Improvement efforts focus on processes
Cross-functionally trained individuals
Enhanced utilization of personnel and/or equipment
Focus on organization mission and objectives
Increased communication with appropriate departments/areas
Shared accountability and responsibility
Organization focus versus department focus
Optimizing organizational performance
Significant gathering and use of feedback from customers
Seamless organization

Figure 11.6   Traits of a Process-Oriented Organization

their activities affect the organization as a whole. More often, they are unable to see the connection between what they do on a day-to-day basis and what the overall mission and objectives of the organization are. This lack of alignment between the mission, objectives, and day-to-day activities of the organization can create a suboptimal organization. Remember, alignment means that if effort is made at one end of the process, the other end responds accordingly. If an organization is to concentrate on customers, alignment must exist within its processes and systems to support this focus. Although functional departments provide for high levels of productivity and control over employees, they do not allow for a clear focus on the customer. Functional departments tend to concentrate on their individuality and on achieving departmental excellence rather than organizational excellence. In relatively small organizations or organizations with few products, these drawbacks can be overcome with good internal communication. However, as an organization or its product and service list grows, this type of functional structure has more difficulty focusing on the customer.

Other drawbacks become apparent as a functionally structured organization grows. In a functional organization, employees are highly skilled at their particular job, and, as a result, they lack the flexibility to react with agility to changing market and customer needs. Since people are not cross-trained in or, in many cases, not even aware of the activities in another function, they cannot be easily shifted to other activities as the need arises. Even if they could be, the management of these individuals becomes a problem. In a functional organization, individuals as well as department managers are measured based on their performance for a particular department. The personnel issues that arise in a functional organization are mirrored in the difficulties that occur when functional departments try to share materials, equipment, information, and other supplies across the boundaries of their departments. Because of the structure of the organization, managers must constantly optimize their own department's budget, sometimes to the detriment of the overall organization. Managers' decisions to conserve departmental resources are their reaction to the performance measures used to judge them.

It is easy to see the difficulties that can arise when a product is shuttled from functional department to functional department. However, another, perhaps less obvious, key organizational resource that is often lost in this shuffle is information. A functional departmental setup is a barrier to the free flow of information from one area of an organization to another. Although functional organization leadership never actually discourages employees from working together and sharing information, the very nature of a functional organization, with its departmental lines clearly drawn, prevents or delays the flow of information. The obstacles created by functional boundaries impede feedback between departments. Often, it is difficult for members of an organization to gain information about the product or information they passed on to another department. A functional organization structure encourages departments to think of themselves as islands working independently of each other, or sometimes even despite each other.

WHAT BENEFITS DOES A PROCESS ORIENTATION PROVIDE?

One of the key differences in a process organization is that the process orientation forces people to become aware of the links between the activities in the process. A process orientation enables an organization to achieve its overall mission and objectives more easily because it gets everyone in the organization aligned with the key business processes that absolutely must take place in order for the company to attract and retain customers. Rather than a "Throw it over the wall—it's their problem now" mentality, effort is fixed on improving the links between activities. People understand how the entire process they are working on operates, and they are also much more aware of how their day-to-day activities support this process. They can clearly see the relationships between various activities. They concentrate on supporting these activities rather than their own particular specialty. If a disconnect exists involving people, materials, information, or equipment, it is readily identified and an effort is made to correct the situation. A process focus supports the use of problem-solving and self-directed work teams. Since individuals in the organization can see the links between activities, they are more likely to work together to enhance their activities and the links between them. For this reason, continuous improvement is easier within a process-oriented organization.

HOW DO WELL-MANAGED PROCESSES CREATE VALUE
AND GENERATE CUSTOMER SATISFACTION?

Time waste differs from material waste in that there can be no salvage.

Henry Ford

Anytime that waste occurs in a process, an organization and its customers lose. This wasteful, hidden factory consumes resources that might otherwise have been used to create valuable products or services. These costs create nothing of value yet are often hidden under the term *overhead*. The costs of poor quality, long process lead times, and variability in products or services significantly affect the profitability of an organization. Process management enables organizations to eliminate wasted time, effort, material, money, and manpower. Effective organizations concentrate on value-driven improvements by recognizing that processes should be measured and the results carefully analyzed to identify opportunities for improvement. This chapter discusses processes, while Chapters 12 and 13 provide information about improvement methods.

EXAMPLE 11.2 Process Improvement with a Customer Focus

One of the chief disadvantages of a functional organization that is not focused on the customer is that the processes supporting interaction with the customer are often fragmented and in reality unsupportive. One university began their improvement efforts only to find that their functional mind-set prevented any significant improvement.

A bursar's office functions as a central location for bill payment and banking functions for the university as a whole, as well as serving as a bank for students and faculty. Unfortunately, service lines at this university bursar's office were long both for phone-in and for walk-in customers. The situation had gotten so out of control that parents, students, and faculty members were phoning the university's president to complain. The president was well aware of the problem. Every day he had to wade through stagnating lines of students and faculty members in the hallway to reach his office. When things reached a breaking point, the president convened a team to deal with the problem: People have to wait too long for service. The team divided itself in two, one to deal with the long waits on hold on the telephone, the other to deal with the long lines at the office.

The First Attempt

Right from the start, many clever solutions were suggested for the long lines at the office. These included moving the service desks back from their current positions to allow to lines form inside the bursar's office rather than the hallway; putting up roping or screening to give definition to the line and direct the flow of traffic in the hallway; and the truly creative suggestion that a lookout be set up and when the president was spotted approaching the building, members of the bursar's office would hide the people waiting in line in the restrooms. The bursar's office has a 1–800 number, and the university must pay for the time that customers spend waiting on hold. For the people on hold on the phone, suggestions were made to add a message requesting that people call back later; to direct excess calls to other departments; and to disconnect callers when a maximum number of people were on hold. Although these suggestions lack customer focus (to say the least), in some ways the constraints placed on the teams had instigated the "Who cares about the customer?" attitude. Two of the critical constraints affecting the team were lack of money for process improvement training (people did not know how to improve the process) and lack of funding for more manpower. The first deficiency was caused by management's lack of understanding of the complexities of problem solving, and the second was caused by a budget crunch.

A Second Chance

After several months of ineffective activity and frustration, the president authorized process improvement training, although nothing was done to provide funding for more manpower. In many ways the money spent on training enabled the university as a whole to make greater strides than would have resulted had the president done the opposite and authorized funding for manpower. Because of this training, the second attempt to tackle the problems facing the bursar's office took a totally different, very customer-focused approach. First the problem was redefined to: Customers are not able to access their accounts or gain information from the bursar's office in a timely manner. Second, measures of performance were created, including number of complaints before and after service; average wait time in line before and after service; and average time on hold before and after service. Last, budget expectations were established so as not to exceed current levels of staffing or finance expenditures.

During the training the teams worked directly with their assigned problem in a hands-on training approach. First, they gathered information about the average wait times (24 minutes on hold, 18 minutes in line) and the reasons why people call or visit the bursar's office. They contacted customers to determine what the customer viewed as the most important issues they faced. They held discussion sessions with members of the bursar's office to determine what internal issues they were facing that reduced their effectiveness when dealing with customers. This information provided the teams with a baseline from which to begin their improvement efforts.

Two key pieces of information emerged from their efforts. In general, once customers made contact with an individual in the bursar's office, they felt that they were well taken care of, their questions were answered, and their needs were addressed. These responses were very positive for the previously misaligned bursar's staff. The staff's skills and abilities took dissatisfied customers who were angry about their wait times and turned them into satisfied customers. The customers' acknowledgment of their satisfaction with service was a true morale booster for members of the bursar's office and the teams. It provided them with the desire to attack the real problems, the second key piece of information.

Internal Customer-Focused Process Improvement

For the tellers at the bursar's office, many of the delays in serving customers were the result of antiquated paper systems, lack of access to credit-card swipe machines, lack of overflow telephone assistance for the 1–800 number, and nonessential paperwork requirements. By changing their approach to the problem and utilizing appropriate quality improvement tools, tellers were able to find many areas for improvement.

Investigation revealed that several of the forms filled out on a daily basis were not used by anyone, anywhere in the university. The tellers were dealing with a "We've always filled out that form mentality" without ever pausing to find out what the information was used for. Now they were able to determine which forms provided important information and at what times these forms had to be completed. Eliminating the other forms resulted in a much better workflow during the day. When a review of the daily drawer balance forms revealed how antiquated the forms were, a much simpler form was devised and utilized. These changes freed up time to service customers.

Years ago, to keep costs low, only one credit-card swipe machine had been purchased for the entire office. When presented with a credit card, the teller had to lock his drawer, log off his computer, leave his workstation, go to the other side of the office to the credit-card swipe machine, and often stand in line behind other tellers who had done the same thing. As credit-card use increased, this process became quite cumbersome, wasting everybody's time and generating long lines. Managers of the bursar's office were chagrined when they learned that the credit-card companies provide these machines free of charge. Very quickly, each teller station was equipped with a credit-card swipe machine.

All of these changes provided the tellers with significant additional time to serve customers. Lines virtually disappeared overnight.

External Customer-Focused Process Improvement

For the telephone operators at the bursar's office, check sheets gathered information about why people use the 1–800 number to call the bursar's office. Pareto diagrams created from this information revealed that the overwhelming reason that people called the bursar's office was "other." A full *seventy percent* of the time, people called the bursar's office for reasons unrelated to the functions of the bursar's office. These callers were not calling about bills, account balances, or scholarship or financial aid funding. They were calling to request information about graduate programs, to be connected to other departments such as education or engineering, to be transferred to food services or conference facilities, to handle a parking ticket, or to use the 1–800 number to talk to their children on campus. They were calling about basketball tickets, play performance times, lost articles, and so on, and so on. For each call the telephone operators in the bursar's office politely provided information and directions to the appropriate offices. This enormous backlog of callers unrelated to the functions of the bursar's office resulted in lengthy wait times on hold. Further investigation into the problem revealed that callers were unaware of what the word *bursar* meant. They knew their questions didn't fit into financial aid or admissions, the other two choices on the 1–800 number, so they simply selected the bursar's office.

Unlike the problems of tellers, who were able to work as a problem-solving team to isolate and deal with the issues directly affecting them, the 1–800 number problem wasn't simply a bursar's office problem. Careful analysis of the data revealed that the three existing choices offered by the 1–800 number were insufficient. Callers regularly tried to reach eight separate university business functions through the bursar's office, including admissions, financial aid, food services, security/parking services, the law school, graduate programs, the school of education, and the bursar's office.

The long wait times experienced by people calling the bursar's office could not possibly be reduced without taking a university-wide customer focus. The 1–800 team was reformed to include participants from the affected departments. At first the other departments were very reluctant to participate. After all, for years this problem was perceived as associated with "the slackers in the bursar's office." Encouraging other departments to take ownership in their customer-related issues required a compelling presentation of the data already gathered by the bursar's office team, as well as timely intervention by the president and the vice president of finance. Leadership's mandate to solve the 1–800 number issues as determined by the former team got the new team moving.

Through discussions with the 1–800 number provider, the new team discovered that adding more choices to the existing line would not incur additional costs to the university's overall budget, although the individual departments would be expected to bear the costs of calls to their own departments. These costs had been borne by the bursar's office in the past. The 1–800 number provider worked with the team to study the financial implications of additional lines and other modifications. In the end, the 1–800 number was revamped to include six choices. Graduate programs, the education school, and law school each received their own 1–800 line. For all lines a brief summary of the services offered by each office was given with each selection; for instance, "Financial Aid: for information about scholarships and financial aid, please press 2. Bursar: for information about billing and account balances, please press 3." Because the time callers waited on hold dropped to virtually zero, the performance measures showed that the total combined costs of the new services were actually lower than the costs associated with the original service.

When ownership in the customer's problem was expanded to include the affected departments, true improvements to the process experienced by the customer could be made. Had the bursar's office merely changed their message on the 1–800 number or removed their name from the choices available—both of which were within their power to do—the problem of customers' long wait would not have been eliminated but merely shifted to another department as customers sought answers to their questions. Only through stressing value-added, customer-focused processes was the university truly able to improve its service to its customers and reduce wait time.

HOW ARE PROCESSES IMPROVED?

Processes are improved through value-added process mapping, problem isolation, root cause analysis, and problem resolution. Many processes develop over time, with little concern for whether they are the most effective manner in which to provide a product or service. To remain competitive in the world marketplace, companies must identify wasteful processes and improve them. The processes providing the products and services should be improved with the aim of preventing defects and increasing productivity by reducing process cycle times and eliminating waste. The key to refining processes is to concentrate on the process from the customer's point of view and to identify and eliminate non–value-added activities.

Two step-by-step process improvement methodologies have been introduced in this text: Shewhart's and Deming's Plan-Do-Study-Act cycle and Six Sigma's Define-Measure-Analyze-Improve-Control cycle. A typical process for improving processes is shown in Figure 11.7. Regardless of the specific steps taken, in order to effectively improve processes it is critical to:

1. Determine the objective of the process as it relates to the customer.
2. Determine the boundaries of the processes as the customer sees them.
3. Involve representatives from each major activity associated with the process in the improvement effort. Identify where conflicts exist between the boundaries of the processes as they are related to existing functional departments.
4. Identify the process owner.
5. Create a process map by identifying all of the activities in the process.
6. Separate the non–value-added activities from the value-added activities.
7. Eliminate the non–value-added activities.
8. Identify, analyze, and eliminate variation in the process.
9. Determine whether the remaining value-added activities are truly the "best practice."
10. Redesign the process using the knowledge gained in the first nine steps.

WHY IS IT IMPORTANT TO DETERMINE THE OBJECTIVE OF THE PROCESS AS IT RELATES TO THE CUSTOMER?

Process improvement efforts should concentrate on what is important to the customer, whether that customer is the next workstation in the line or the ultimate end user of the product or service. Without this focus, an organization can go about improving processes without

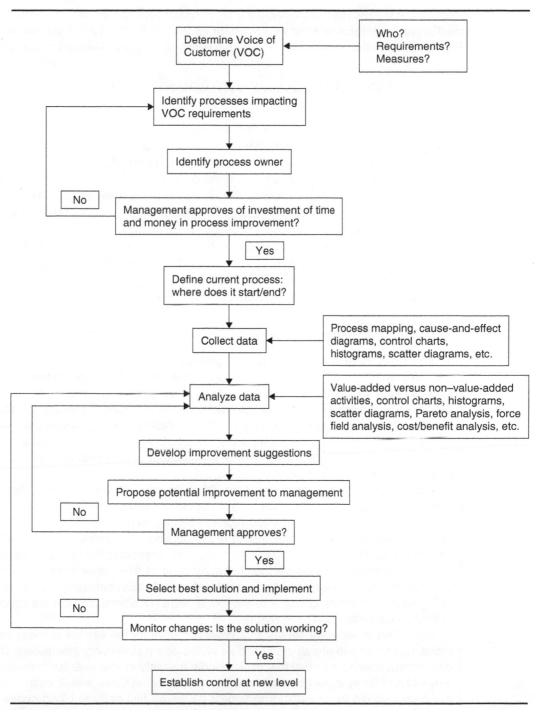

Figure 11.7   A Process for Improving Processes

321

ever improving the right processes, the ones the customer cares about. If the customer doesn't care about it, chances are that improvement efforts will have no effect on the organization's bottom line because these efforts are not related to creating successful customers who return.

WHY IS IT IMPORTANT TO IDENTIFY THE BOUNDARIES OF THE PROCESS?

The process should be studied from the customer's point of view. True process improvement comes from a knowledge of what adds value or meaning for a customer. Not looking at the process from a customer viewpoint often leads to a narrow definition of the process. This narrow definition limits improvement efforts because it fails to study the customer experience. For example, if a hotel views the check-in process internally—that is, not as a customer—it will probably identify check-in as the actual process of being assigned a room and receiving a key. Customers may take a broader view of where check-in begins, seeing where it begins as the moment they reach the front door of the hotel, and where it ends as the moment they are seated in their room with their shoes off. Improving simply the time spent receiving a room key fails to take into account the customer's overall impression or experience, thus missing the opportunity to increase customer perceived value.

EXAMPLE 11.3   Cross-Functional External Customer Focus

A department store is facing a significant problem. The monthly billing statements it sends out are cryptic at best, resulting in numerous calls to the billing department's customer service office for clarification. Customer service representatives report that many of the callers begin the conversation with the words, "I have a question about my bill," after which they hear the person tearing open the bill. The representatives confirmed their suspicions by asking people if they had opened their bills before calling. The callers admitted that they were so used to not understanding their bill that they would call first, knowing that they were going to have a question.

A team, composed of members of the billing department, was created to deal with the problem. At first, the answer to the problem appeared straightforward: Change the bill format. It wasn't until the team applied problem-solving techniques that they realized there were actually two problems: the format of the bill and the timeliness of information from the departments where the charges were incurred. Improving the format of the bill would be relatively straightforward. Various billing statements from other stores and credit-card companies could be studied, with the best features of each combined into the store's new billing format. However, changes to the format were not attempted until the second issue of timely information had been resolved.

The billing department serves as a collection point for the various charges customers make. Customer bills are as many and as varied as the customers themselves. The billing department needed a format that could handle a variety of charges, but more importantly they needed timely input from the departments assessing those fees. It wasn't unusual for a charge incurred by a customer to appear on the bill two or three billing cycles later. By then, the customers would not always remember what the charge was for, so they called the billing department. Naturally, the customer service representatives couldn't answer any questions about why the charge existed because their office did not originate the charge. The representatives would try to determine the source of the fees and then place a return call to the customer with the explanations. Although this process satisfied the customer, it was fraught with hassles and was wasteful and time-consuming for everyone concerned.

Itemizing the fees on the bill required only a formatting change. Gaining timely information about the charges required a store-wide process viewpoint change. The team was re-formed to include representatives of the affected departments. Reluctance was overwhelming. The affected departments weren't concerned about the problem because "It's the duty of the billing department to collect the money. Let them worry about it." Store leadership had to mandate the involvement of the affected departments.

When ownership in the customer's problem was expanded to include the affected departments, true improvements to the process experienced by the customer could be made. Had the billing department merely changed the format of the bill, the real problem of customers' not understanding their bill would not have been eliminated. Only through focus on value-added customer processes was the store truly able to improve its service to its customer, reduce errors, provide timely information, and collect the money owed them in a timely fashion.

WHY IS IT IMPORTANT TO INVOLVE INDIVIDUALS FROM KEY ACTIVITIES IN THE PROCESS IMPROVEMENT EFFORT?

Buy-in: You want the people who are going to have to live with the new process to be the ones who fix it. If they are involved in identifying, creating, and making the necessary changes, chances are very good that they will live with those changes and work to make them permanent. Involvement from all key activities also breaks down barriers between existing departments and provides everyone with a clearer understanding of how work gets done in the organization. Once the key processes have been identified, it is important to determine whether there are conflicts between the existing functional organization structure and the newly emphasized process approach. By identifying these conflicts, leadership can takes steps to minimize the difficulties associated with the transition.

WHAT IS MEANT BY PROCESS OWNERSHIP?

Process ownership refers to identifying who is ultimately responsible for seeing that a process is completed in a manner that results in customer satisfaction. These individuals are in a position to make, and have the necessary power to make, changes to the process.

WHY IS IT NECESSARY TO CREATE A PROCESS MAP BY IDENTIFYING ALL OF THE ACTIVITIES IN THE PROCESS?

In most organizations very few people truly understand the myriad required activities in a process that creates a product or service. Process maps are powerful communication tools that provide a clear understanding of how business is conducted within the organization. Identifying and writing down the process in pictorial form helps people understand just how they do the work that they do. Process maps have the ability to accurately portray

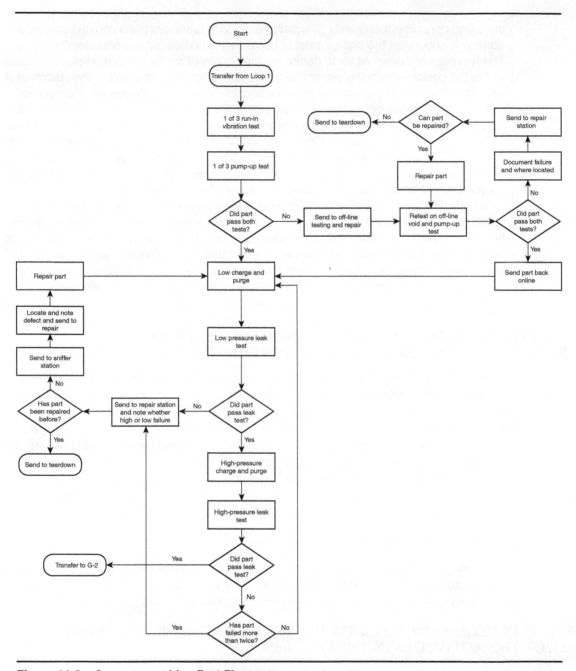

Figure 11.8   Compressor Line Part Flow

current operations, and they can also be used to evaluate these operations. In addition, a process map identifies the activities that have been added to a process over time in order to adapt older processes to changes in the business. Once changes have been proposed, process maps are equally powerful for communicating the proposed changes in the process.

Process maps are known by many names, including *flowcharts*, *process flowcharts*, and *process flow diagrams*. A **process map** *is a graphical representation of all of the steps involved in an entire process or a particular segment of a process.* An example of a process map is shown in Figure 11.8. Diagramming the flow of a process or system aids in understanding it. Flowcharting is effectively used in the first stages of problem solving because the charts enable those studying the process to quickly understand what is involved in a process from start to finish. Problem-solving team members can clearly see what is being done to a product or provided by a service at the various stages in a process. Process flowcharts clarify the routines used to serve customers. Problems or non–value-added activities nested within a process are easily identified by using a flowchart.

The construction of process maps is fairly straightforward. The steps to creating such charts are the following:

1. Define the process boundaries. For the purpose of the chart, determine where the process begins and ends.
2. Define the process steps. Use brainstorming to identify the steps for new processes. For existing processes actually observe the process in action.
3. Sort the steps into the order of their occurrence in the process.
4. Place the steps in appropriate flowchart symbols (Figure 11.9) and create the chart.
5. Evaluate the steps for completeness, efficiency, and possible problems such as non–value-added activities.

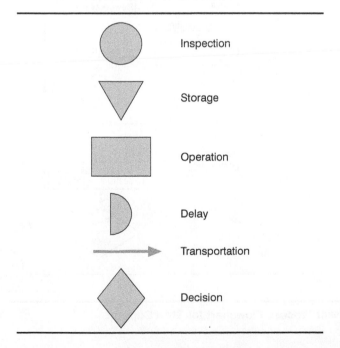

Figure 11.9   Flowchart Symbols

Process flowcharts can be constructed with symbols like those shown in Figure 11.9. Because processes and systems are often complex, in the early stages of flowchart construction, removable 3-by-5-inch sticky notes placed on a large piece of paper or board allow creators greater flexibility when creating and refining a flowchart. When the chart is complete, a final copy can be made utilizing the correct symbols. Either the symbols can be placed next to the description of the step, or they can surround the information.

A variation on the traditional process flowchart is the deployment flowchart. When a deployment flowchart is created, job or department titles are written across the top of the page, and the activities of the process that occur in that job or department are written

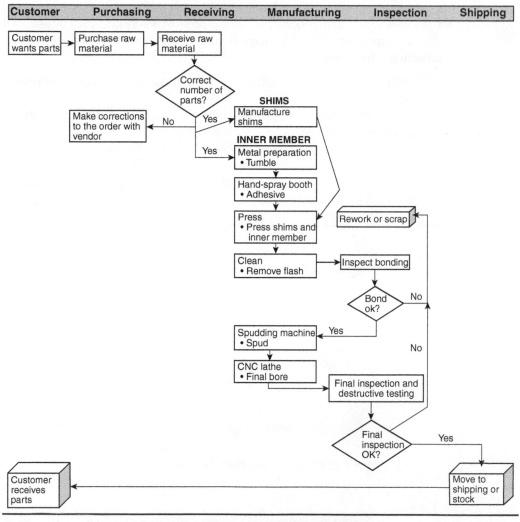

Figure 11.10   Deployment Process Flowchart for Strut Support

below that heading (Figure 11.10). Flowcharts can also be constructed with pictures for easier understanding (Figure 11.11). When process flowcharts are used as routing sheets, they often appear as shown in Figure 11.12, which includes additional details such as process activities, operator self-inspection notes, and specifications.

Value stream process mapping is a technique that adds process information to a flow diagram. To create a value stream map, the steps are the same as a process map or flow diagram except that during the investigation, details like cycle time, changeover time, uptime, and process waste are recorded. These details are displayed on the diagram using the value stream map symbols shown in Figure 11.13. A completed value stream map for JQOS is shown in Figure 11.14.

EXAMPLE 11.4   Creating a Process Map

A process improvement consultant for a university was asked by the leadership of the university to assist in the problem of scheduling students. Students complained about the length of time the process took and the large number of errors that occurred. The team had already decided that a computerized system would significantly improve the process. However, the consultant wanted to be sure that the process being computerized was the best process for registering and scheduling students. She asked the team to map the process. They came up with the process map shown in Figure 11.15.

Because this simple process flow diagram did not mesh with the numerous student complaints about the process complexity, the consultant and the team selected five students to shadow during their registration process. The shadowing began on a Monday, the first day to begin the process, and none of the five were completely registered before 14 days had passed. The team stifled numerous complaints of their own about how much effort it took to truly shadow these students in their process. Instead, they focused on creating a process flow map that reflected the reality of the process. Utilizing 3-by-5-inch sticky notes, the team members listed each of the steps in the process as they had experienced them; then they affixed the notes to the wall in the conference room.

By sorting out the notes, they combined their experiences into one large process flow map. The resulting map contained over 32 steps (Figure 11.16). By carefully studying the actual process, the team was able to identify a myriad of non–value-added activities as well as confusing instructions and odd requirements. By the end of the team's work as a committee, the process had changed dramatically. Information was made available for students concerning course availability, course requirements, and course closings. All but one approval was removed. Scheduling holds (due to nonpayment of current-term bills) were simplified. In the end, only four steps remained in the process. Once the students have selected their courses, the entire registration process can take place online in the advisor's office. The only paperwork generated is a printout of the student's schedule.

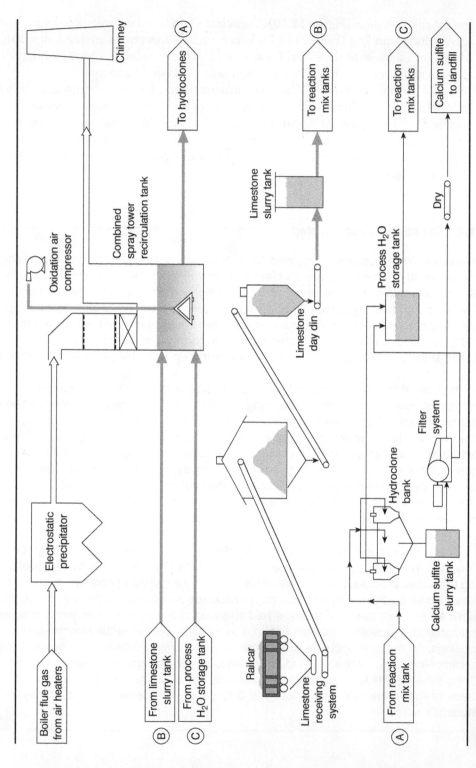

Figure 11.11   Flue Gas Desulfurization Process

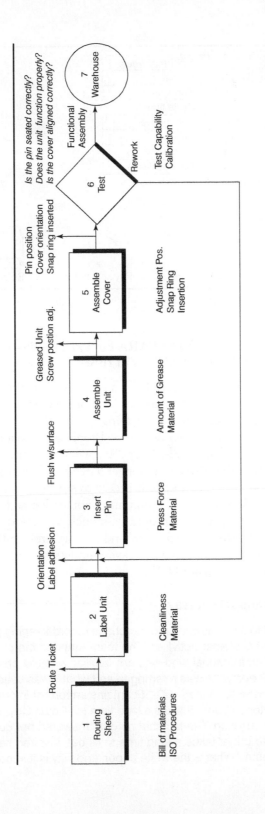

Figure 11.12   Process Flowchart with Instructions and Specifications

Variable	Target	Upper Spec	Lower Spec
Press Force	12 lbs	15 lbs	10 lbs
Amount of Grease	2 cc's	1.5 cc's	2.5 cc's

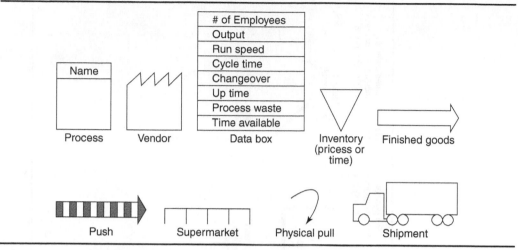

Figure 11.13   Value Stream Map Symbols

WHY SHOULD AN EFFORT BE MADE TO SEPARATE THE NON–VALUE-ADDED ACTIVITIES FROM THE VALUE-ADDED ACTIVITIES?

Processes evolve over time. What may have been a necessary activity in the past may no longer be needed because of changes in technology or changes in the desires of the customer. Because people are so accustomed to doing the job that they do, very few even realize that these extra steps are no longer necessary. People get into a routine. Over time, traditions are established for work methods, and people will comment, "We've always done it that way," without realizing that there is no longer a need to do it that way. By identifying these extra activities on the process map, it is easy to see that some of these activities may not be necessary and can be removed. Eliminating non–value-added activities gets rid of waste in the process, resulting in savings of time, money, and effort. As with any problem-solving or improvement activity, the 5 'whys' and 2 'hows' are questions that should be asked. The answers to these questions provide a clearer understanding of the process under study and lead to improvements (Figure 11.17)

EXAMPLE 11.5   We've Always Done It That Way

A local company decided to study its customer purchase record-keeping process to ensure that it accurately reflected customer activities. The team mapped the process. During this activity the team leader, an individual who was not familiar with the process, asked what the "white paper" was that everyone was referring to and what it was used for. The answers she received were, "We have to fill that out"; "It contains important information"; "We have one of those for each customer"; and "It takes a long time to fill out." Obviously these replies did not really answer her question. The fact that no one answered her question, combined with the fact that the white paper takes a long time to fill out, inspired her to be persistent and ask over and over again: "What is the white paper, and why is it important?" Finally, an

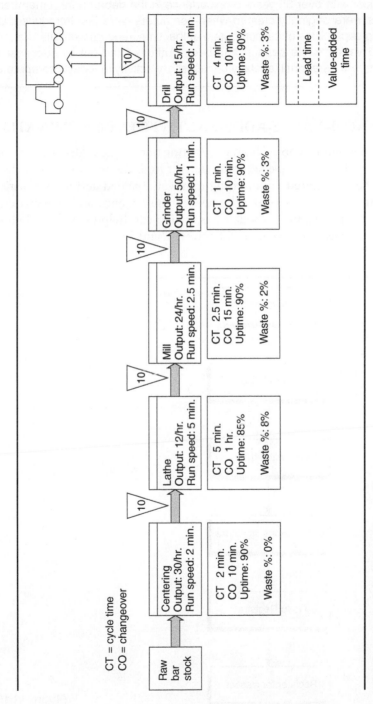

CT = cycle time
CO = changeover

	Centering		Lathe		Mill		Grinder		Drill
Raw bar stock	Output: 30/hr. Run speed: 2 min.	10	Output: 12/hr. Run speed: 5 min.	10	Output: 24/hr. Run speed: 2.5 min.	10	Output: 50/hr. Run speed: 1 min.	10	Output: 15/hr. Run speed: 4 min.
	CT 2 min. CO 10 min. Uptime: 90% Waste %: 0%		CT 5 min. CO 1 hr. Uptime: 85% Waste %: 8%		CT 2.5 min. CO 15 min. Uptime: 90% Waste %: 2%		CT 1 min. CO 10 min. Uptime: 90% Waste %: 3%		CT 4 min. CO 10 min. Uptime: 90% Waste %: 3%

Lead time

Value-added time

Figure 11.14   JQOS Value Stream Map

331

individual with over 35 years of experience in the department remembered that the white papers were originally created when computers were first introduced. Due to the limited backup capabilities of the computers at that time, records were also kept on "white papers." For 35 years "white papers" had been dutifully filled out by successive generations of department personnel, a totally unnecessary activity in today's computerized world.

WHY SHOULD NON–VALUE-ADDED ACTIVITIES BE ELIMINATED?

As the previous examples show, eliminating non–value-added activities saves time, money, and effort. Because these activities detract from the main business, they are wasteful and should be eliminated. Eliminating non–value-added activities clarifies the process and allows it to focus on meeting the needs, requirements, and expectations of the customer. Process improvement methodologies, which can help identify and eliminate non–value-added activities, are covered in Chapters 12 and 13.

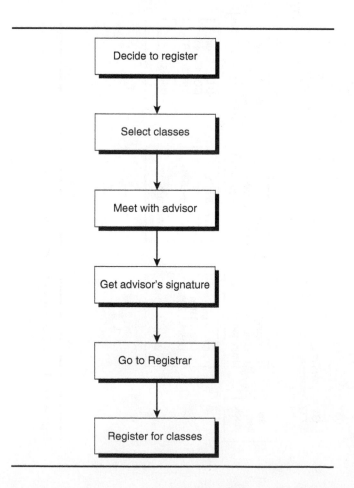

Figure 11.15   First Try at a Process Map (Example 11.4)

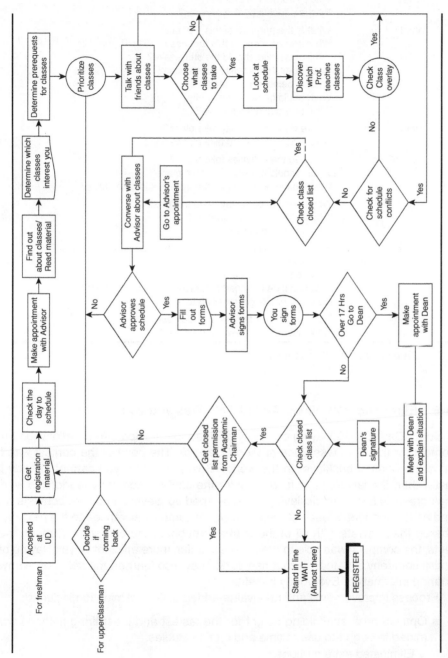

Figure 11.16   Registration Flowchart

333

	Questions to Be Asked
Who?	Who performs the process?
	Who is affected by this process?
What?	What is the purpose of this process?
	What are the steps in this process?
	What sequence should the steps take?
	What does this process accomplish?
	What is being done better?
	What could be done differently?
	What purpose does it serve?
Where?	Where does the activity take place?
	Where does it need to take place?
`When?	When do the activities take place?
	When should the activities take place?
	When is the right time for the activities to take place?
Why?	Why does the company need this process?
	Why is this process important?
	Why must it be done?
How?	How does this process relate to other company processes?
	How is the work being done?
	How could it be done differently?
	How can it be changed to match or exceed the best?
	How will results get measured?
How much?	How much does the old method cost?
	How much will the new method cost?

Figure 11.17   The Five Why's and Two How's

EXAMPLE 11.6   Eliminate Non–Value-Added Activities and Win

A South Carolina bricklaying company decided to participate in a nationwide bricklaying contest for the *Guinness Book of World Records.* The point of the contest was to determine the "fastest bricklayer in the world." Believing that if you cannot already lay brick faster than the fastest person, why try for the contest, company leadership worked with their crews to see how bricklaying could be sped up. Several in-house contests were held, and as the winners of each of these individual contests advanced to the next round, they shared their secrets with all of the teams throughout the company. Through these contests, the company was able to develop new, safer, more efficient ways of laying brick. The entire company saw their output rise 30%. They also learned the value of teamwork and sharing information: Everybody benefits.

Process improvements and non–value-added activity elimination included:

- Optimal brick scaffolding height for the fastest and least tiring motions was determined through the use of time and motion studies.
  - Eliminated extra motions.
  - Improved the quality of workmanship.

- New scaffolding designs allow the scaffolding to be raised two inches at a time through the use of a foot pedal.
  - Eliminated non–value-added activities related to setting and resetting scaffolding height.
- New method was developed to set the brick height line using a pole with marker holes and a pin attached to a string.
  - Improved quality by eliminating height variation in brick courses.
  - Eliminated non–value-added activities related to setting and resetting brick line height.
- New arrangements were created for bricks on the scaffold, including optimum stack height, new angled positioning, and new distance between stacks for bricks.
  - Increased effectiveness of bricklayer by aligning work with job.
  - Decreased work-related injuries related to falling brick and awkward movements.
- Decreased product waste by allowing for handgrip space.

The original record in the *Guinness Book of World Records* was 1,024 bricks, laid in one hour. By eliminating non–value-added activities and implementing process improvements, the winning team member laid 1,493 bricks.

WHAT IS VARIATION, AND WHY SHOULD IT BE ELIMINATED?

Because variation is present in any natural process, including any process that produces a product or provides a service, no two products or occurrences are exactly alike. When a product is manufactured, variation is often identified as the difference between the actual part dimensions of one part versus another. In service industries, variation may be the difference between the type of service received and the type of service expected. Variation may be caused by one of four factors: common causes, special causes, tampering, or structural difference. Companies interested in providing a quality product or service use statistical process control techniques to carefully study the variation present in their processes. Determining why differences exist between similar products or services and then removing the causes of these differences from the processes that produce them enable a company to more consistently provide a high-quality product or service. Think of it this way: If you are carpooling with an individual who is sometimes late, sometimes early, and sometimes on time, it is difficult to plan when you should be ready to leave. If, however, the person is always five minutes late, you may not like it, but you can plan around it. The first person exhibits a lot of variation—you never know when to expect him or her. The second person, although late, has very little variation in his or her process; hence you know that if you need to leave at exactly 5 P.M., you had better tell that person to be ready at 4:55. The best situation would be to be on time, every time.

Companies seek to eliminate or reduce the variation in a process. Variation can be present in a process in a variety of ways. There may be variations in lead times, quality, processing time, inventory availability, and so on. Effective organizations focus on both time and quality as important metrics in process improvement. Reducing process lead times and the variation present in the amount of time it takes to complete a process is just as critical as improving product or service quality. Slow processes are costly in terms of inventory that must be moved, counted, stored, or retrieved. Low process lead times reduce overhead cost and inventory and may prevent damage to inventory or inventory obsolescence. Variation should be investigated for any critical-to-quality issue. Any activities that cause critical-to-quality issues for the customer or cause long time delays in any process offer the greatest opportunity for improvement in cost, quality, capital, and lead time.

WHY IS IT CRITICAL TO DETERMINE WHETHER THE REMAINING VALUE-ADDED ACTIVITIES ARE TRULY THE "BEST PRACTICE"?

For the same reason that it is important to remove the non–value-added activities in a process, it is critical to revisit the value-added process steps and check to see whether better ways to do the work do exist. The "we've always done it that way" mentality carries over into the value-added activities just as easily as it did the entire process. Determining best practices enables an effective organization to eliminate poor work methods, out-of-date practices, wasteful activities, and unnecessary steps. Doing so makes the organization highly competitive.

WHY SHOULD THE PROCESS BE REDESIGNED USING THE KNOWLEDGE GAINED DURING THE IMPROVEMENT EFFORT?

Process improvement focuses on eliminating waste—the waste of time, effort, material, money, and manpower. It is the combined knowledge gained during the improvement efforts that enables an organization to develop its own best practices and reach a new level of performance, resulting in delight for its customers.

How Do We Know It's is Working?

Prior to new leadership at JQOS, the activity of turning a quote into a finished part had no uniform process. This led to incorrect quotes, lost quotes or orders, past-due orders, incorrect orders, improperly priced orders, and other problems. Knowing that standardized processes reduce the chance of errors, a team was formed with the objective of defining order processing. The boundaries of the process were set at customer requests quote to customer receives parts. The team included the order entry clerk, the scheduler, and the process engineer. The order entry clerk was designated the process owner since the clerk is responsible for a majority of the initial interaction with the quote/order.

By tracking the flow of an actual order through the shop, the team developed a list of activities that must take place. As they studied the existing process, they used the five Ws and two Hs from Figure 11.17 to help them see the whole picture. Following a number of orders through the shop also helped. Since the original shop layout (Figure 12.6) resembled a spider web of parts moving here and there in the plant, order tracking was not easy. Further complications arose when incorrect routers accompanied the orders. These were modified when necessary, before, it was hoped, incurring significant costs of quality (see Chapter 9) from producing the parts incorrectly.

The first thing that the team noticed was that there were no checks or balances in the process. Quotes were rarely checked to ensure correct material costs, routings, or part print revisions. Those converting a quote into an order did not verify up-to-date material price, shop schedules, or delivery availability information. Promised due dates often did not correspond to true shop uptime availability. This frequently was the cause of late orders (see Chapter 13). So, while the activities in the process were value added, they did not create value because of the variation in processing technique.

As the team analyzed the process, it became clear that a better order processing method was needed. The team readily considered the idea of computerizing the process; however, they recognized that computerizing a nonfunctional process failed to help company performance. Their efforts focused on redesigning the process. Figure 11.18 shows the new process map. The new steps in the process are shown in bold italics. Errors have declined significantly. Computerization of the new process speeds the quoting/order process even more.

HOW DO THE MALCOLM BALDRIGE NATIONAL QUALITY AWARD CRITERIA SUPPORT PROCESS MANAGEMENT?

Section 6.0 of the Malcolm Baldrige National Quality Award (MBNQA) criteria, Process Management, invites organizations to examine their process management efforts by looking at their key product, service, support, and business processes. Effective organizations design and manage these processes in a manner that creates customer value and enables the organization to achieve business success and growth. The criteria are interested in how an organization identifies and manages its key processes. Key to these efforts is how an organization's processes are improved in order to achieve better performance. The criteria require that organizations use performance measures for the control and improvement of processes. They encourage organizational alignment by emphasizing the use of input from internal and external customers when designing and improving processes.

The task of leadership is to design and manage effective processes that support an unwavering focus on the customer and what the customer values. Key product, service, support, and business processes are found throughout an organization. When evaluating how effective an organization's processes are, ask these questions based on the MBNQA criteria:

Leadership

- Are the organization's leaders converting policies into actions related to the key business processes?
- Is leadership focused on processes rather than activities?

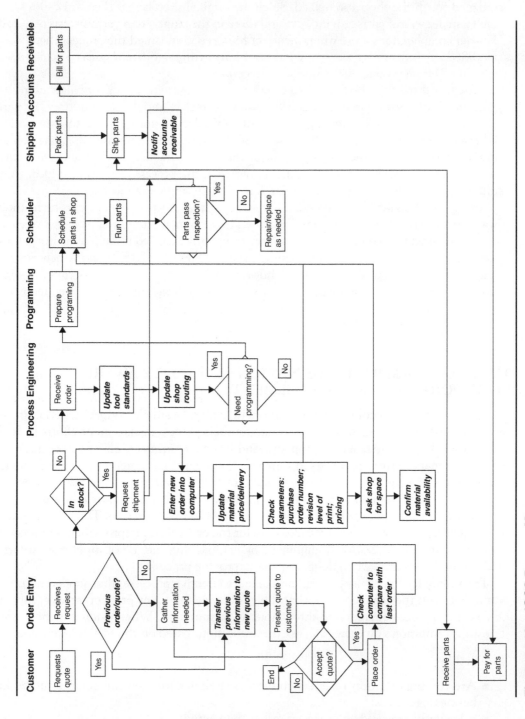

Figure 11.18 JQOS Process Map

Strategic Planning

- Has the strategic plan identified the key processes that the organization absolutely must perform well in order to attract and retain customers and stay in business?
- Has the strategic plan mapped out a strategy for improving each key process?

Customer and Market Focus

- Is the organization constantly checking to determine whether the customer and market information indicates a need to change a key process?

Measurement, Analysis, and Knowledge Management

- Is the organization gathering, analyzing, disseminating, and using information related to its key processes?
- Is there a communication system that links everyone in the organization with the customers?

Workforce Focus

- Do employee policies and reward systems support improvement activities related to the key processes?
- Is there a reward and recognition process that is supportive of the needs of the employees while contributing to the organization's success?

Process Management

- Have the key processes been identified?
- Are the key processes being continuously monitored for improvement opportunities?
- How is the organization measuring key process performance?
- Does the organization have corrective and preventive action processes in place?
- How does the organization design its work systems to support core competencies?
- How are the key work processes designed, improved, and innovated in order to increase organizational performance and customer value?
- How does the organization prepare for emergencies and continuity of operation?
- How does the organization manage and improve its key work processes?
- How does the organization reduce variability in its key work processes?
- How does the organization keep its key processes current with business needs and market directions?

Results

- Does the organization's skill at managing key business processes translate to improvements in the financial performance of the organization?
- Are business results, combined with customer information, being used to guide changes in organizational processes?
- Is customer and market information being integrated with business results information in order to make changes to key processes?

LESSONS LEARNED

Organizations trying to compete on a world-class level cannot afford to waste time or effort. Effective organizations use process management methods to enhance their overall organizational performance. By concentrating on eliminating non–value-added activities from their existing processes, and then improving those processes based on input from their customers and employees, organizations eliminate many forms of waste. The resulting processes provide customer delight because they display consistent high performance from beginning to end. Properly managed processes focus on the activities critical for achieving high customer-perceived value and satisfaction ratings.

1. Processes take inputs and perform value-added activities on those inputs to create an output.
2. Key processes are the business processes that have the greatest impact on customers' value perceptions about the product or service and the greatest impact on customer retention.
3. Functional organizations are structured so that functional activities are grouped together and managed as separate entities.
4. Process organizations are organized around processes or product lines.
5. Well-managed processes create value and generate customer satisfaction.
6. Processes are improved through value-added process mapping, problem isolation, root cause analysis, and problem resolution.
7. Processes should be designed to support customer requirements.
8. Process boundaries should be established before beginning process improvement.
9. Process owners should be identified before beginning process improvement.
10. A process map is a graphical representation of all the steps involved in an entire process or a particular segment of a process.
11. Non–value-added activities should be identified and removed from any process.
12. Variation causes differences in the products and services customers receive.

Are You a Quality Management Person?

It's the little day-to-day activities that make a quality management person. We'd all like to think that we are, but how do you know? Your life is full of processes that you must complete. Are those processes effective and efficient? Check out these questions and find out whether you are a quality management person. How do you answer them on a scale of 1 to 10?

Can you identify the key processes that you use?

| 1 | 2 | 3 | 4 | 5 | 6 | 7 | 8 | 9 | 10 |

Do you understand how your key processes affect your overall goal and objective achievement?

| 1 | 2 | 3 | 4 | 5 | 6 | 7 | 8 | 9 | 10 |

Do you understand how your key processes affect your customers?

| 1 | 2 | 3 | 4 | 5 | 6 | 7 | 8 | 9 | 10 |

Do you take ownership of your processes?

| 1 | 2 | 3 | 4 | 5 | 6 | 7 | 8 | 9 | 10 |

Do you constantly strive to improve your skills and knowledge?

| 1 | 2 | 3 | 4 | 5 | 6 | 7 | 8 | 9 | 10 |

Do you use your knowledge and skills to improve the processes you interact with on a daily basis?

| 1 | 2 | 3 | 4 | 5 | 6 | 7 | 8 | 9 | 10 |

Have you identified the value-added and non–value-added activities that make up your key processes?

| 1 | 2 | 3 | 4 | 5 | 6 | 7 | 8 | 9 | 10 |

Have you removed non–value-added activities from your processes?

| 1 | 2 | 3 | 4 | 5 | 6 | 7 | 8 | 9 | 10 |

Have you redesigned your processes to make them effective?

| 1 | 2 | 3 | 4 | 5 | 6 | 7 | 8 | 9 | 10 |

Chapter Questions

1. What is the difference between a functionally structured organization and a process-focused organization?
2. How does a focus on processes and process improvement help an organization become more effective?
3. How would you recognize that a process needed to be improved?
4. What is meant by *value-added operation*?

5. How would you recognize a non–value-added activity?

6. Why is process mapping an important tool for improving processes?

7. Why is it important to determine the process boundaries?

8. Why is it important to study the process from the customer's point of view?

9. What is meant by *process ownership*?

10. Who should be involved in process improvement efforts? How should the team be structured?

11. Describe a process that you are familiar with. Who is the process owner?

Flowcharts

12. Create a flowchart for registering for a class at your school.

13. Create a flowchart for ordering a meal at a restaurant.

14. WP Uniforms provides a selection of lab coats, shirts, trousers, uniforms, and outfits for area businesses. For a fee, WP Uniforms will collect soiled garments once a week, wash and repair these garments, and return them the following week while picking up a new batch of soiled garments.

    At WP Uniforms, shirts are laundered in large batches. From the laundry, these shirts are inspected, repaired, and sorted. To determine whether the process can be done more effectively, the employees want to create a flowchart of the process. They have brainstormed the following steps and placed them in order. Create a flowchart with their information. Remember to use symbols appropriately.

Shirts arrive from laundry.
Pull shirts from racks.
Remove shirts from hangers.
Inspect.
Ask: Does shirt have holes or other damage?
Make note of repair needs.

Ask: Is shirt beyond cost-effective repair?
Discard shirt if badly damaged.
Sort according to size.
Fold shirt.
Place in proper storage area.
Make hourly count.

Section 6.0: Process Management

The Process Management section of the MBNQA criteria addresses how an organization designs, manages, measures, and improves key product and service processes. This section also focuses on how an organization designs, manages, measures, and improves key business and support processes.

The key goals of this section are to:

1. Examine the key processes of Remodeling Designs, Inc. and Case Handyman Services, specifically:
   a. Product and service processes
      i. Design processes, production processes, and delivery processes
   b. Business processes
      i. Nonproduct/nonservice processes
   c. Support processes
2. Examine how Remodeling Designs and Case Handyman improve their product and service processes, business processes, and support processes.

This section was created by Luke Parks, Chris Dolan, Erich Eggers, Mike Cordonnier, and Donna Summers.

QM Process Management Questions

1. How would you define *process management*?
2. What are your key processes?
3. Can you name the key processes involved in Remodeling Designs/Case Handyman, from the initial contact with the customer to the finished project and the customer's final billing?
4. How do you and/or your associates establish and retain an effective communication channel between yourself and the customer?
5. How do you measure your key processes to ensure that they are on schedule? During these processes are there any particular "milestones" established in order to ensure that everything is headed in the proper direction?
6. If an error occurs in one of your key processes mentioned earlier, what is your process to handle the error?
7. Does your organization use any particular strategies to facilitate continuous improvement? Any particular methods?
8. What is your process to set your job schedule? Do you use standard time data from past jobs?

9. What is your process to receive feedback from your employees, such as on-the-job problems? Do you use surveys?
10. Does your organization have a well-developed strategic plan for process improvement set for the future? If so, how do you fit in the picture as a leader, and how do you keep all members of the company aligned with this plan?

6.0 Process Management

6.1 VALUE CREATION PROCESS

Remodeling Designs and Case Handyman concentrate on the areas of customer satisfaction, communication with customers, communication with employees, hiring exceptional employees, and technological advances. However, current business practices do not allow for continuous improvement because in most cases the company's key processes are not clearly defined. When a process is not well defined, it cannot be repeated and the cycle times cannot be tracked, preventing Remodeling Designs and Case Handyman from reaching their fullest potential.

Remodeling Designs and Case Handyman put customer satisfaction above maximizing profits. This approach is both a strength and a weakness. With a better understanding of their processes and how to manage them, they will be able to enhance customer satisfaction while increasing profits.

6.1a Design Process

The design process begins with the sales call and continues through preliminary plan development, preliminary budget preparation, project cost estimation, design finalization, material selection, scheduling, final budget preparation, and construction (Figure 1). Leaders of Remodeling Designs and Case Handyman realize that although a basic process can be carried over from job to job, every opportunity is unique. Due to the very nature of their business, customer requirements are different for each new job. The single most important purpose of the design process is to please the customer. The companies believe in taking a customer's dreams and turning it into reality. Remodeling Designs and Case Handyman concentrate on this goal and take advantage of opportunities to improve the design process. Recent software acquisitions have changed business practices for everything from labor cost estimation (Case Handyman) to interior decoration (Remodeling Designs). The new software also facilitates customer communication.

Remodeling Designs and Case Handyman can improve their design process by mapping the process, determining where non–value-added activities exist, and removing those activities. By comparing cycle time information from project to project, they can learn which methods are effective and which are not. Currently, because of their lack of a clearly defined process, cycle time information is not used to enhance process performance.

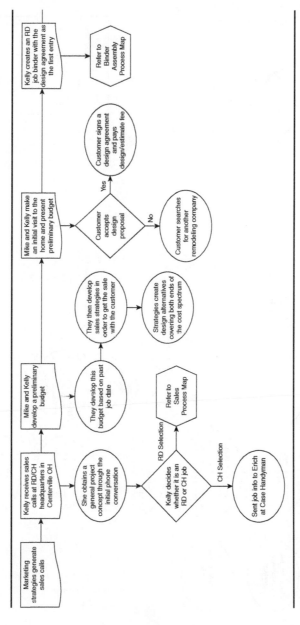

Figure 1   Design Phase Process Map

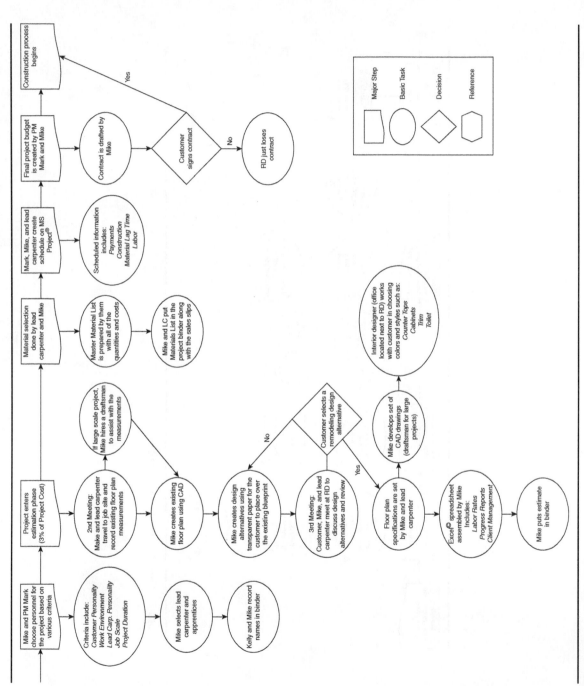

Figure 1 (continued)

6.1b Production/Delivery Processes

Key production and delivery processes include the assembly of the production binder, communication with the customer, budgeting and cost estimating, and scheduling. The key performance requirement is customer satisfaction. Several methods are in place to ensure the performance of these processes. Two-way communication with the customer takes place through the use of daily logs kept at the job site. These log books have space for the lead carpenter and the customer to leave notes to each other about questions and concerns (Figure 2). All employees are issued a cell phone and thus can be easily contacted during business hours. Effort is taken to contact the customer daily to keep lines of communication open.

Lead carpenters inspect the job site daily. Each Tuesday, members of the production team meet to discuss their project to ensure that it is on track. One area where Remodeling Designs can make improvements is in communication between lead carpenters. Since these individuals can learn a lot from one another about methods improvement, current business practices should be changed to allow for greater communication between lead carpenters working on different jobs.

6.2 SUPPORT PROCESSES

6.2a Business Processes

Business processes are one of the many strengths of Remodeling Designs and Case Handyman. Key processes include sales and marketing, billing, and warranty. Their sales and marketing techniques have enabled them to expand their influence in the market. The marketing of Case Handyman is done through two main avenues: the Yellow Pages™ and customer referral. Marketing of Remodeling Designs is primarily through referral. The key to the sales of both companies is communication with the customer. The customer is treated with courtesy, respect, and attentiveness. Marketing effectiveness is measured with surveys asking how the customers heard of Remodeling Designs or Case Handyman. The survey provides feedback as to which marketing techniques are effective and which are not. The marketing and sales processes have been modified based on this information.

The billing and warranty processes are not mapped and have experienced some difficulties. Customers pay one third of the estimated cost of the project before the project begins (when the contract is signed), one third when the project begins, and one third when the project has been completed. Completion of the project refers to satisfying the customer and interwining the billing and warranty processes (Figure 3). Remodeling Designs goes to great lengths to make the customer happy, sometimes at the expense of their profit.

6.3a Support Processes

The main support processes are the communication between employees and the hiring of new employees. Remodeling Designs and Case Handyman are proficient in these two support processes. The communication process has one basic requirement: It must be a short process, with as few steps as possible. This requirement was voiced by both employees and

Remodeling Designs, Inc.   Daily Log

Name:_____

Job Name:_____

Date:_____

Inspections: (P=Pass F=Fail)

Plumbing_____   Electric_____   Framing_____   Insulation_____   Foundation_____   Final_____

Other_____

Conditions: Fair___Overcast___Rain___Snow___

Temp.: 0–30___30–40___40–50___50–60___60–70___70+___

Other Observations:_____

Labor on Site: Company	Name	Time	Total Hours	Work Completed
Subcontractors:_____	_____	____to____	_____	_____
	_____	____to____	_____	_____
	_____	____to____	_____	_____
RD/CH Employees	_____	____to____	_____	_____
	_____	____to____	_____	_____

Daily Safety Inspections:

Hazards Identified:	Corrective Action Taken:

Calls Made	Discussion/Results

Homeowner Communications: (selections, additional work authorizations, changes, etc.)

_____

_____

_____

_____

_____

_____

Figure 2   Daily Log

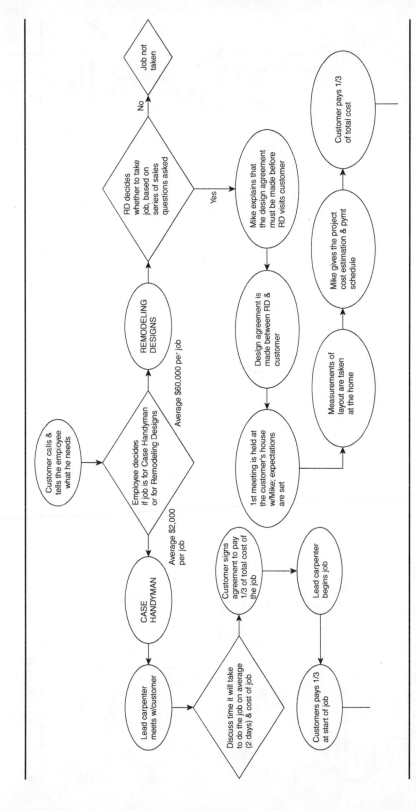

Figure 3 Customer Billing Process Map

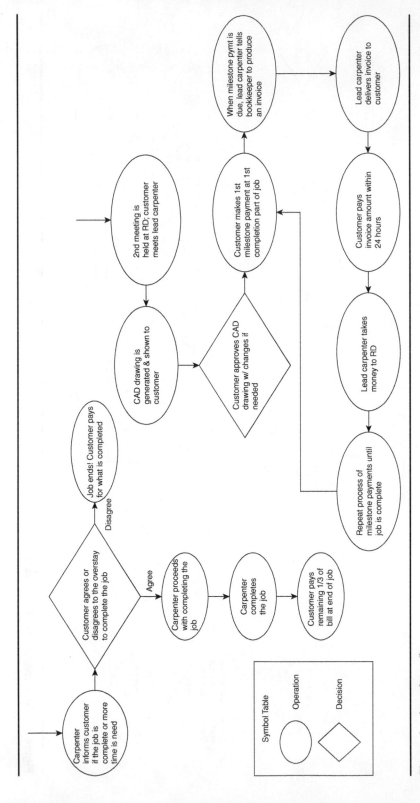

Figure 3 (continued)

customers. Employees utilize cell phones so that they can be easily contacted by fellow employees, the lead carpenter, or customers. The use of on-site log books also enhances the communication process.

Human resources is in charge of the hiring process, a critical process because employees at Remodeling Designs and Case Handyman have great autonomy in their jobs. The quality of the skills and the knowledge of these individuals is key to a great remodeling or repair job, and thus customer satisfaction. The hiring process is fairly standardized, but it has not been mapped or studied for improvement opportunities. A new employee "pitch" book is used during the hiring process. This pitch book explains the direction in which the company wants to head and the performance measures they use to evaluate their progress. It also provides succinct information about what Remodeling Designs and Case Handyman expect from their employees. Current employees participate in the hiring process. During interviews they are asked to evaluate the potential employee for both technical and social skills. Since the company is not large, everyone must get along. Each new employee has a two-month probationary period during which their skills and their ability to fit into the organization are judged. The hiring process is shown in Figure 1 of the Case Study in Chapter 7.

12 Lean Tools and Techniques

What is lean thinking?

What is value stream process mapping?

What is kaizen?

What are the five Ss?

What is kanban (pull inventory management)?

What is error proofing (poka-yoke)?

What is productive maintenance?

What is setup time reduction?

What are reduced batch sizes?

What is line balancing?

What is schedule leveling?

What is standardized work?

What is visual management?

Lessons Learned

Learning Opportunities

1. To become familiar with the tools and techniques related to lean organizations

2. To explore how the concepts of lean relate to quality management

On shop floors, the plane makers have also learned from car makers. At an Airbus factory in Wales, where it builds wings, production teams used to walk far to the stockroom for bags of bolts and rivets, and frequently left them scattered about—a wasteful and unsafe practice—because they lacked nearby storage. Using work-analysis methods developed by the auto industry, project teams studied which fasteners were needed where, and when, and then organized racks on the shop floor. Now carefully labeled bins contain tidy sets of supplies needed for specific tasks. The change has sped up work and saved over $100,000 in rivets and bolts at the Welsh factory alone.

Boeing, Airbus Look to Auto Companies for Production Tips,
The Wall Street Journal, April 1, 2005

WHAT IS LEAN THINKING?

Henry Ford said *"Time waste differs from material waste in that there can be no salvage."* With this in mind, he focused on ways to speed up the manufacturing process without sacrificing quality. He was the first to utilize a moving assembly line on a grand scale. On Ford Motor Company's final assembly line, car bodies slid down a ramp onto waiting chassis. This moving assembly line approach, so different from previous manufacturing methods, regularly broke daily production records.

In his books *The Machine That Changed the World* and *Lean Thinking*, Jim Womack presents the principles and practices related to lean manufacturing. A lean system provides what is needed, in the amount needed, when it is needed. The principal focus of lean thinking concentrates on value-added process flow. For a process to have value-added flow, there must be an uninterrupted adding of value to a product or service as it is being created. Interruptions or non–value-added activities found in the process, such as downtime, rework, waiting, and inspection, must be eliminated. In a lean process, the value-added time in the process equals more than 25% of the total lead time of that process.

Lean thinking is a mindset best described as a relentless war on waste. Lean tools and techniques attack the activities, behaviors, and conditions that lead to waste. When used, these tools identify and help correct excessive setup time, incapable processes, poor maintenance, variable processes and work methods, redundant activities, and lack of communication. Some of the benefits of lean thinking include shorter lead times, less handling, lower costs for storage and subsequently floor space, and fewer customer service activities. Companies implementing lean thinking have reported significant reductions in cycle times, lead times, floor space usage, and inventory while at the same time they see a significant improvement in quality, inventory turns, profit margins, and customer responsiveness.

Effective organizations realize that devising new methods to cut production costs, improve quality, speed assembly, and increase throughput are vital to staying competitive. Lean techniques enhance the effectiveness of quality management. One of the critical aspects is the realization that working hard to keep things simple saves money. Six Sigma strengthens company performance by concentrating on reducing process variation. Lean thinking enhances company performance by focusing on the reduction of waste. Effective quality management process improvement efforts involve both (Figure 12.1).

Lean thinking focuses on eliminating wasted time, effort, and material, particularly time, time wasted waiting for value to be added, or time wasted waiting in inventory for a customer or time wasted waiting for the next step in the process, and so on. There are many sources of waste. Tadamitsu Tsurouka, a Honda process engineer, identified seven sources of waste:

> Overproduction waste
> Idle time waste (waiting time/queue time)
> Delivery waste (transport/conveyance waste)
> Waste in the work itself
> Inventory waste
> Wasted operator motion
> Waste of rejected parts

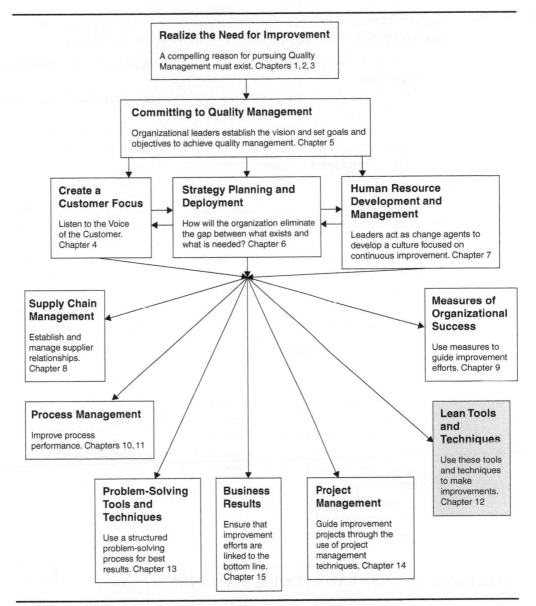

Figure 12.1 Achieving Organizational Success

Intellect can also be wasted. Lean thinking tackles the causes of these wastes. Lean projects focus on inadequate processes, inadequate tools/equipment, inefficient layouts, lack of training, inadequate suppliers, lack of standardization, poor management decisions, inadequate communication, mistakes by the operator, inadequate scheduling, and so on. Lean thinking generates process improvement by following five key steps:

1. Study the process by directly observing the work activities, their connections, and flow.
2. Study the process to systematically eliminate wasteful activities, their connections, and flow.
3. Establish agreement among those affected by the process in terms of what the process needs to accomplish and how the process will accomplish it.
4. Attack and solve problems using a systematic method.
5. Integrate the above approach throughout the organization.

Three key performance measures related to lean thinking are cycle time, value-creating time, and lead time. Cycle time represents how often a product is completed by a process or the time it takes an operator to complete all the steps in their work cycle before repeating them. Value-creating time is the part of the cycle time during which the work activities actually transform the product in a way the customer is willing to pay for. Lead time is the time it takes for one piece to move all the way through a process from start to finish.

Lean thinking essentially takes a diverse set of tools, techniques, and practices and combines them into a system. Lean thinking tools include:

Value stream process mapping
Kaizen
The five Ss
Kanban (pull inventory management)
Error proofing (Poka-yoke, pronounced "poke-a-yoke")
Preventive and predictive maintenance
Setup time reduction (single minute exchange of dies [SMED])
Reduced lot sizes (single piece flow)
Line balancing
Schedule leveling
Standardized work
Visual management

WHAT IS VALUE STREAM PROCESS MAPPING?

Processes provide customers with goods and services and wherever there is a process, there is a value stream. Value streams, which include both value-added and non–value-added activities, are the actions required to create a product or service from raw material until it reaches the customer. Since lean thinking focuses on needs as perceived by the end customer, value stream maps should focus on what is important to meeting customer requirements. Value stream process mapping makes the current production situation clear by drawing the material and information flows related to the process. It provides a clearer understanding of the process by allowing users to visualize the process, recognize sources of waste, and eliminate non–value-added activities.

When doing value stream process mapping, accurately specify the value desired by the customer. Go to where the action is taking place and identify every step in the value stream. Collect current information by walking along the actual pathways of material and information flow. Really do the work required. This may require several walkthroughs, beginning with a quick walk along the entire value stream and continuing with subsequent visits to gather more detailed information. Process mapping is often easier if the process is worked backwards. Working backwards reduces the probability of missing an activity because it takes place more slowly, without jumps to the conclusion of "I know what is going to happen next." Process mapping is best done with pencil, paper, and stopwatch. Though many computer programs exist to help with process mapping, it is difficult to get them to where the action is taking place. After all, the point of creating the map is to understand the flow of information and material, not in creating a map. Once the map is complete, remove the waste. Based on the requirements of the customer, make value flow from the beginning to the end of the process.

Typical data gathered during the value stream mapping process includes cycle time and changeover time information, machine uptime, production batch sizes, number of operators, lead times, number of products and their variations, pack sizes, working time, scrap rates, rework rates, and others. Cycle time refers to the actual time it takes to complete a product, part, or service. It is the time it takes for an operator to go through their work activities one full cycle. Value creation time is rarely equal to cycle time. Value creation time is the time it takes to complete those work activities that actually transform the product into what the customer wants. Lead time is the time it takes to move one piece, part, product, or service all the way through the process. All of this information can be captured on the value stream map (Figure 12.2). When the value stream map is complete, it will show the areas that are in need of improvement. From this, create an improvement plan establishing what needs to be done and when, outlining measurable goals and objectives, complete with checkpoints, deadlines, and responsibilities laid out. Process mapping and process improvement are covered in detail in Chapter 11.

EXAMPLE 12.1 Value-Added Process Mapping

The chocolate candy manufacturing line at Yummy Chocolates is a fully automated line, requiring little human intervention. However, when a Six Sigma team reviewed cost of quality information related to production cost overruns, they found excessive scrap and rework rates, and high inspection and overtime costs. To tackle the cost overruns, they wanted to isolate and remove waste from the process, thus preventing defectives by removing the sources of variation. In order to enable them to better understand the production line activities, they mapped the process (Figure 12.2).

The process map revealed several non–value-added activities. Chocolate is melted and mixed until it reaches the right consistency, then it is poured into mold trays. As the chocolates leave the cooling chamber, two workers reorganize the chocolate mold trays on the conveyor belt. This prompted the Six Sigma team to ask: why are the trays bunching up on the conveyor in the first place? This non–value-added activity essentially wastes the time of two workers. It also could result in damaged chocolates if the trays were to flip over or off the conveyor.

Yummy Chocolates prides itself on quality product. To maintain high standards, before packaging, the process map reveals that four workers inspect nearly every piece of chocolate as it emerges from the wrapping machine. A full 25% of the chocolate production is thrown out in a large garbage can. Though this type of inspection prevents defective chocolates from reaching the customer, this is a very high internal failure cost of quality.

This huge waste of material, manpower, and production time prompted the Six Sigma team to ask: why are the chocolates being thrown away? The simple answer was that the chocolates didn't meet standards. They persevered and discovered that the rejected chocolates were improperly wrapped. Despite the large amount of chocolate being thrown away, no one had suggested that the wrapper machine should be repaired.

If the team hadn't studied the process carefully using value-added process mapping, these two enormous sources of waste and their associated costs would have gone unnoticed. Up until this study, Yummy Chocolates' employees considered these costs part of doing business. During a subsequent kaizen event, modifications made to the production flow through the cooling chamber and to the wrapping machine resulted in a manpower savings of five people (who were moved to other areas in the plant) and rejected chocolates went from 25% to 0.05%. Savings were evidenced in decreases in in-process inspection, scrap, rework, production cost overruns, overtime, inefficient and ineffective production, and employee lost time.

WHAT IS KAIZEN?

Effective organizations create conditions that promote creative improvements. Improvements to current operations can come about from several sources, including innovation and continuous improvement, or kaizen. Innovation involves coming up with great new products or services or ideas or technology. Innovation comes about when someone looks at an existing product or service differently or discovers an unmet need in the marketplace. While innovation delivers excitement, kaizen focuses on improving how organizations meet customers' basic needs. Kaizen supports effective organizations as they build and keep a solid foundation of quality products and services. Kaizen encourages all employees to participate in the development and implementation of improvements to the systems affecting them the most. Through these efforts, they develop skills that enable them to improve their performance on the job. Kaizen supports quality management because it involves the constant and continual daily effort to improve a product or service.

Kaizen's guiding words are: combine, simplify, eliminate. Kaizen seeks to standardize processes while eliminating waste. Waste in a process or system can be removed by combining steps or activities, simplifying steps or activities, and eliminating any waste in a system or process. Kaizen practitioners go to the actual work area, work with the actual part or service, and learn the actual activities required in the work situation. Kaizen activities may

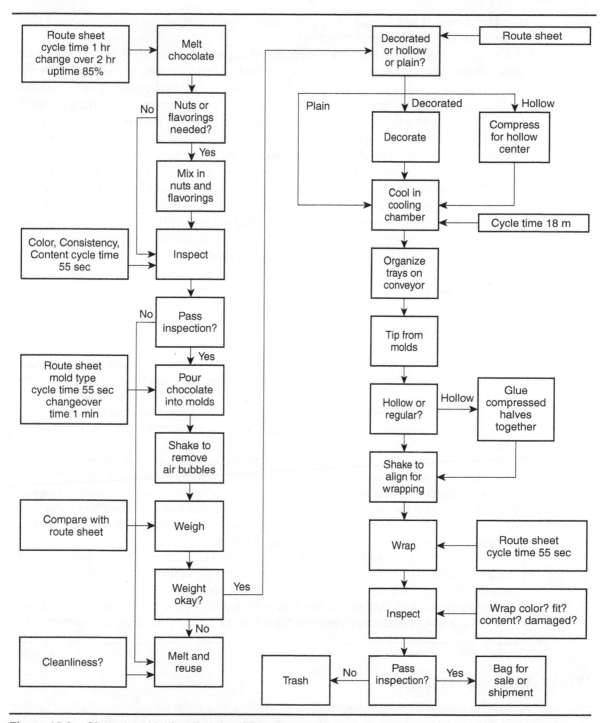

Figure 12.2   Chocolate Making Process Map

take two forms: flow kaizen, focusing on value stream improvement, and process kaizen, focusing on the elimination of waste.

Kaizen activities are typically a concentrated improvement event that lasts for about a week. Teams meet on the first day for training and problem identification. From there they analyze the process by documenting activities, processes, and cycle times. This information enables them to discuss process improvement options. These improvement options are implemented and tested for their ability to solve the problem. This may involve rearranging existing equipment, creating mock-ups, or going through the motions of the newly designed process. The improvements are refined, tested again, and the results presented to management.

At Kaizen events, lean thinking and Six Sigma tools and techniques often blend together to enable participants to maximize the process, product, or service. As we have learned, Six Sigma tools and techniques emphasize root cause analysis through the use of a standardized problem-solving technique (DMAIC) in combination with statistical analysis and performance measures. Many of the examples used in this text are from kaizen events. Lean thinking integrates lean tools into the kaizen process.

EXAMPLE 12.2 Kaizen Event

At PLC, Inc. a Six Sigma kaizen team has been investigating the process of returning defective parts to suppliers. The original fourteen step process takes 175 hours, nearly four weeks, for a defective part to be returned to a supplier.

Recognizing that their kaizen team's purpose is to combine, simplify, and eliminate while creating a standardized defective part return process, the team looked for waste that could be removed by combining or simplifying steps and activities. The Six Sigma team went through the entire process to learn the actual activities required. During four meetings, the team developed a new process. The new process included nine steps and reduced the time from 175 hours to 69 hours.

When the team presented their proposed process changes to upper management, the managers asked the team to begin the exercise again, starting with a clean slate, keeping only what absolutely had to take place in order to get the job done. The team was given one hour to report their new process. The new process has four steps and takes one hour to complete. This process was achieved by focusing on what the customer absolutely had to have and eliminating all non–value-added activities.

Though they were able to develop a new, simplified process map for the defective part return process, while in their meeting, it dawned on the team members that, in the big scheme of things, why were they allowing defective parts at all?

The team members learned two things from this kaizen experience: it doesn't pay to spend time fixing a bad process and sometimes the reason for the process needs to be eliminated.

WHAT ARE THE FIVE Ss?

Lean thinking practitioners recognize that the five Ss are the foundation of a lean facility. The five Ss are activities that focus on creating orderliness in a facility, thus supporting error proofing, setup time reduction, single piece flow, line balancing, visual management, and preventive maintenance. These five words serve as a reminder that process improvement lies in the basics.

Seiri	Separate
Seiton	Straighten/orderliness
Seiso	Scrub, cleanliness
Seiketsu	Standardize repeat seiri, seiton, and seiso at regular intervals
Shitsuke	Systematize, discipline

The five Ss mean more than just tidying up. They make up an organizational culture where all employees, regardless of position, are expected to be organized, clean, neat, and disciplined. They are expected to follow standardized procedures. The five Ss continually improve the work and the workplace by focusing on creating and maintaining an ever more organized, neat, and clean work space. It's more like a lifestyle than a one-time cleanup event. Employees practice the five Ss in everything they do.

When work processes are arranged according to the five Ss, workers can expect to have everything they need to perform their work right at their fingertips. Whether it is a receptionist taking a phone message needing pen and paper or a computer program or a surgeon needed particular scalpels for an operation, their tools are right at hand, ready for use. Proper arrangement also includes having work surfaces and equipment designed for ease of use.

Seiri refers to separating needed tools, parts, and instructions from unnecessary items. To define an unnecessary item, consider the 24-hour rule: If an item is not used or touched within 24 hours, it should be stored elsewhere. Keep only what is essential at the workplace. Begin by considering what work is being done at the workplace, then remove anything that isn't needed or that is not used very often. Critical to this step is to also bring to the work area anything that *is* needed to do the work.

Seiton, or orderliness, is quite simply "a place for everything and everything in its place." No work processes can be performed well if the work area is in disarray. Straightening up the work space means putting things where they can most easily be used. This step isn't about just making the work space neat by putting things away. Work spaces that have been straightened have everything necessary to do the job within easy reach in a logical position. While straightening, ask the question, Where do items need to be placed in order to improve flow? Straightening may mean that the equipment, measuring devices, tooling, and even machines need to be laid out differently. Readily accessible may mean that very little is kept in drawers or cabinets. Work centers utilizing five Ss often have taped or labeled locations for everything that is needed to perform the work. After each use, the operator returns the item to its convenient and marked location.

Seiso, cleanliness, focuses on the need to reduce the clutter. The only things kept at a work area should be those things that are needed to perform the work. Cleanliness also means that every day the work place should be straightened up, order restored, tools and equipment cleaned, and anything unnecessary disposed of properly. Seiso, cleanliness, is often associated with scrubbing, cleaning up, and painting. Seiso goes beyond this to find the root causes of waste and uncleanliness. Cleaning and painting spruce the place up, but if the sources of the problems are not eliminated, the untidiness will return. Ask questions like, How did the dirt get there? and How do we keep it from coming back?

Seiketsu, or standardize, refers to establishing a new way of working. Now that tools and equipment necessary to do the job are in the right place to support the flow of value-added work and the sources of waste and dirt have been eliminated, it needs to be maintained. *Seiketsu* reminds people to conduct seiri, seiton, and seiso at frequent intervals in order to maintain the workplace in pristine condition. It refers to regularly picking up, returning things to their proper location, and eliminating unneeded items.

Shitsuke, or discipline, is the self-control needed to continuously implement the other four Ss. When no one is looking, does disorder return? Does the work area get dirtier? Do unneeded items creep back in? The five Ss can fail if there is no effort to make a habit of the first four Ss. Management must be sure through audits and reward systems that the five Ss become ingrained in the organization's work practices.

How Do We Know It's Working?

JQOS management applied the five Ss to the entire facility within a few weeks of the purchase.

Seiri: Separate

One Monday, large trash cans appeared at every workstation. Employees were rather taken aback when at a short meeting all employees were given two directives: make a list of what you absolutely must have at your workstation and throw the rest out. This meant filling their trash cans by the end of the week.

The owner quickly dispelled disbelief by filling his own trash can with a multitude of unnecessary files from his own office. Slowly, others did the same. Soon, front-office desktops contained only what was needed for their function. As they dug out from the mounds of unnecessary files and papers, the office seemed bigger and lighter. Some separating was easy, like throwing out a phone book from 1998 or old lunch wrappers. Other separating took time, determining which files to keep and which to dispose of.

In the shop, employees held many discussions concerning the optimum places to store tools and tooling. There weren't enough tools for each workstation, so setups frequently involved walking around looking for the right tool. Management invested in the needed tools so that each workstation had a complete set. To ensure that these tools didn't migrate around the plant, maintenance constructed boards drawn with outlines of each essential tool to hang vital tools and tooling on for each workstation. Involving all the operators, supervisors, setup, and maintenance employees in preparing these boards meant that their acceptance and use is ensured.

Seiton: Straighten

The most visible examples of straightening also took place in the inventory area and in tooling. Cooperative education students were hired to create a tooling area complete with labeled bins for all tools, gages, and spare parts. They also created tooling sheets listing all the tools and gages required to start a job. Setups used to be a process of having to walk around the shop looking for things. Now, the employee prints the job information, takes a basket, and walks around the small room gathering what is needed. Strict rules are in place for replacing what has been used. One individual has been assigned the task of daily receiving new tooling inventory and keeping the room organized.

The inventory storage area received special attention from the co-ops. Outdated and unneeded material was sold for scrap. New shelving units got material up off the floor and out of harm's way. Changes made to material ordering policies prevent overstocking. These changes resulted in a net space gain of 50%.

Seiso: Scrub

Making metal shafts means making metal chips. Chips attached themselves to shoes and got tracked everywhere. Chips overflowed in work areas, sometimes damaging parts during processing. Chip control became a key focus of cleanup efforts. Investigation revealed that chips, when compressed into hockey puck–sized units, could be sold quite profitably for reuse. This fit nicely into management's concern for the environment, the need to recycle, and the need to eliminate waste. Management invested in a chip collection system. At each machine, vacuums placed on 55-gallon drums collect the chips. When full, these barrels are taken to a machine that compresses the chips into pucks. Once a week, a scrap dealer collects the chips. The $50,000 system paid for itself in four months. The shop is cleaner, and the root cause of many quality problems and machine maintenance headaches has been removed.

As machining areas employed the five Ss successfully, the machines in those areas were painted. Lathes sport a blue and white paint scheme. Mills are gray, and grinders are burnt orange. This visual statement also served as an incentive for other machinists to separate, straighten, and scrub their work areas.

Seiketsu: Standardize

Maintaining known and agreed-on procedures and conditions takes effort. The new tooling boards support faster setups, but more improvements could be made. The co-ops videotaped a number of different machine setups, and then the employees discussed a variety of ways to standardize the setups. The new methods are more effective, reducing some setup times by as much as 80%. In other areas, standards have also been set. Shipping procedures changed to ensure safer travel of the shafts to customers. The challenge in shipping has been to keep the area clean and organized with the increased volume of parts flowing through.

> **Shitsuke: Systematize**
>
> Discipline to maintain the new environment is key. This discipline must occur on a personal level. Each employee must agree to follow standards and procedures necessary to maintain organizational neatness and cleanliness. JQOS management has put into place rewards and recognition that help support their employees as they make a habit of the five Ss.

WHAT IS KANBAN (PULL INVENTORY MANAGEMENT)?

Kanban improves process management by focusing on visual control of the process. Whether the kanban cards are physically present or digital on an Internet-based system, their purpose is to order the creation of a product or service on an as needed basis. Kanban cards follow the product through its stages of production. Information on the card includes the name of the part, the number of part, instructions related to the part, the creation date of the card, the due date of the parts, and other pertinent information. Kanban cards keep track of inventory. To be genuinely called a pull inventory system, parts must not be produced or conveyed without a kanban. The number of parts must correspond with the number of parts needed as listed on the kanban.

EXAMPLE 12.3   Kanban

Through previous process improvements, QRC, Inc. reduced the amount of in-process inventory significantly. What little remained still needs to be managed. Controlling inventory means that it is easy to identify what has been made and when, as well as where it is and in what order it should be used. QRC, Inc. came up with a simple system. A small portion of their manufacturing floor was marked in bright yellow squares matching the size of the carts used to hold inventory awaiting processing. Each square had a corresponding label painted in it, A-1, B-1, C-1, A-2, B-2, etc. For each of these squares, small discs labeled with the square location were created.

When a cart needs to be parked, a disc is selected from the rack. The number on this disc represented the location in which to park the cart. After parking the cart, the disc is inserted into the top of the appropriate tube, each type of product being represented by a different tube. Since different tubes represent different types of inventory, the disc being placed in the tube matches the inventory on the cart. The more in-process inventory, the more discs in the tube, a visual representation of the in-process inventory for each type of product. This tube also serves to make sure that a first-in, first-out inventory control system is in place. When a cart is needed, the operator pulls a disc from the bottom of the tube. This tells the operator which cart needs to be taken next.

WHAT IS ERROR PROOFING (POKA-YOKE)?

To eliminate waste at its source, lean thinking seeks to develop simple methods of preventing errors from occurring in a process. This concept of preventive action known as errors proofing or poka-yoke follows five principles: elimination, replacement, facilitation, detection, and mitigation. Elimination refers to the need to design work and processes that eliminate the potential for error. Replacement means replacing a faulty process with another, more reliable process, that has less of a potential for error. When a process is facilitated, it is easer for the operator to perform without error. Detection encourages the use of methods that enable an error to be easily spotted, either at the original workstation or at the very next operation. This prevents errors from compounding as the work progresses. Mitigation, the final choice, refers to minimizing the effect of the error if it does occur.

Error proofing may be accomplished through the design and use of fail-safe devices, counts, redundancy, magnifying the senses, or special checking and control devices. Fail-safe devices may fool-proof an action by preventing the work from being done any other way. Mechanisms or work-holding devices may signal the operator when the work has been done correctly. Limiting mechanisms on tools may be used to prevent a tool from exceeding a certain position or amount of force. Counts or count-downs can help an operator keep track of where they are in a process. Redundancy can be effective when identifying parts, for instance, labeling a part both with color and a bar code. Double checks can be used to ascertain if the work has been completed correctly. Humans have five senses, seeing, hearing, feeling, tasting, and smelling. Tools that provide feedback in a variety of ways can alert an operator to whether or not their process is operating the right way. Lastly, special checking and control devices of varying levels of complexity may be designed to help an operator detect whether or not the work they are performing is correct.

EXAMPLE 12.4   Error Proofing

The quality specialist at RQM, Inc. received the following information concerning warranty claims.

Condition/Symptom Customers have been requesting warranty service for inoperative and incorrectly reading fuel gages for recently purchased automobile models R and Q.

Probable Cause Incorrect fuel gages or incorrect fuel pumps installed at the factory.

Immediate Corrective Action Replace fuel units (pump, gages, tank) with correct parts depending on model.

Market Impact To date there have been 39 warranty claims associated with fuel systems on models R and Q. Each claim costs $3,000 in parts and labor.

During a kaizen event, the quality specialist visited the Fuel System Creation workstation to study its layout (Figure 12.3). He created a process flow map (Figure 12.4). He was able to determine that the creation of the fuel system requires that the operator visually identify the handwritten designation for the fuel tank type and install the correct fuel system (pump and gage combination). Visiting the related workstations, he determined that the labeling of the tank is currently done by hand at the station where the part is unloaded from the paint rack. The part sequence, date, shift, and model type are written in a location visible to the person at the Fuel System Installation workstation, but not easily read by the operator at the Fuel System Creation workstation. The pumps and gages are not labeled, but arrive at the workstation on separate conveyors. The operator is responsible for designating the tank type and sequencing the tank into the correct production line for either model R or model Q. The unit then proceeds to the final assembly line.

Models R and Q are produced on the same final assembly line. Similar in size, they require fuel pumps, gages, and tanks which on the outside look remarkably similar. A variety of opportunities for error exist:

An R pump and gage combination may be installed on a model Q tank
A Q pump and gage combination may be installed in a model R tank
An R pump with a Q gage may be installed in a model Q tank
An R pump with a Q gage may be installed in a model R tank
A Q pump with an R gage may be installed in a model Q tank
A Q pump with a Q gage may be installed in a model R tank
The pump may be inoperable or substandard, repeating the same permutations above
The gage may be inoperable or substandard, repeating the same permutations above

With only 55 seconds in which to select and assemble a pump, a gage, and a tank, the possible errors are numerous. An operator might misread the handwritten identification on the fuel tank, select the wrong pump, select the wrong gage, or place the completed fuel system on the wrong assembly line.

Countermeasures In order to eliminate errors and foolproof the process, the following countermeasures were put into place:

Label all components with a barcode
Install barcode readers on the conveyors feeding the Fuel System Creation workstation
Install gates on all incoming and outgoing part conveyors
Utilize computer software that reads the type of tank and releases only the correct fuel pump and gage from material storage. This system also selects the appropriate conveyor line leaving the workstation.

This kaizen activity prevents future errors by more clearly identifying the incoming parts and limiting the choices the operator can make.

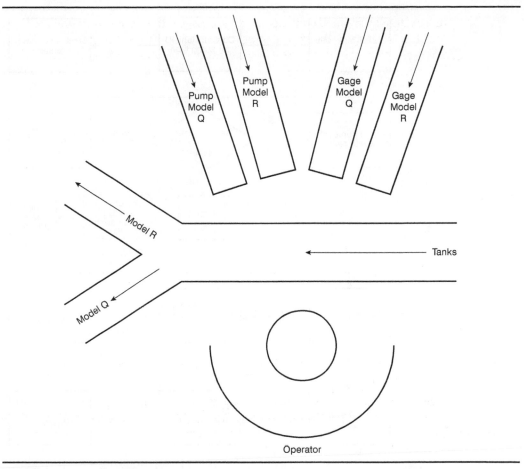

Figure 12.3   Workstation Layout for Fuel Systems

WHAT IS PRODUCTIVE MAINTENANCE?

Productive maintenance efforts, by preventing problems, seek to increase overall equipment effectiveness and machine reliability. Productive maintenance includes the concepts of preventive and predictive maintenance. Overall equipment effectiveness is a function of available time for value-added activities. Total available equipment time is reduced by planned downtimes and setup losses. Unfortunately, this is further reduced by unplanned downtimes, idling and minor stoppages for adjustments, speed losses, spoilage, and rework caused by faulty equipment. Unscheduled machine downtime, whether from parts shortages, setup time, absenteeism, reduced speeds, minor stoppages, or major machine breakdowns, represents wasted time and money. Preventive and predictive maintenance programs attack equipment failures as a source of waste. Preventive maintenance strives to maintain the equipment at the peak of its condition eliminating the waste of reduced speeds and

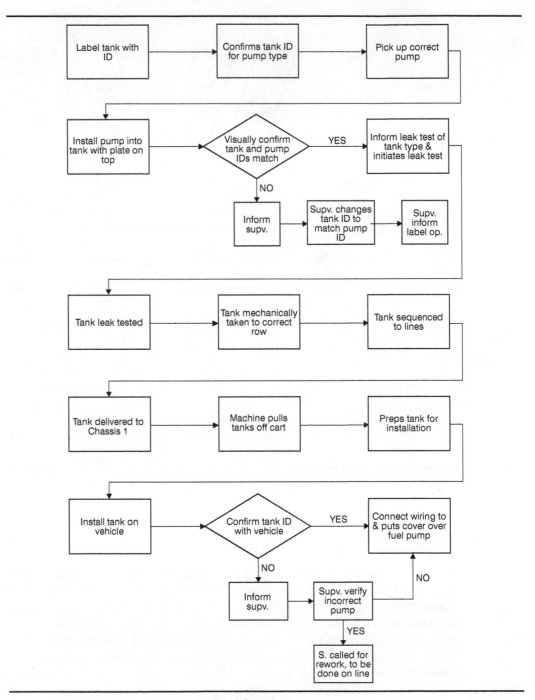

Figure 12.4 Process Flow Map for Fuel Systems

368

minor stoppages. The benefits of practicing preventive and predictive maintenance include reduced costs, lower inventories, shorter lead times, and fewer breakdowns, jams, standbys, speed losses, and startup losses; shorter changeover times; improved productivity; increased flexibility; improved quality, and so on. Productive maintenance prevents disasters too. Or, as in the following example, lack of maintenance can be deadly.

EXAMPLE 12.5

Productive maintenance could have prevented two disasters at British Petroleum (BP) facilities. In 2005, in Texas City, Texas, a refinery explosion and fire killed several employees. In 2006, a pipeline failure resulted in the leakage of more than 200,000 gallons of oil into an environmentally fragile area. Death, lost production, equipment and facility damage, wasted raw materials, and environmental impact—could these have been prevented by a sound productive maintenance program? According to a presentation made on May 16, 2007, in Washington, D.C., by the Chemical Safety Board (CSB) Chairperson Carolyn Merritt, the answer is yes. Chairperson Merritt told the House of Representatives subcommittee that "virtually all of the seven root causes identified for the Prudhoe Bay incidents have strong echoes in Texas City." These included the "significant role of budget and production pressures in driving BP's decision making—ultimately harming safety." The Prudhoe Bay pipeline suffered extensive corrosion due to lack of maintenance over several years. At the Texas City refinery, "abnormal start-ups were not investigated and became routine, while critical equipment was allowed to decay. By the day of the accident, the distillation equipment had six key alarms, instruments, and controls that were malfunctioning." Clearly, no one took the initiative to maintain, repair, or replace equipment as needed. As operating conditions deteriorated in both situations, the end result, disaster, was inevitable. A sound productive maintenance program could have averted disaster.

Preventive maintenance requires that maintenance of equipment be planned, scheduled, and then executed to performance standards. These performance standards denote the plan for scheduled maintenance including the interval, how the downtime will be scheduled, and what type of work is to be done (basic check or major overhaul). This plan will also have procedures in place to make sure that appropriate parts/materials have been procured and labor scheduled. Because there is a distinct maintenance plan for each piece of equipment, there is also a budget in place to cover the costs. Also important are the maintenance records documenting what has taken place, machine performance, inspections, and lubrications.

Predictive maintenance monitors equipment performance. It seeks to measure, recognize, and use signals from the process to diagnose the condition of the equipment and determine when maintenance will be required. Many of the records kept for preventive maintenance are used in predictive maintenance.

How Do We Know It's Working?

Leaders at JQOS turned to productive maintenance in order to increase overall equipment effectiveness and machine reliability. In the past, as machines wore out, employees developed all sorts of work-arounds. As a result, the machines in the plant were unreliable and prone to breakdowns. High costs of quality existed due to machines that were unable to hold tolerance and exhibited poor repeatability. One machine squealed so badly when operating that the sound level in the plant approached OSHA's 90-decibel maximum level.

Management at JQOS worked together to create a schedule of maintenance for the machines. A new employee skilled in machine repair was hired. Money was spent to scrape the ways and replace the bearings, switches, and handles. A worn-out lathe was sold for parts and replaced with a significantly more capable machine. Because of its larger motor and newer CNC controls, this machine can handle shafts a foot longer.

Unscheduled machine downtime, minor stoppages for adjustments, speed losses, spoilage, and rework have all decreased significantly since the productive maintenance program was put in place. There are fewer breakdowns, jams, standbys, and changeover times. Training is easier too. The quirks have been removed from the machines, making them easier to operate. These improvements have enabled the plant to increase production throughput by improving quality and productivity, increasing flexibility, and shortening lead times. The end result: JQOS has been able to double sales in two years without a major investment in new equipment, employees, or plant size.

WHAT IS SETUP TIME REDUCTION?

In the broad sense, setup time can be defined as the time between the production of the last good part in one series of parts and the production of the first good part in the next series of parts. Setup times encompass taking an existing setup apart, preparing for the next setup, installing the next setup, and determining whether or not the new setup can create good parts.

Lean thinking encourages the reduction of setup time. Though necessary, setup times do not add value and therefore are a source of waste. Tooling and equipment should be designed to allow easy changes from one tool to another. Setup processes should be studied for wasteful activities. Without quick changeovers single minute exchange of dies [SMED], lead times remain long. Since the cost of changing over tooling is high, small batch sizes cannot be justified. Lengthy changeovers limit an organization's flexibility to respond quickly to customer needs.

EXAMPLE 12.6   Setup Time Reduction

QRC Corporation faces a dilemma. On the one hand, they are woefully short of the capacity required to take on a new, rather lucrative job from a customer. On the other hand, they are short of the funds needed to purchase a new stamping press vital to the new order. Management has approached the operators of the equipment to discuss their concerns. Together, they have decided to conduct a kaizen event focused on reducing the setup times on their existing machines. Nearly everyone, from engineers to operators, expressed concern that there was not any room for improvement at all.

On the first day, the first two hours of the morning is spent discussing the machines, how they operate, the importance of certain steps in the setup process, and the concerns of the operators and setup people. Following the meeting, the team returned to their work-stations with one small change. Each operator and setup person was videotaped while performing their normal duties. After a few moments of discomfort, they forgot they were on camera.

On the morning of the second day, each of the four videos were watched by the team. Two things were immediately apparent to all who viewed the films. Operators often stopped their machines to make minor adjustments, and during machine changeovers the setup people were always walking off in search of tools. When questioned, the response was "that's the way we've always done it."

Quickly changes were made. Setup personnel identified tools necessary for their jobs. Management invested in tool carts to contain the necessary tools. Setup personnel also identified minor machine and tooling modifications that could be made to greatly simplify their work. These changes, modest in scale, were rapidly implemented, especially when it was discovered that the improvements to setup would eliminate most, if not all, of the work stoppages for minor adjustments.

At the end of their kaizen event, the company was proudly able to report that the changes resulted in a capacity increase equal to one-third more production time. This exceeded the capacity increase from a new machine by more than two times. The company accepted the new order, rewarded their employees monetarily for their help, and scheduled future kaizen events for other areas of the plant.

WHAT ARE REDUCED BATCH SIZES?

Reduced lot sizes result in shorter lead times, a shorter period of time between when a raw material arrives and is paid for and when a product or service reaches the customer and is paid for. Shorter lead times increase the number of inventory turns. Part of lean thinking's ultimate goal is to create continuous flow wherever possible. In a continuous flow environment, as each piece is created, it flows immediately to the next activity with no delays, storages, or work-in-process inventories. Kanban or pull systems work well with single piece flow because only what is needed is being made. Smaller lot sizes reveal problems quickly. The process of going from large batches to single piece flow requires much problem resolution.

In some situations, it won't be possible to meet single piece flow, however, as downtimes due to changeovers are reduced and smaller in-line equipment utilized, batch sizes will get smaller and smaller, providing many of the benefits of single piece flow. Single piece flow does not work well when processes are too unreliable to be linked closely to other processes. It can also happen when raw materials or components have to travel a great distance. Also, some processes, such as injection molding, stamping, or forging, achieve better economies of scale by batch processing. Still, where continuous single piece flow is not possible, batches should be linked with the same pull system, making only what is needed. Safety stock or buffers should be temporary, not a crutch to avoid improvement. Kaizen efforts, using the five Ss and other lean tools, should focus on working toward SMED, single minute exchange of dies.

EXAMPLE 12.7   Single Piece Flow

In its article *"Boeing, Airbus Look to Auto Companies for Production Tips,"* the *Wall Street Journal,* April 1, 2005, reported that "Boeing, though, made one of the most dramatic production changes yet in 2001 when it began putting together planes on a huge moving line—a la Henry Ford and his Model T. The motion 'lent a sense of urgency to the process that we really didn't have when the planes were sitting still,' says Carolyn Corvi, the executive who oversaw the change. Ms. Corvi and other top Boeing executives made multiple visits to Toyota when they were first beginning to study how to convert the production process to a moving line. When many workers initially balked at the production line and unions filed complaints, Boeing took extra pains to win them over. The change paid off: Boeing halved the time it takes to assemble a single-aisle 737, and has started putting its other planes—including its oldest and largest product, the 747—on moving lines."

WHAT IS LINE BALANCING?

When systems are in balance, the work is performed evenly over time with no peaks and valleys placing undue burdens on employees or machines. Each machine and each operator makes what is needed timed to match when it is needed. Unbalanced lines are evident when workstations make more than is needed or have to wait for production from the previous work center to reach them. Overproduction represents significant waste. Making more than the customer wants or more than can be sold means that valuable resources of time, material, and money have been spent to no benefit for the organization.

Schedule leveling and line balancing create a thread that links customer needs with production. Lean thinking focuses on getting processes to make only what is needed by their customer, either internal or external. All processes should be linked, from the raw material to the final customer. This smooth flow will provide the shortest lead time, highest quality, lowest cost, and least waste.

Value stream mapping and kanban systems will reveal where lines need to be balanced. Takt times, how often a single part should be produced, provide the starting point for line balancing.

$$\text{Takt time} = \frac{\text{Available working time per day}}{\text{Customer demand rate per day}}$$

Takt times are used to synchronize the pace of production to the demands of the customer. Because it is based on the customer requirements, this measure enables organizations to balance their lines. No operation should run faster than the takt time, just as no operation should produce more than what has been ordered by the customer.

Takt times do not leave a margin for error. To meet takt times, organizations have to remove sources of waste from their processes. The organization must be able to respond quickly to problems and eliminate possible causes of unplanned downtime.

EXAMPLE 12.8   Line Balancing

At QRC Inc., a kaizen event studied the packaging of driver's side airbags. This three step process takes a completed driver's side airbag, folds it properly for instantaneous expansion, wraps it for insertion into the steering wheel, and inserts it into a metal bracket that will later hold the airbag to the interior of the steering column. The existing design resulted in an unbalanced line as evidenced by the percent load chart (also called a Yamhzumi chart) in Figure 12.5. Operators 1 and 2 are overburdened, and they are unable to keep Operator 3 busy.

Takt time calculations require that one driver's side airbag package must be created every minute. With this knowledge, the entire work area was redesigned to accommodate three operators for folding, two operators for wrapping, and one operator for insertion.

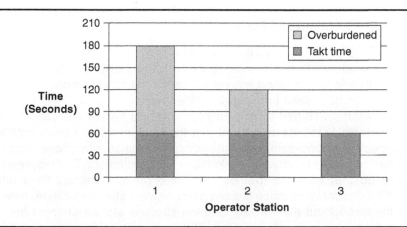

Figure 12.5   Percent Load Chart

WHAT IS SCHEDULE LEVELING?

Much of the strength of a kanban system comes from level scheduling. Though level scheduling becomes more difficult the closer the work is to the customer's final product or service, being able to predict the amount of work to be performed in a day reduces the demands on the processes. Fluctuating customer demand can overwhelm the system's ability to respond appropriately. Uneven demand can result in waste associated with unpredictable downtime for setups or an excessive number of setups. Waste can also be found in the unpredictability in the mix and volume of customer demand.

Level scheduling enables an organization to link the customer schedule with the pace of manufacturing, thus it requires cooperation from the customer. The need to expedite disappears because the organization is neither ahead nor behind. The drawback is that responding to customer requirement changes becomes more complicated. To remain flexible, lean organizations develop the ability to make every part every day. Their processes are very flexible.

WHAT IS STANDARDIZED WORK?

In some organizations, standardized work is referred to as standard operating procedures. These standard operating procedures refer to the activities that must happen in order to complete a process. They mean that everyone doing that job does it exactly the same way each time. There would be no difference between the way operator A performs the work versus operator B. There would also be no difference between the fifth time they did the work or the 1,000th time they performed the work. Kaizen improvement teams often create standard operating procedures as they make improvements to an area. In order to become a learning organization, when changes prove to be useful in one work area, these same countermeasures should be adopted for other similar areas.

EXAMPLE 12.9   Standardized Work

XYZ Corporation asked a consultant to visit their plant to tackle noise-related issues at workstation 18. While walking through the plant, the consultant noticed that other workstations contained equipment nearly identical to workstation 18. Sound pressure level comparisons between the other workstations and workstation 18 showed that workstation 18 had significantly higher noise levels. The consultant asked to see the other workstations and found that the equipment had been outfitted with three very effective sound-dampening devices. She asked why these same countermeasures had not been adopted to workstation 18's equipment. The response: we never thought to do that because that part of the plant is managed by a different supervisor. XYZ Corporation, since it didn't practice work standardization, failed to reap the benefits throughout their facility. Sharing success stories enables the entire organization to improve.

How Do We Know It's Working?

Flow velocity is a term used to describe the movement of parts, people, or information through any system. At JQOS, flow velocity ran at a snail's pace. The chief problem related to how machines were placed in the plant. The machine arrangement arose over time; as new machines were added, the previous owners found places to put them without being concerned about material flow through the plant. This resulted in a confusing flow and lots of backtracking.

JQOS is a job shop making parts for specific customer orders. Using shop routings, cooperative education students performed an analysis of the flow of parts through the plant. They divided the parts into three part families: small, medium, and large. All raw bar stock is received and sent to be centered. Following centering, all bars proceed to the lathes to be turned to size. The next step is the mills, where keyways are cut. From there, differences arise. Approximately 40% of the shafts go to be drilled and tapped on their ends before going to the grinding area. The remaining 60% go directly to the grinders to be precision ground. Following this, the finished parts are inspected, packed, and shipped.

Changing the layout of a plant while product is being made is challenging. The realization that the parts traveled over 1,000 feet in a spider-web pattern throughout the plant (Figure 12.6) encouraged management to take the plunge. Management and employees worked together to create a new layout. The new layout emphasizes U-shaped flow. It places all the lathes and millwork centers in one area. Each work center combines two lathes with a mill. In the new lathe/mill machining centers, one operator runs parts simultaneously on all three machines. Process studies showed that two shafts could easily be processed while the operator ran a part on the mill. For specialty products, two end mills have been combined in one work area staffed by a single employee. Having two loading docks enabled receiving and shipping to be separated. Additional floor space resulted from combining quality, maintenance, and scheduling offices together. A reorganization of the incoming materials area also gained floor space.

These changes significantly reduced material handling and therefore material damage. The new flow takes just 360 feet from receiving through the plant to shipping (Figure 12.7). Visual pull signals have been implemented because the work areas can see each other. Capacity planning and line balancing are simplified because productive maintenance and setup time reduction activities have made machines more capable. Changes have been made to make operators more flexible too. Unlike the previous system with one man to one machine, when work backs up in one area, cross-trained workers using standardized work procedures can move from work area to work area as needed. Closer tracking of customer orders using a computerized system has reduced batch sizes, inventory requirements, and work-in-progress. Many orders approach single piece flow. Visual management is easier too, because the shop is cleaner and the more organized product flow more visible.

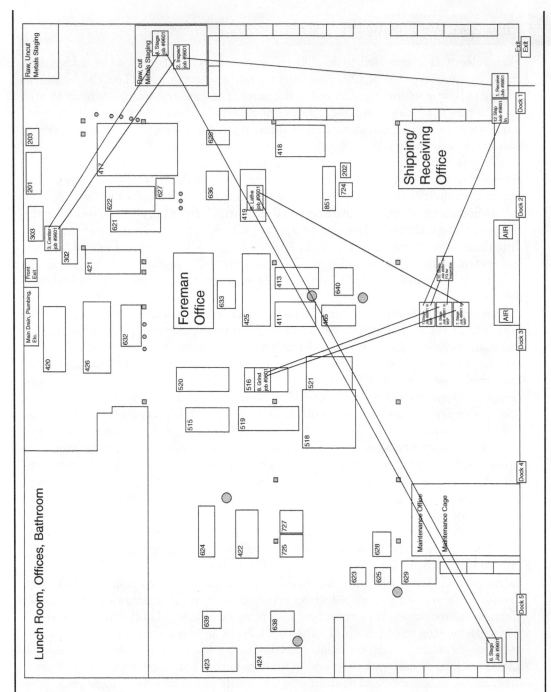

Figure 12.6   JQOS Original Layout with One Part Shown

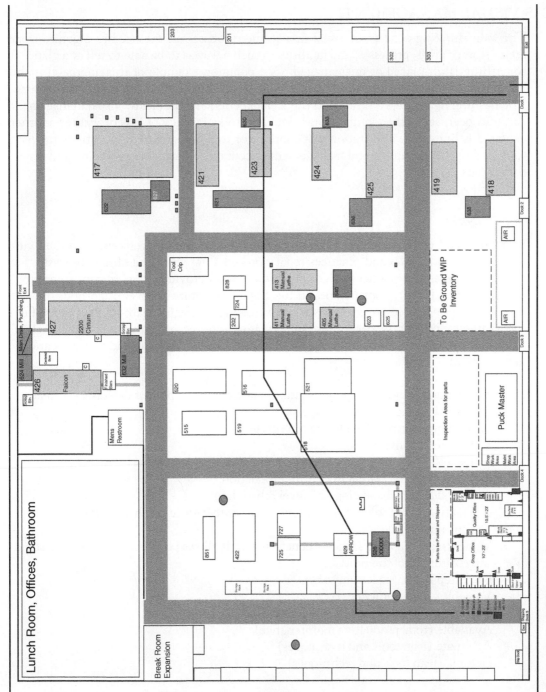

Figure 12.7 JQOS New Layout

WHAT IS VISUAL MANAGEMENT?

The lean thinking concept of visual management focuses on the need to organize work areas, storage areas, processes, and facilities in such a way as to be able to tell at a glance if something is misplaced or mismanaged. One common example of this approach is the lights that flash at a workstation if it is unable to meet its production quota or takt time. The flashing light draws attention to the area experiencing problems. Visual management can also been seen when kanban cards are used to keep track of inventory. Yet another example is lines on a workstation work surface or tool board marking locations for each tool used in the process. Visual management encourages a place for everything and everything in its place.

LESSONS LEARNED

Lean tools and techniques, enable organizations to become more responsive to their customers. Lean principles and practices reduce process flow times and eliminate waste. Six Sigma principles and practices reduce process variation providing a more consistent product or service. Users of both lean and Six Sigma recognize that lean methods are not designed to bring process under statistical control and Six Sigma methods are not as effective at improving process speed. Successful organizations recognize that any and all helpful tools, techniques, and practices, when used together as a system, will lead to organizational improvement.

1. Six Sigma tools work to reduce variation and eliminate process defects while lean tools work to increase process speed.
2. Lean improvements often follow this cycle:
   a. Practice the five Ss.
   b. Develop a continuous flow that operates based on takt time.
   c. Establish a pull system to control production.
   d. Introduce line balancing and level scheduling.
   e. Practice kaizen to continually eliminate waste, reduce batch sizes, and create continuous flow.
3. Value stream mapping's biggest benefit is that it allows people to see and understand the entire flow of the process.
4. Each step in the process must be:
   Valuable (customer focused)
   Capable (Six Sigma tools and techniques)
   Available (total productive maintenance)
   Adequate (lean tools and techniques)
   Flexible (lean tools and techniques)

Are You a Quality Management Person?

Lean-thinking people use a variety of tools, and techniques to improve their daily activities. Check out the quality management person questions and find out whether you are a quality management person. How do you answer them on a scale of 1 to 10?

Can you identify sources of waste from overproduction or idle time in your life?

| 1 | 2 | 3 | 4 | 5 | 6 | 7 | 8 | 9 | 10 |

Can you identify wasted motion or travel time in your life?

| 1 | 2 | 3 | 4 | 5 | 6 | 7 | 8 | 9 | 10 |

Have you made changes to reduce the waste in your life so that you can get more work done than you were 12 months ago?

| 1 | 2 | 3 | 4 | 5 | 6 | 7 | 8 | 9 | 10 |

Do you observe your work activities and their connections and flow?

| 1 | 2 | 3 | 4 | 5 | 6 | 7 | 8 | 9 | 10 |

Do you strive to eliminate non–value-added activities from your life?

| 1 | 2 | 3 | 4 | 5 | 6 | 7 | 8 | 9 | 10 |

Have you ever done a kaizen event on a problem in your life?

| 1 | 2 | 3 | 4 | 5 | 6 | 7 | 8 | 9 | 10 |

Have you taken the time to separate the necessary from the unnecessary in your life?

| 1 | 2 | 3 | 4 | 5 | 6 | 7 | 8 | 9 | 10 |

Do you make a habit of the five Ss, including straightening and scrubbing?

| 1 | 2 | 3 | 4 | 5 | 6 | 7 | 8 | 9 | 10 |

Have you standardized certain activities or procedures in your life in order to make things easier for you?

| 1 | 2 | 3 | 4 | 5 | 6 | 7 | 8 | 9 | 10 |

Have you error-proofed aspects of your life?

| 1 | 2 | 3 | 4 | 5 | 6 | 7 | 8 | 9 | 10 |

Do you perform productive maintenance on key equipment in your life, like your car, computer, or cell phone?

| 1 | 2 | 3 | 4 | 5 | 6 | 7 | 8 | 9 | 10 |

Have you found ways to reduce your setup times for key activities like getting ready for work, washing clothes, or making dinner?

| 1 | 2 | 3 | 4 | 5 | 6 | 7 | 8 | 9 | 10 |

Chapter Questions

1. Describe the goals of lean manufacturing.
2. How does lean support quality management?
3. Review the seven sources of waste provided in the chapter. Consider where you work, provide examples of each of these types of wastes in your organization.
4. Describe the five steps of lean process improvement in your own words.
5. Describe the kaizen concept.
6. What is value stream process mapping? How does it go beyond traditional process mapping? Why is value stream process mapping such an important tool when trying to make organizational improvements?
7. Describe each of the five Ss. Go to an area in your organization or home and apply the five Ss. What changes did you make? What results did you see?
8. How do kanban cards work?
9. Describe a process, system, or product that you have error proofed.
10. Why is preventive and predictive maintenance critical to overall system or process performance?

11. What tasks do you perform daily, either at home or at work? How would you go about reducing the setup times involved in these tasks? Select a task and make some changes to reduce the setup times. How much of a reduction did your changes provide?

12. What are the benefits of producing smaller batch sizes?

13. Describe what is meant by the term *line balancing*.

14. What is meant by the term *takt time*?

15. Describe a situation where you use visual management to monitor.

13 Problem-Solving Tools and Techniques

How does an effective organization make value-driven improvements?

What are typical steps in a problem-solving process?

What happens during the "Plan" phase?

What happens during the "Do" phase?

What happens during the "Study" phase?

What happens during the "Act" phase?

Lessons Learned

Learning Opportunities

1. To become familiar with the tools and techniques needed to make value-driven improvements

2. To become familiar with using a structured problem-solving process

What would happen if your boss walked in and said, **"Solve this problem"?**
What would you do?

HOW DOES AN EFFECTIVE ORGANIZATION MAKE VALUE-DRIVEN IMPROVEMENTS?

In many companies, people spend their time attacking problems in a firefighting mode. As each new crisis arises, they run to put the new fire out. Effective organizations employ fire marshals, those people who are able to detect potential problems and design solutions to prevent them. Effective organizations provide the problem-solving tools, training, and incentives to change their employees from firefighters to fire marshals.

Problem solving, the isolation and analysis of a problem and the development of a permanent solution, is an integral part of process improvement in an effective organization (Figure 13.1). An effective organization uses the problem-solving tools and techniques introduced in Chapter 2 to create value through process improvement (Figure 13.2). Problem-solving tools and techniques are essential to effective process improvement because they help teams uncover the root causes of problems and help them develop solutions to eliminate the problems. Well-structured solutions attack the root cause of a problem rather than the symptoms. To solve problems effectively and therefore make value-driven improvements, people need to be trained in correct problem-solving procedures. Problem-solving efforts should be objective and focused on finding root causes. Proposed solutions should prevent a recurrence of the problem. Controls should be present to monitor the solution. Teamwork, coordinated and directed problem solving, problem-solving techniques, and statistical training are all part of ensuring that problems are isolated, analyzed, and corrected. In this chapter, a structured problem-solving process is introduced along with five of the seven quality improvement tools including Pareto charts, cause-and-effect diagrams, check sheets, histograms, and control charts. The seventh quality tool, flowcharts, was covered in Chapter 11. WHY-WHY diagrams and brainstorming are also covered. The concepts, tools, and techniques taught in this chapter are integral to the continuous improvement and Six Sigma methodologies.

Problem-solving teams, utilizing their problem-solving tools and techniques, are the foundation of process improvement. Teams who follow a structured problem-solving process can identify and fix costly, inefficient, and ineffective processes in an organization. Problem-solving efforts should be focused on problems that relate closely to customer-value–driven issues, key organizational goals, or business priorities. Problems can fall into several categories: reaction, improvement, innovation, and invention. Problems in the reaction category focus on reacting to immediate concerns, creating a short-term fix. The reaction is followed by a long-term fix, that is, correcting mistakes by removing the root causes of the problem to prevent its return. Improvement goes a step further by looking at processes that are not necessarily experiencing an immediate problem or customer issue yet have faced difficulties in the past. These problems become part of a continual improvement of methods and processes effort on which all effective organizations embark. The improvement activities concentrate on products, processes, systems, and activities making up the day-to-day activities of an organization. Some problems require breakthroughs and innovations to current methods, processes, and systems. For these innovation problems, technological advances are key. The final category, invention, assumes that the current approach or a modification to it will never

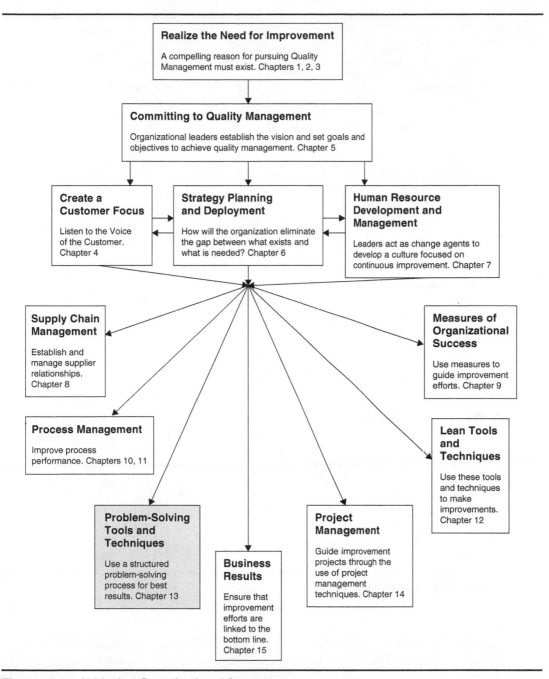

Realize the Need for Improvement

A compelling reason for pursuing Quality Management must exist. Chapters 1, 2, 3

Committing to Quality Management

Organizational leaders establish the vision and set goals and objectives to achieve quality management. Chapter 5

Create a Customer Focus

Listen to the Voice of the Customer. Chapter 4

Strategy Planning and Deployment

How will the organization eliminate the gap between what exists and what is needed? Chapter 6

Human Resource Development and Management

Leaders act as change agents to develop a culture focused on continuous improvement. Chapter 7

Supply Chain Management

Establish and manage supplier relationships. Chapter 8

Measures of Organizational Success

Use measures to guide improvement efforts. Chapter 9

Process Management

Improve process performance. Chapters 10, 11

Lean Tools and Techniques

Use these tools and techniques to make improvements. Chapter 12

Problem-Solving Tools and Techniques

Use a structured problem-solving process for best results. Chapter 13

Business Results

Ensure that improvement efforts are linked to the bottom line. Chapter 15

Project Management

Guide improvement projects through the use of project management techniques. Chapter 14

Figure 13.1   Achieving Organizational Success

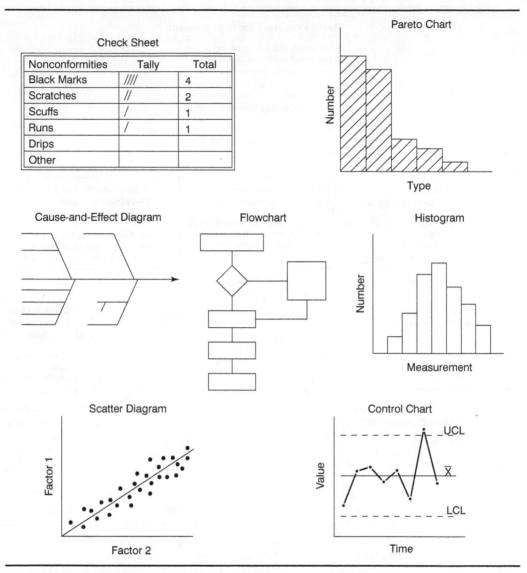

Figure 13.2 Seven Tools of Quality

satisfy the customer requirements and thus a new system, process, or product will have to be developed. The new system, process, or product will be more effective; will perform at a higher level; and will deliver greater levels of customer satisfaction. Newly invented systems, processes, and products should exceed current and future customer expectations. As teams approach various problems existing in the workplace, they should keep in mind the type of problem they are dealing with and apply appropriate tools and techniques to solving it.

EXAMPLE 13.1   Root Cause Analysis

RQM Inc., an automobile manufacturer, purchases most of its parts from suppliers and assembles the parts at its final assembly plant. Over the past six months, RQM has been experiencing intermittent smoke/odor problems when the completed vehicles are started for the first time to be driven off the assembly line. The problem does not occur with any pattern. Sometimes it reappears several times in a shift; sometimes it will not happen for days. The smoke is thick and white, and the worse the smoke, the worse the odor. The odor carries a strong scent of ammonia and is causing the operators to complain of violent sudden headaches. Several operators have requested transfers to other positions.

Since this is a serious safety concern, a variety of Band-Aid solutions were immediately proposed, such as starting all of the vehicles in a curtained area or inserting the tailpipe into a hose vented to the outside. Management recognized that these suggestions would merely cover up the problem rather than tackle its root cause. Instead, they formed a corrective-action, problem-solving team to isolate the root cause of the problem and determine a solution. Cognizant that action must be taken for the short term, the operators in the area were required to wear respirators, and exhaust fans were temporarily added to the area.

While searching for a root cause of the problem, team members spoke with operators to see whether they had any idea what the source of the white smoke might be. Several operators suggested that the team check the exhaust pipes. In some shipments excess oil has been found in the exhaust pipes. If the pipes are soaked in oil, they are rejected, but perhaps some are getting through to assembly. Several of the people questioned didn't think the oil would cause a smell as bad as what was occurring.

At the same time that the team was discussing possible sources of the smoke with the operators, they also requested that air samples be taken during vehicle start-ups. Analysis of the air samples revealed that ammine was present in large amounts when the vehicles smoked during start-up. If the vehicles did not smoke on start-up, ammine was not present. The presence of ammine accounted for the ammonia smell.

Since the engine, manifolds, and catalytic converter parts come from several suppliers, the team set out to investigate the types of materials present, either in the finished parts or during their assembly, that might release an ammonia-type smell. At first, the team suspected that adhesives containing ammine were used during assembly. However, further investigation revealed that no adhesives containing ammine were used during vehicle or part assembly, thus ruling out adhesives as the root cause or source of the ammine.

While RQM was talking with its suppliers, it came to light that the catalytic converter was coated with palladium in an ammine solution. Since this was the only source of ammine located, the team studied the process of manufacturing the converters. In the current process the converters are coated with palladium in an ammine solution and then conveyed to an oven for drying. A gap is left on the conveyor between different lots of converters. When the converters are in the oven, hot air is forced up from the bottom to dry the converters. Team members determined that the gap between lots allows an escape path for the heated air. Escaping hot air lowered the operating temperature to 250 degrees Celsius, not the required 300 degrees. The airflow and lower temperature dried the converters inconsistently, thus the reason for the intermittent nature of the problem. If the palladium-coated converters were not dried entirely, the heat created by the car engine during operation completed the drying, causing the white smoke and ammonia odor upon start-up. Once the converter dried, the smoke and smell disappeared. So this problem was never experienced by the customer—only by RQM during initial vehicle start-up.

Having determined the root cause of the problem, the team, working with the converter supplier, made the following changes. A jig is now placed between the different converter lots. This jig separates the lots but does not allow an escape path for hot air out of the oven. Other minor modifications were made to the oven to allow it to hold a temperature of 300 degrees Celsius.

Once the root cause of the problem was located and the changes were made to the process, the smoke/odor problem disappeared. Although the team was disbanded, a representative from the team was assigned the duty of monitoring the situation to ensure that the problem did not reappear.

WHAT ARE TYPICAL STEPS IN A PROBLEM-SOLVING PROCESS?

In Chapter 2, Dr. Deming's Plan-Do-Study-Act (PDSA) cycle was introduced. Since its inception, effective organizations have used it as a guide to develop their own problem-solving methodologies. Motorola divided their Six Sigma problem-solving methodology into five phases: Design, Measure, Analyze, Improve, and Control (DMAIC). Figure 13.3 shows how PDSA compares with DMAIC. Figure 13.4 shows the problem-solving methodology and associated tools presented in this chapter.

WHAT HAPPENS DURING THE "PLAN" PHASE?

Many problem solvers are so eager to do something that they are tempted to reduce the amount of time spent on planning, and they immediately propose a solution. The best solutions truly solve the problem and are found only after the root cause of the problem has been identified.

In problem solving, the PDSA cycle places a strong emphasis on determining the current conditions and planning how to approach a problem. In the Plan phase, problem investigators are looking at the processes, products, or services involved to determine how they are presently performing. This gives the team a benchmark against which to judge the success of their improvements. Planning is the most time-consuming, yet most important, part of the PDSA cycle. The following sections describe the steps of the Plan phase.

Step 1. Plan: Recognizing That a Problem Exists

Information concerning the problem(s) may come from a number of different sources, including but not limited to manufacturing, assembly, shipping, or product design departments, or from employees or customers. To emphasize the importance of solving problems, management should participate in the recognition and identification of problems.

During the problem recognition stage, the problems are outlined in very general terms. At this point in the problem-solving process, management has recognized or identified that a problem or problems exist, but as yet the specifics of the problem(s) have not been clearly defined.

Define, Measure, and Analyze	Plan
1. Select appropriate metrics Key process output variables	1. Recognize a problem exists
	2. Form a quality improvement team
2. Determine how these metrics will be tracked over time	3. Clearly define the problem
3. Determine the current baseline performance of project/process	4. Develop performance measures
4. Determine key process input variables that drive the key process output variables	5. Analyze problem/process
	6. Determine possible causes
5. Determine what changes need to be made to the key process input variables in order to positively affect the key process output variables	
Improve	**Do**
6. Make the changes	7. Select and implement the solution
Control	**Study and Act**
7. Determine whether the changes have positively affected the key process output variables	8. Evaluate the solution
	9. Ensure permanence
8. If the changes made result in performance improvements, establish control of the key process input variables at the new levels.	10. Continuous improvement

Figure 13.3   Comparing PDSA with DMAIC

Step 2. Plan: Forming Quality Improvement Teams

Once a problem situation has been recognized and before the problem is attacked, an *interdisciplinary problem-solving* or *quality improvement team* must be created. *This team will be given the task of investigating, analyzing, and finding a solution to the problem situation within a specified time frame. The problem-solving team consists of people who have knowledge of the process or problem under study.* Management gives the project teams a mandate to focus on a particular process, area, or problem. Generally, this team is composed of those closest to the problem as well as a few individuals from middle management with the power to effect change. Teams, including their behavior and motivations, were covered in greater detail in Chapter 7.

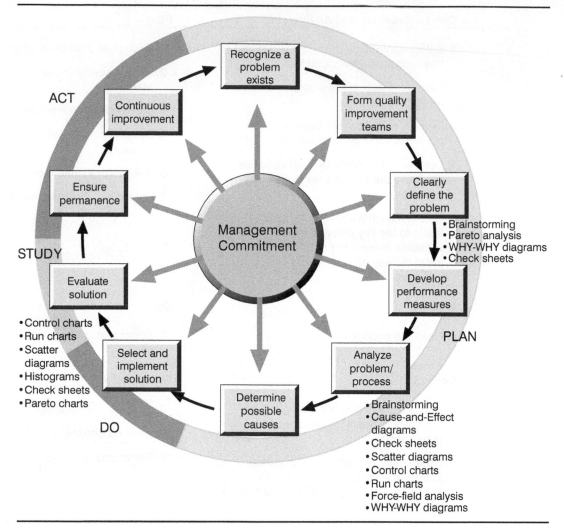

Figure 13.4   Problem-Solving Steps with Tools

EXAMPLE 13.2   Steps 1 and 2: Recognizing the Problem and Forming a Quality Improvement Team

Plastics and Dashes Inc. supplies instrument panels and other plastic components for automobile manufacturers. Recently their largest customer informed them that there have been an excessive number of customer complaints and warranty claims concerning the P&D instrument panel. The warranty claims have amounted to over $200,000, including the cost of parts and labor. In response to this problem, the management of Plastics and Dashes has initiated a corrective action request (Figure 13.5) and has formed an improvement team to investigate. The steps they will take to solve this problem are detailed in the examples throughout this chapter.

CORRECTIVE/PREVENTIVE ACTION REQUEST

TO DEPARTMENT/VENDOR: INSTRUMENT PANEL

DATE: 8/31/2004   ORIGINATOR: R. SMITH

FINDING/NONCONFORMITY: CUSTOMER WARRANTY CLAIMS FOR INSTRUMENT PANEL 360ID ARE
EXCESSIVE FOR THE TIME PERIOD 1/1/04–8/1/04

APPARENT CAUSE: CLAIMS ARE HIGH IN NUMBER AND CITE ELECTRICAL PROBLEMS AND
NOISE/LOOSE COMPONENTS

ASSIGNED TO: M. COOK   DATE RESPONSE DUE: 10/1/04
ASSIGNED TO: Q. SHEPHERD   DATE RESPONSE DUE: 10/1/04

IMMEDIATE CORRECTIVE ACTION:

ROOT CAUSE:

PREVENTIVE ACTION:

EFFECTIVE DATE:

ASSIGNEE   DATE   ASSIGNEE   DATE

QUALITY ASSURANCE   DATE   R. SMITH   8/3/2004
   ORIGINATOR   DATE

COMMENTS/AUDIT/REVIEW: SATISFACTORY   UNSATISFACTORY

NAME   DATE

Figure 13.5   P&D Corrective Action Request Form (Example 13.2)

Step 3. Plan: Defining the Problem

Once established, the quality improvement team sets out to clearly define the problem and its scope. Several techniques exist to help team members determine the true nature of their problem. The most basic of these is the check sheet.

Technique: Check Sheets

A check sheet is a data-recording device and is essentially a list of categories. As events occur in these categories, a check or mark is placed on the check sheet in the appropriate category. Given a list of items or events, the user of a check sheet marks down the number

EXAMPLE 13.3   Check Sheets

The problem-solving team at Plastics and Dashes has a great deal of instrument panel warranty information to sort through. In order to gain a better understanding of the situation, they have decided to investigate each warranty claim from the preceding six months. A check sheet has been chosen to record each type of claim and the number of each type.

To create a check sheet, they first brainstormed a list of potential warranty problems. The categories they came up with include loose instrument panel components, noisy instrument panel components, electrical problems, improper installation of the instrument panel or its components, inoperative instrument panel components, and warped instrument panel. The investigators will use the check sheet created from this list to record the types of warranty problems. As the investigators make their determination, they will make a mark in the appropriate category of the check sheet (Figure 13.6). Once all of the warranty information has been reviewed, these sheets will be collected and turned over to the team to be tallied. The information from these sheets will help the team focus their problem-solving efforts.

Loose instrument panel components	///// ///// ///// ///// ///// ///// //
Noisy instrument panel components	///// ///// ///// /
Electrical problems	///// ///
Improper installation of the instrument panel or its components	///// ///// /////
Inoperative instrument panel components	///// ///// //
Warped instrument panel	/////
Other	

Figure 13.6   Partially Completed Data Recording Check Sheet for Warranty Panel Information (Example 13.3)

of times a particular event or item occurs. A check sheet has many applications and can be custom-designed by the user to fit the particular situation. Check sheets are often used in conjunction with other quality assurance techniques. Be careful not to confuse a check sheet with a checklist. The latter lists all of the important steps or actions that need to take place, or things that need to be remembered.

Technique: Pareto Analysis

*The **Pareto chart** is a graphical tool for ranking causes of problems from the most significant to the least significant.* Named after the Italian economist Vilfredo Pareto, Pareto charts are a graphical display of the 80–20 rule. During his study of the Italian economy, Pareto found that 80% of the wealth in Italy was held by 20% of the people, thus the name "80–20 rule." In 1950 Dr. Joseph Juran applied this principle to quality control when he noticed that 80% of the dollar loss due to quality problems was found in 20% of the quality problems. Since then, the 80–20 rule, through Pareto charts, has been applied to a wide variety of situations, including scrap rates, sales, and billing errors.

Pareto charts are a helpful tool for problem analysis. Problems and their associated costs are arranged according to their relative importance in bar-chart form. Although the split is not always 80–20, the chart is a visual method identifying which problems are most significant. Pareto charts allow users to separate the vital few problems from the trivial many. The use of Pareto charts also limits the tendency of people to focus on the most recent problems rather than on the most important problems.

A Pareto chart is constructed using the following steps:

1. Select the subject for the chart, for example, a particular product line exhibiting problems, a department, or a process.
2. Determine what data need to be gathered. Determine whether numbers, percentages, or costs are going to be tracked. Determine which nonconformities or defects will be tracked.
3. Gather data related to the quality problem. Be sure that the time period during which data will be gathered is established.
4. Use a check sheet to gather data. Record the total numbers in each category. Categories will be the types of defects or nonconformities.
5. Determine the total number of nonconformities and calculate the percent of the total in each category.
6. Determine the costs associated with the nonconformities or defects.
7. Select the scales for the chart. The y-axis scale is typically the number of occurrences, number of defects, dollar loss per category, or percent. The x-axis usually displays the categories of nonconformities, defects, or items of interest.
8. Draw a Pareto chart by organizing the data from the largest category to the smallest. Include all pertinent information on the chart.
9. Analyze the chart or charts. The largest bars represent the vital few problems. If there do not appear to be one or two major problems, recheck the categories to determine whether another analysis is necessary.

EXAMPLE 13.4   Constructing a Pareto Chart

At Plastics and Dashes, the team members working on the instrument panel warranty issue first discussed in Example 13.2 have decided to begin their investigation by creating a Pareto chart.

Step 1. *Select the subject for the chart.*   The subject of the chart is instrument panel warranty claims.

Step 2. *Determine what data need to be gathered.*   The data to be used to create the chart are the different reasons customers have brought their cars in for instrument panel warranty work. Cost information on instrument panel warranty work is also available.

Step 3. *Gather the data related to the quality problem.*   The team has determined that it is appropriate to use the warranty information for the preceding six months. Copies of warranty information have been distributed to the team.

Step 4. *Make a check sheet of the gathered data and record the total numbers in each category.*   As noted in Example 13.3, based on the warranty information, the team has chosen the following categories for the x-axis of the chart: loose instrument panel components, noisy instrument panel components, electrical problems, improper installation of the instrument panel or its components, inoperative instrument panel components, and warped instrument panels (Figure 13.7).

Step 5. *Determine the total number of nonconformities and calculate the percent of the total in each category.*   From the six months of warranty information, they also have the number of occurrences for each category:

1. Loose instrument panel components	355	41.5%
2. Noisy instrument panel components	200	23.4%
3. Electrical problems	110	12.9%
4. Improper installation of the instrument panel or its components	80	9.4%
5. Inoperative instrument panel components	65	7.6%
6. Warped instrument panel	45	5.2%

Warranty claims for instrument panels total 855.

Step 6. *Determine the costs associated with the nonconformities or defects.*   The warranty claims also provided cost information associated with each category.

1. Loose instrument panel components	$115,000
2. Noisy instrument panel components	$25,000
3. Electrical problems	$55,000
4. Improper installation of the instrument panel or its components	$10,000
5. Inoperative instrument panel components	$5,000
6. Warped instrument panel	$1,000

Step 7. *Select the scales for the chart.*   The team members have decided to create two Pareto charts, one for number of occurrences and the other for costs. On each chart, the x-axis will display the warranty claim categories. The y-axis will be scaled appropriately to show all of the data.

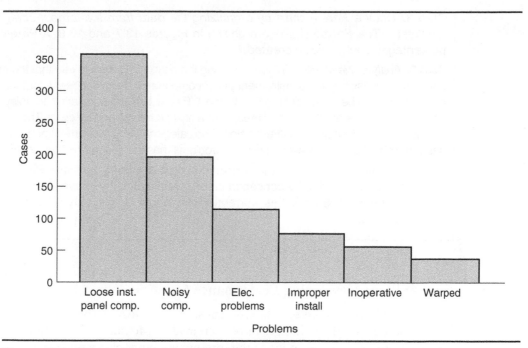

Figure 13.7   Pareto Chart of Problems Related to the Instrument Panel

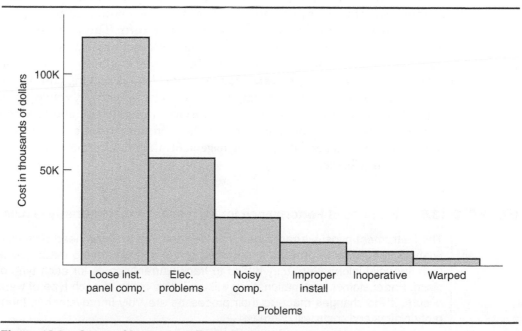

Figure 13.8   Costs of Instrument Panel Problems

Step 8. *Draw a Pareto chart by organizing the data from the largest category to the smallest.* The Pareto charts are shown in Figures 13.7 and 13.8. A Pareto chart for percentages could also be created.

Step 9. *Analyze the charts.* When analyzing the charts, it is easy to see that the most prevalent warranty claim is loose instrument panel components. It makes sense that loose components might also be noisy, and the Pareto chart (Figure 13.7) does reflect this: Noisy instrument panel components are the second most frequently occurring warranty claim. The second chart, in Figure 13.8, tells a slightly different story. The category "loose instrument panel components" has the highest costs; however, "electrical problems" has the second-highest costs.

At this point, although all the warranty claims are important, the Pareto chart has shown that efforts should be concentrated on investigating the causes of loose instrument panel components. Solving this warranty claim would significantly affect warranty numbers and costs.

Step 4. Plan: Developing Performance Measures

Measures of performance enable problem solvers to answer the question "How will we know whether the right changes have been made?" Measures may be financial in nature, customer oriented, or pertinent to the internal workings of the organization. Examples of financial measures are costs, return on investment, value added, and asset utilization. Financial measures usually focus on determining whether the changes made will enhance an organization's financial performance. Companies use customer-oriented measures to determine whether their plans and strategies keep the existing customers satisfied, bring in new customers, and encourage customers to return. These measures may include response times, delivery times, product or service functionality, price, quality, or other intangible factors. Measures pertinent to the internal workings of an organization concentrate on the business processes that are critical for achieving customer satisfaction. These measures focus on process improvement and productivity; employee and information system capabilities; and employee satisfaction, retention, and productivity. Once developed, measures should be used to develop cost/benefit scenarios to help sell improvement recommendations to management. Performance measures were covered in more detail in Chapter 9.

EXAMPLE 13.5   Measures of Performance for Instrument Panel Warranty Issues

The instrument panel warranty team has decided that they will need both customer and financial measures in order to know whether the changes being made are working. To gain financial information, they intend to track warranty costs for each type of warranty claim. For customer information, they will track how many of each type of warranty claim occurs. If the changes made to their processes are truly improvements, then both warranty claims and costs will decrease.

Step 5. Plan: Analyzing the Problem/Process

Once the problem has been defined, the problem and its processes are investigated to identify the potential constraints and to determine the sources of difficulties. Investigators are seeking a deeper understanding of the problem. Information gathered at this stage will help determine potential solutions. The analysis must be thorough to uncover any intricacies involved in or hidden by the problem. To understand an involved process, problem solvers often utilize flowcharts (Figure 13.9). Flowcharts were covered in greater detail in Chapter 11.

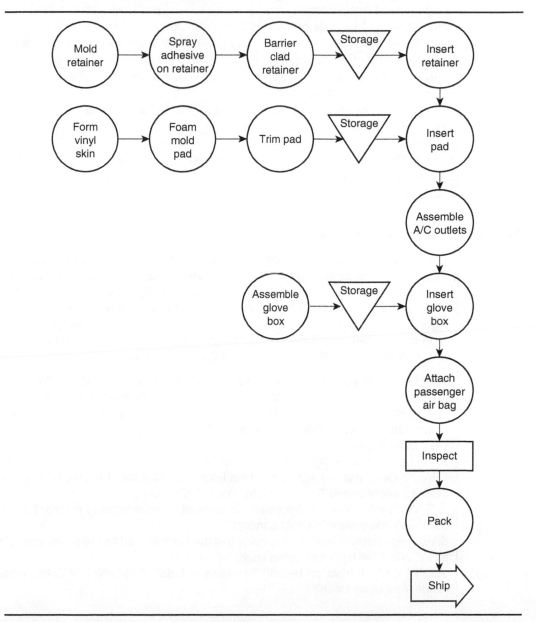

Figure 13.9   Plastics and Dashes: Glove Box Assembly Flowchart

Step 6. Plan: Determining Possible Causes

Determining the possible causes of a problem requires that the problem be clearly defined. A flowchart gives the problem solvers a greater understanding of the processes involved. Now the problem statement can be combined with knowledge of the process to isolate potential causes of the problem. An excellent technique to begin to determine causes is brainstorming.

Technique: Brainstorming

The purpose of **brainstorming** *is to generate a list of problems, opportunities, or ideas from a group of people.* Everyone present at the session should participate. The discussion leader must ensure that everyone is given an opportunity to comment and add ideas. Critical to brainstorming is that no arguing, no criticism, no negativism, and no evaluation of the ideas, problems, or opportunities take place during the session. It is a session devoted purely to the generation of ideas.

The length of time allotted to brainstorming varies; sessions may last from 10 to 45 minutes. Some team leaders deliberately keep the meetings short to limit opportunities to begin problem solving. A session ends when no more items are brought up. The result of the session will be a list of ideas, problems, or opportunities to be tackled. After being listed, the items are sorted and ranked by category, importance, priority, benefit, cost, impact, time, or other considerations.

EXAMPLE 13.6   Brainstorming

The team at Plastics and Dashes Inc. conducted a further study of the causes of loose instrument panel components. Their investigation revealed that the glove box in the instrument panel was the main problem area (Figure 13.10). They were led to this conclusion when further study of the warranty data allowed them to create the Pareto chart shown in Figure 13.11. This figure displays problems related specifically to the glove box.

In order to better understand why the glove box might be loose, the team assembled to brainstorm the variables associated with the glove box.

JERRY: I think you all know why we are here today. Did you all get the opportunity to review the glove box information? Good. Well, let's get started by concentrating on the relationship between the glove box and the instrument panel. I'll list the ideas on the board here, while you folks call them out. Remember, we are not here to evaluate ideas. We'll do that next.
SAM: How about the tightness of the latch?
FRANK: Of course the tightness of the latch will affect the fit between the glove box and the instrument panel! Tell us something we don't know.
JERRY: Frank, have you forgotten the rules of a brainstorming session? No criticizing. Sam, can you expand on your concept?
SAM: I was thinking that the positioning of the latch as well as the positioning of the hinge would affect the tightness of the latch.
JERRY: Okay. [Writes on board.] Tightness of Latch, Positioning of Latch, Positioning of Hinge. Any other ideas?

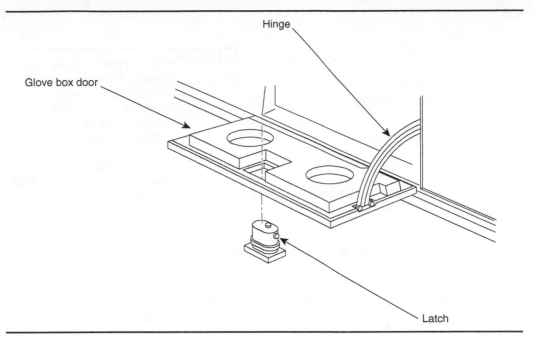

Hinge

Glove box door

Latch

Figure 13.10   Glove Box

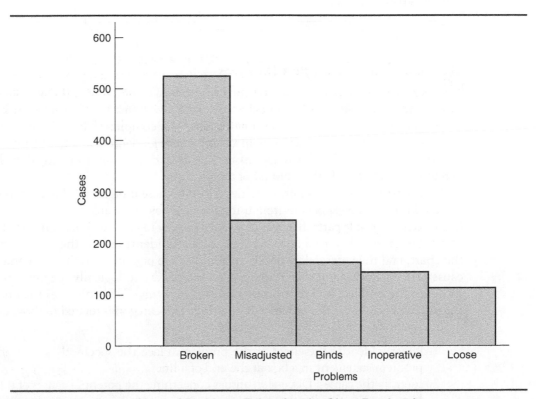

Figure 13.11   Pareto Chart of Problems Related to the Glove Box Latch

Positioning of the Glove Box
Strength of the Glove Box
Tightness of the Latch
Positioning of the Latch
Strength of the Latch
Positioning of the Hinge
Strength of the Hinge
Glove Box Handle Strength
Glove Box Handle Positioning
Glove Box Construction Materials

Figure 13.12   Variables Associated with the Glove Box

SUE: What about the strength of the hinge?
JERRY: [Writes on board.] Strength of Hinge.
SHARON: What about the glove box handle strength?
FRANK: And the glove box handle positioning?
JERRY: [Writes on board.] Glove Box Handle Strength. Glove Box Handle Positioning.

The session continues until a variety of ideas have been generated (Figure 13.12). After no more ideas surface, or at subsequent meetings, discussion and clarification of the ideas can commence.

Technique: Cause-and-Effect Diagrams

Another excellent method of determining root causes is the cause-and-effect diagram. The **cause-and-effect diagram** is also called the *Ishikawa diagram* after Kaoru Ishikawa, who developed it, and the *fish bone diagram* because the completed diagram resembles a fish skeleton. A chart of this type will help *identify causes for nonconforming or defective products or services*. Cause-and-effect diagrams can be used in conjunction with flowcharts and Pareto charts to identify the cause(s) of the problem.

This chart is useful in a brainstorming session because it organizes the ideas that are presented. Problem solvers benefit from using the chart by being able to separate a large problem into manageable parts. It serves as a visual display to aid the understanding of problems and their causes. The problem or effect is clearly identified on the right-hand side of the chart, and the potential causes of the problem are organized on the left-hand side. The cause-and-effect diagram also allows the session leader to logically organize the possible causes of the problem and to focus on one area at a time. Not only does the chart permit the display of causes of the problem; it also shows subcategories related to those causes.

To construct a cause-and-effect diagram:

1. Clearly identify the effect or the problem. Place the succinctly stated effect or problem statement in a box at the end of a line.
2. Identify the causes. Discussion ensues concerning the potential causes of the problem. To guide the discussion, attack just one possible cause area at a time. General

topic areas are usually methods, materials, machines, people, environment, and information, although other areas can be added as needed. Under each major area, subcauses related to the major cause should be identified. Brainstorming is the usual method for identifying these causes.

3. Build the diagram. Organize the causes and subcauses in diagram format.
4. Analyze the diagram. At this point, solutions will need to be identified. Decisions will also need to be made concerning the cost-effectiveness of the solution as well as its feasibility.

EXAMPLE 13.7   Constructing a Cause-and-Effect Diagram

As the Plastics and Dashes Inc. instrument panel warranty team continued their investigation, they determined that defective latches were causing most of the warranty claims associated with the categories of loose instrument panel components and noise.

Step 1. *Identify the effect or problem.*   The team identified the problem as defective latches.

Step 2. *Identify the causes.*   Rather than use the traditional methods, materials, machines, people, environment, and information, this team felt that the potential areas to search for causes were related directly to the latch. For that reason, they chose these potential causes: broken, misadjusted, binds, inoperative, loose.

Step 3. *Build the diagram.*   The team brainstormed root causes for each category (Figure 13.13).

Step 4. *Analyze the diagram.*   The team discussed and analyzed the diagram. After much discussion they came to the following conclusions. Latches that are broken, misadjusted, or inoperable or those that bind had two root causes in common: improper alignment and improper positioning. Latches that are loose or broken had a root cause of low material strength (materials supporting the latch were low in strength). From their findings the team determined that three root causes were associated with defective latches: improper alignment, improper positioning, and low material strength.

Technique: WHY-WHY Diagrams

An excellent technique for finding the root cause(s) of a problem is to ask "Why?" five times. This is also an excellent method for determining what factors must be in place in order to respond to an opportunity. **WHY-WHY** *diagrams organize the thinking of a problem-solving group* and *illustrate a chain of symptoms leading to the true cause of a problem.* By asking "Why?" five times, the problem solvers are stripping away the symptoms surrounding the problem and are getting to its true cause. At the end of a session it should be possible to make a positively worded, straightforward statement defining the true problem to be investigated.

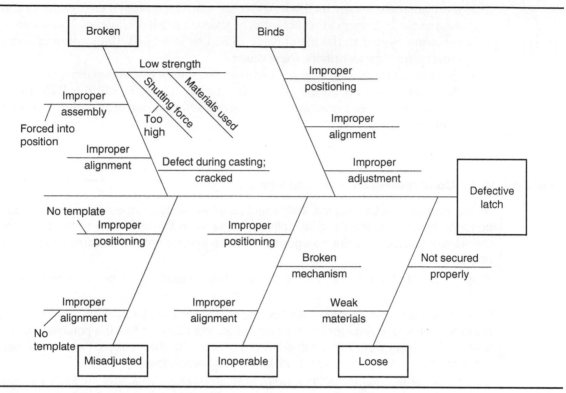

Figure 13.13   Cause-and-Effect Diagram

Developed by group consensus, the WHY-WHY diagram flows from left to right. The diagram starts on the left with a statement of the problem to be resolved. Then the group is asked why this problem might exist. The responses will be statements of causes that the group believes contribute to the problem under discussion. There may be only one cause or there may be several. Causes can be separate or interrelated. Regardless of the number of causes or their relationships, the causes should be written on the diagram in a single, clear statement. "Why?" statements should be supported by facts as much as possible and not by hearsay or unfounded opinions. Figure 13.14 shows a WHY-WHY diagram the Plastics and Dashes problem-solving team completed for instrument panel warranty costs.

This investigation is continued through as many levels as needed until a root cause is found for each of the problem statements, original or developed during the discussions. Frequently five levels of "Why?" are needed to determine the root cause. In the end, this process leads to a network of reasons the original problems occurred. The ending points indicate areas that need to be addressed to resolve the original problem. These become the actions the company must take to address the situation. WHY-WHY diagrams can be expanded to include notations concerning who will be responsible for action items and when the actions will be completed.

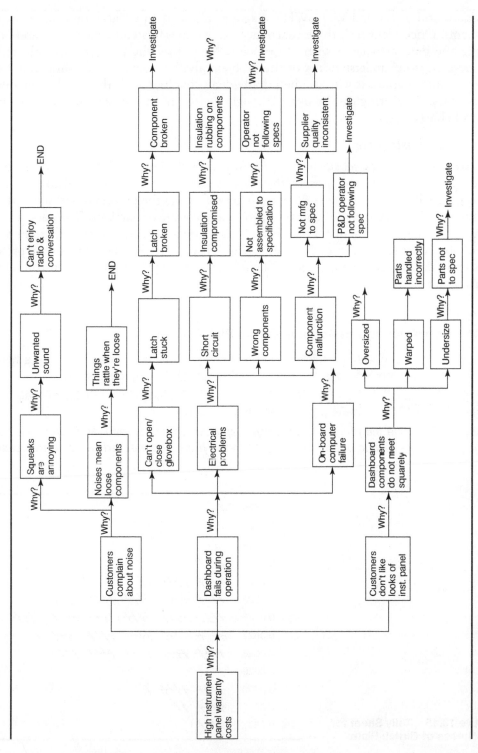

Figure 13.14 WHY-WHY Diagram for Plastics and Dashes Instrument Panel, Work-in-Progress

403

Cause-and-effect and WHY-WHY diagrams allow us to isolate potential causes of problems. Once identified, these causes need to be investigated by measuring and organizing the data associated with the process. Measuring the process will help refine the investigators' understanding of the problem and help sort out relevant from non-relevant information. It will also help individuals involved with the problem maintain objectivity. Measuring can be done through the use of histograms, scatter diagrams, control charts, and run charts.

Technique: Histograms

A **histogram** is a graphical summary of the frequency distribution of the data. When measurements are taken from a process, they can be summarized by use of a histogram. Data are organized in a histogram to allow those investigating the process to see any patterns in the data that would be difficult to see in a simple table of numbers. The data are separated into classes in the histogram. Each interval on a histogram shows the total number of observations made in each separate class. Histograms display the variation present in a set of data taken from a process. The following example details the construction of a histogram.

EXAMPLE 13.8   Constructing a Histogram: Clutch Plate Thickness

Engineers working with the thickness of a clutch plate have decided to create a histogram to aid in their analysis of the process. They are following these steps:

Step 1. *Collect the data and construct a tally sheet.*   The engineers will use the data collected (Table 13.1) to create a tally sheet (Figure 13.15).

Step 2. *Calculate the range.*   The **range**, *represented by the letter R, is calculated by subtracting the lowest observed value from the highest observed value.* In this case,

0.0620	/
0.0621	//
0.0622	//
0.0623	////
0.0624	₴₴₴ ₴₴₴ //
0.0625	₴₴₴ ₴₴₴ ₴₴₴ ₴₴₴ ₴₴₴ /
0.0626	₴₴₴ ₴₴₴ ₴₴₴ ₴₴₴ ₴₴₴ ₴₴₴
0.0627	₴₴₴ ₴₴₴ ₴₴₴ ₴₴₴ ₴₴₴ //
0.0628	₴₴₴ ₴₴₴ ₴₴₴ ₴₴₴ ///
0.0629	/
0.0630	₴₴₴ ₴₴₴ /
0.0631	₴₴₴ //
0.0632	////

Figure 13.15   Tally Sheet for Thickness of Clutch Plate

Table 13.1 Clutch Plate Grouped Data for Thickness (in Inches)

Subgroup 1	0.0625	0.0626	0.0624	0.0625	0.0627
Subgroup 2	0.0624	0.0623	0.0624	0.0626	0.0625
Subgroup 3	0.0622	0.0625	0.0623	0.0625	0.0626
Subgroup 4	0.0624	0.0623	0.0620	0.0623	0.0624
Subgroup 5	0.0621	0.0621	0.0622	0.0625	0.0624
Subgroup 6	0.0628	0.0626	0.0625	0.0626	0.0627
Subgroup 7	0.0624	0.0627	0.0625	0.0624	0.0626
Subgroup 8	0.0624	0.0625	0.0625	0.0626	0.0626
Subgroup 9	0.0627	0.0628	0.0626	0.0625	0.0627
Subgroup 10	0.0625	0.0626	0.0628	0.0626	0.0627
Subgroup 11	0.0625	0.0624	0.0626	0.0626	0.0626
Subgroup 12	0.0630	0.0628	0.0627	0.0625	0.0627
Subgroup 13	0.0627	0.0626	0.0628	0.0627	0.0626
Subgroup 14	0.0626	0.0626	0.0625	0.0626	0.0627
Subgroup 15	0.0628	0.0627	0.0626	0.0625	0.0626
Subgroup 16	0.0625	0.0626	0.0625	0.0628	0.0627
Subgroup 17	0.0624	0.0626	0.0624	0.0625	0.0627
Subgroup 18	0.0628	0.0627	0.0628	0.0626	0.0630
Subgroup 19	0.0627	0.0626	0.0628	0.0625	0.0627
Subgroup 20	0.0626	0.0625	0.0626	0.0625	0.0627
Subgroup 21	0.0627	0.0626	0.0628	0.0625	0.0627
Subgroup 22	0.0625	0.0626	0.0628	0.0625	0.0627
Subgroup 23	0.0628	0.0626	0.0627	0.0630	0.0627
Subgroup 24	0.0625	0.0631	0.0630	0.0628	0.0627
Subgroup 25	0.0627	0.0630	0.0631	0.0628	0.0627
Subgroup 26	0.0630	0.0628	0.0629	0.0628	0.0627
Subgroup 27	0.0630	0.0628	0.0631	0.0628	0.0627
Subgroup 28	0.0632	0.0632	0.0628	0.0631	0.0630
Subgroup 29	0.0630	0.0628	0.0631	0.0632	0.0631
Subgroup 30	0.0632	0.0631	0.0630	0.0628	0.0628

0.0620 is the lowest value and 0.0632 is the highest:

$$\text{Range} = R = X_h - X_l$$

where

R = range
X_h = highest number
X_l = lowest number
$R = 0.0632 - 0.0620 = 0.0012$

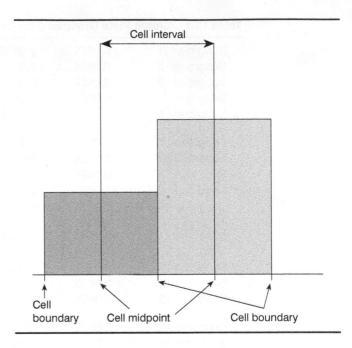

Figure 13.16   Histogram Cell Description

Step 3. *Create the cells.* In a histogram, data are combined into cells. Cells are composed of three components: cell intervals, cell midpoints, and cell boundaries (Figure 13.16). **Cell midpoints** *identify the centers of cells. A* **cell interval** *is the distance between the cell midpoints. The* **cell boundary** *defines the limits of the cell.*

Cell Intervals

Odd-numbered cell intervals are often chosen for ease of calculation. For example, if the data were measured to one decimal place, then the cell intervals could be 0.3, 0.5, 0.7, or 0.9. If the gathered data were measured to three decimal places, then the cell intervals to choose from would be 0.003, 0.005, 0.007, 0.009. (The values of 1, 0.1, 0.001 are not chosen for a histogram because they result in the creation of a frequency diagram.) For this example, because the data were measured to four decimal places, the cell interval could be 0.0003, 0.0005, 0.0007, or 0.0009.

Cell interval choice plays a large role in determining the size of the histogram created. To determine the number of cells, the following formula is used:

$$h = \frac{R}{i} + 1$$

where

h = number of cells
i = cell interval
R = range

Since both i, the cell interval, and h, the number of cells, are unknown, creators of histograms must choose values for one of them and then solve for the other. For our example, if we choose a cell interval of 0.003, then

$$h = \frac{0.0012}{0.0003} + 1$$
$$h = 5$$

and the histogram created will contain 5 cells.

For a cell interval value of 0.0005,

$$h = \frac{0.0012}{0.0005} + 1$$
$$h = 3$$

For a cell interval value of 0.0007,

$$h = \frac{0.0012}{0.0007} + 1$$
$$h = 3$$

As the cell interval gets larger, the number of cells necessary to hold all the data and make a histogram decreases. When deciding the number of cells to use, it is sometimes helpful to follow this rule of thumb:

> For fewer than 100 pieces of data, use 4 to 9 cells.
> For 100 to 500 pieces, use 8 to 17 cells.
> For 500 or more pieces, use 15 to 20 cells.

Another helpful rule of thumb exists for determining the number of cells in a histogram. Use the square root of n ($\sqrt{n}$), where n is the number of data points, as an approximation of the number of cells needed.

For this example we will use a cell interval of 0.0003. This will create a histogram that provides enough spread to analyze the data.

Cell Midpoints

When constructing a histogram, it is important to remember two things: (1) Histograms must contain all of the data; (2) one particular value cannot fit into two different cells. Cell midpoints are selected to ensure that these problems are avoided. To determine the midpoint values that anchor the histogram, use either one of the following two techniques:

1. The simplest technique is to choose the lowest value measured. In this example the lowest measured value is 0.0620. We determine other midpoint values by adding the cell interval of 0.0003 to 0.0620 first and then adding it to each successive new midpoint. If we begin at 0.0620, we find the other midpoints at 0.0623, 0.0626, 0.0629, and 0.0632.
2. If the number of values in the cell is high and the distance between the cell boundaries is not large, the midpoint is the most representative value in the cell.

Cell Boundaries

The cell size, set by the boundaries of the cell, is determined by the cell midpoints and the cell interval. Locating the cell boundaries, or the limits of the cell, allows the user to place values in a particular cell. To determine the lower cell boundary, divide the cell interval by 2 and subtract that value from the cell midpoint. To calculate the lower cell boundary for a cell with a midpoint of 0.0620, divide the cell interval by 2:

$$0.0003 \div 2 = 0.00015$$

Then subtract 0.00015 from the cell midpoint:

$$0.0620 - 0.00015 = 0.06185, \text{ the first lower boundary}$$

To determine the upper cell boundary for a midpoint of 0.0620, add the cell interval to the lower cell boundary:

$$0.06185 + 0.0003 = 0.06215$$

The lower cell boundary of one cell is the upper cell boundary of another. Continue adding the cell interval to each new lower cell boundary calculated until all of the lower cell boundaries have been determined.

Note that the cell boundaries are a half decimal value greater in accuracy than the measured values. This helps ensure that values can be placed in only one cell of a histogram. In our example the first cell will have boundaries of 0.06185 and 0.06215. The second cell will have boundaries of 0.06215 and 0.06245. Where would a data value of 0.0621 be placed? Obviously in the first cell. Cell intervals with midpoint values starting at 0.0620 are shown in Figure 13.17.

Step 4. *Label the axes.* Scale and label the horizontal axis according to the cell midpoints determined in step 3. Label the vertical axis to reflect the amount of data collected, in counting numbers.

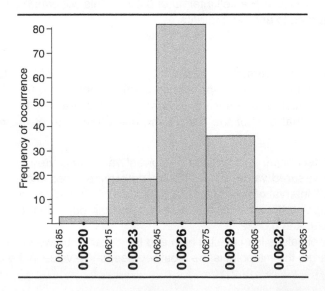

Figure 13.17   Clutch Plate Thickness Histogram

Step 5. *Post the values.* The final step in the creation of a histogram is to post the values from the check sheet to the histogram. The x-axis is marked with the cell midpoints and, if space permits, the cell boundaries. The histogram creator uses the cell boundaries as a guide when posting the values to the histogram. On the y-axis, the frequency of the values within a particular cell is shown. All of the data must be included in the cells (Figure 13.17).

Step 6. *Interpret the histogram.* As we can see in Figure 13.17, the data are grouped around 0.0626 and are somewhat symmetrical. In the following sections we will study histogram shapes, sizes, and locations compared to a desired target specification. We will also utilize measures such as means, modes, and medians to create a clear picture of where the data are grouped (the central tendency of the data). Standard deviations and ranges will be used to measure how the data are dispersed around the mean. These statistical values will be used to fully describe the data comprising a histogram.

Analysis of Histograms

When analyzing a distribution, it is important to remember that it has the following characteristics: shape, location, and spread (Figure 13.18). These three characteristics combine to give us the ability to describe a distribution.

Shape: Symmetry, Skewness, Kurtosis *Shape refers to the form that the values of the measurable characteristics take on when plotted or graphed.* Identifiable characteristics include **symmetry** or, in the case of lack of symmetry, **skewness** of the data; **kurtosis**, or *peakedness of the data;* and **modes**, *the number of peaks in the data.*

When a distribution is **symmetrical**, *the halves are mirror images of each other.* The halves correspond in size, shape, and arrangement. When a distribution is not symmetrical, it is considered to be skewed (Figure 13.19). *With a* **skewed distribution**, *the majority of the data are grouped either to the left or to the right of a center value, and on the opposite side a few values trail away from the center. When a distribution is* **skewed to the right**, *the majority of the data are found on the left side of the figure, with the tail of the distribution going to the right.* The opposite is true for a distribution that is **skewed to the left**.

Kurtosis describes the peakedness of the distribution. *A distribution with a high peak is referred to as* **leptokurtic;** *a flatter curve is called* **platykurtic** (Figure 13.20). Typically, the

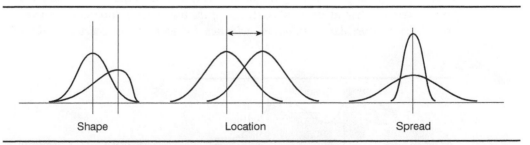

Shape   Location   Spread

Figure 13.18   Shape, Location, and Spread

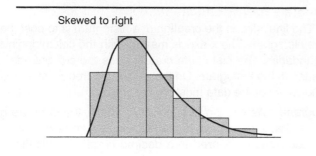

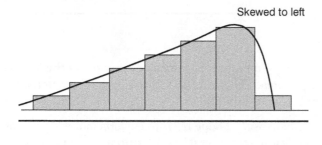

Figure 13.19   Skewness

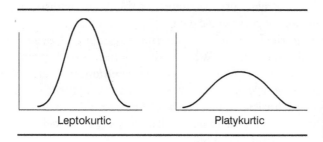

Figure 13.20   Leptokurtic and Platykurtic

kurtosis of a histogram is discussed by comparing it with another distribution. As we will see later in the chapter, skewness and kurtosis can be calculated numerically. Occasionally distributions will display unusual patterns. *If the distribution displays more than one peak, it is considered **multimodal**. Distributions with two distinct peaks are called **bimodal*** (Figure 13.21).

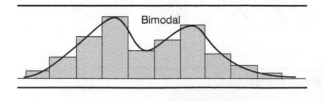

Figure 13.21   Bimodal Distribution

EXAMPLE 13.9   Analyzing the Histogram for the Clutch Plate Data

An analysis of Figure 13.17 based on the three characteristics of shape, location, and spread reveals that the clutch plate thickness data are fairly consistent. The **shape** of the distribution is relatively symmetrical, although skewed slightly to the right. The data are unimodal, centering on 0.0626 inch. Since we have no other distributions of the same type of product, we cannot make any comparisons or comments on the kurtosis of the data. **Location,** or where the data are located or gathered, is around 0.0626. If the engineers have specifications of 0.0625 ± 0.0003, then the center of the distribution is higher than the desired value. Given the specifications, the **spread** of the data is broader than the desired 0.0622 to 0.0628 at 0.0620 to 0.0632.

Technique: Control Charts

A **control chart** is a chart with a centerline showing the average of the data produced. It has upper and lower control limits that are based on statistical calculations (Figure 13.22). It is used to determine process centering and process variation and to locate any unusual patterns or trends in the data.

Analysis of process performance often begins with the construction of a histogram and the calculation of ranges, averages, and standard deviations. The only shortcoming of this type of analysis is its failure to show process performance over time. Let's take a closer look at the clutch plate thickness data in Table 13.2 and see why knowing process performance over time may be important.

When the averages of Table 13.2 are graphed in a histogram, the result closely resembles a normal curve, as shown in Figure 13.23. Graphing the averages by subgroup number, according to when they were produced, gives a different impression of the data, as shown in Figure 13.23b. From the chart, it appears that the thickness of the clutch plate is increasing as production continues. This was not evident during the creation of the histogram or during the analysis of the average, range, and standard deviation.

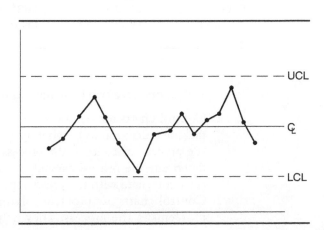

Figure 13.22   A Control Chart, Showing Centerline, Upper Control Limits, and Lower Control Limits

Table 13.2   Clutch Plate Thickness: Sums and Averages

						ΣX_i	$\overline{X}$	$\overline{R}$
Subgroup 1	0.0625	0.0626	0.0624	0.0625	0.0627	0.3127	0.0625	0.0003
Subgroup 2	0.0624	0.0623	0.0624	0.0626	0.0625	0.3122	0.0624	0.0003
Subgroup 3	0.0622	0.0625	0.0623	0.0625	0.0626	0.3121	0.0624	0.0004
Subgroup 4	0.0624	0.0623	0.0620	0.0623	0.0624	0.3114	0.0623	0.0004
Subgroup 5	0.0621	0.0621	0.0622	0.0625	0.0624	0.3113	0.0623	0.0004
Subgroup 6	0.0628	0.0626	0.0625	0.0626	0.0627	0.3132	0.0626	0.0003
Subgroup 7	0.0624	0.0627	0.0625	0.0624	0.0626	0.3126	0.0625	0.0003
Subgroup 8	0.0624	0.0625	0.0625	0.0626	0.0626	0.3126	0.0625	0.0002
Subgroup 9	0.0627	0.0628	0.0626	0.0625	0.0627	0.3133	0.0627	0.0003
Subgroup 10	0.0625	0.0626	0.0628	0.0626	0.0627	0.3132	0.0626	0.0003
Subgroup 11	0.0625	0.0624	0.0626	0.0626	0.0626	0.3127	0.0625	0.0002
Subgroup 12	0.0630	0.0628	0.0627	0.0625	0.0627	0.3134	0.0627	0.0005
Subgroup 13	0.0627	0.0626	0.0628	0.0627	0.0626	0.3137	0.0627	0.0002
Subgroup 14	0.0626	0.0626	0.0625	0.0626	0.0627	0.3130	0.0626	0.0002
Subgroup 15	0.0628	0.0627	0.0626	0.0625	0.0626	0.3132	0.0626	0.0003
Subgroup 16	0.0625	0.0626	0.0625	0.0628	0.0627	0.3131	0.0626	0.0003
Subgroup 17	0.0624	0.0626	0.0624	0.0625	0.0627	0.3126	0.0625	0.0003
Subgroup 18	0.0628	0.0627	0.0628	0.0626	0.0630	0.3139	0.0627	0.0004
Subgroup 19	0.0627	0.0626	0.0628	0.0625	0.0627	0.3133	0.0627	0.0003
Subgroup 20	0.0626	0.0625	0.0626	0.0625	0.0627	0.3129	0.0626	0.0002
Subgroup 21	0.0627	0.0626	0.0628	0.0625	0.0627	0.3133	0.0627	0.0003
Subgroup 22	0.0625	0.0626	0.0628	0.0625	0.0627	0.3131	0.0626	0.0003
Subgroup 23	0.0628	0.0626	0.0627	0.0630	0.0627	0.3138	0.0628	0.0004
Subgroup 24	0.0625	0.0631	0.0630	0.0628	0.0627	0.3141	0.0628	0.0006
Subgroup 25	0.0627	0.0630	0.0631	0.0628	0.0627	0.3143	0.0629	0.0004
Subgroup 26	0.0630	0.0628	0.0629	0.0628	0.0627	0.3142	0.0628	0.0003
Subgroup 27	0.0630	0.0628	0.0631	0.0628	0.0627	0.3144	0.0629	0.0004
Subgroup 28	0.0632	0.0632	0.0628	0.0631	0.0630	0.3153	0.0631	0.0004
Subgroup 29	0.0630	0.0628	0.0631	0.0632	0.0631	0.3152	0.0630	0.0004
Subgroup 30	0.0632	0.0631	0.0630	0.0628	0.0628	0.3149	0.0630	0.0004
						9.3990		

Control charts serve two basic functions:

1. Control charts are decision-making tools. They provide an economic basis for making a decision as to whether to investigate for potential problems, to adjust the process, or to leave the process alone. Control chart information is used to determine the process capability. Samples of completed product can be statistically compared with the process specifications.

2. Control charts are problem-solving tools. They assist in the identification of problems in the process. They help to provide a basis on which to formulate

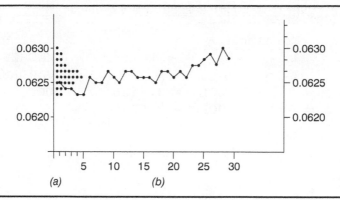

Figure 13.23   Chart with Histogram

improvement actions. Control chart information can be used to help locate and investigate the causes of the unacceptable or marginal quality. Control charts enhance the analysis of a process by showing how that process is performing over time. By combining control charts with an appropriate statistical summary, those studying a process can gain an understanding of what the process is capable of producing.

To create a control chart, samples, arranged into subgroups, are taken during the process. The averages of the subgroup values are plotted on the control chart. *The **centerline** ($\overline{\overline{X}}$) of the chart shows where the process average is centered, or the central tendency of the data. The **upper control limit (UCL)** and the **lower control limit (LCL),** calculated based on ± 3 sigma, describe the spread of the process.* Once the chart has been constructed, it presents the user with a picture of what the process is currently capable of producing. In other words, we can expect future production to fall between these ± 3 sigma limits 99.73 percent of the time, provided that the process does not change and is under control.

Two categories of control charts exist: variables charts and attribute charts. Control charts for variables are covered in this chapter, and control charts for attributes are covered in Appendix 5.

$\overline{X}$ and R Charts

Variables are the measurable characteristics of a product or service, for example, the height, weight, or length of a part. One of the most commonly used variables chart combinations in statistical process control consists of the $\overline{X}$ (also called X bar) and R charts. Typical $\overline{X}$ and R charts are shown in Figure 13.24. $\overline{X}$ and R charts are used together to determine the distribution of the subgroup averages of sample measurements taken from a process. The importance of using these two charts in conjunction with each other will become apparent shortly.

The $\overline{X}$ chart is used to monitor the variation of the subgroup averages that are calculated from the individual sampled data. Averages, rather than individual observations, are used on control charts because average values will indicate a change in the amount of variation much

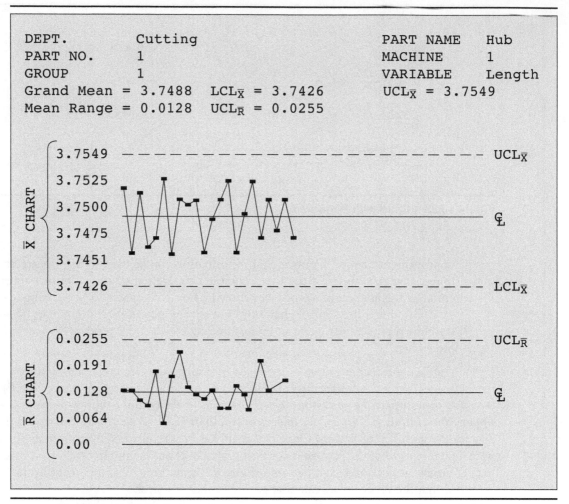

Figure 13.24   Typical X̄ and R Chart

faster than will individual values. Control limits on this chart are used to evaluate the variation from one subgroup to another.

The following steps and examples explain the construction of an $\overline{X}$ chart.

1. ***Define the problem and select characteristic to measure.***   In any situation it is necessary to determine what the goal of monitoring a particular quality characteristic or group of characteristics is. To merely say, "Improve quality," is not enough. Nor is it sufficient to say, "We would like to see fewer parts out of specification." Out of which specification? Is it total product performance that is being affected or just one particular dimension? Sometimes several aspects of a part are critical for part performance; occasionally only one is. It is more appropriate to say, "The length of these parts appears to be consistently below the lower specification limit. This causes the parts to mate incorrectly.

Why are these parts below specification, and how far below are they?" In the second statement we have isolated the length of the part as a critical dimension. From here, we can place control charts on the process to help determine where the true source of the problem is located.

EXAMPLE 13.10 Defining the Problem

An assembly area has been experiencing serious delays in the construction of computer printers. As quality assurance manager, you have been asked to determine the cause of these delays and fix the problems as soon as possible. To best utilize the limited time available, you convene a meeting involving those closest to the assembly problems. Representatives from production, supervision, manufacturing engineering, industrial engineering, quality assurance, and maintenance have been able to generate a variety of possible problems. During this meeting a cause-and-effect diagram was created, showing the potential causes for the assembly difficulties (Figure 13.25). Discussions during the meeting revealed that the shaft that holds the roller in place could be the major cause of assembly problems.

As the troubleshooting meeting continues, further investigation reveals that the length of the shaft is hindering assembly operations. The characteristic to measure has been identified as piece-to-piece variation in the length of the shafts. To begin to study the situation, measurements of the lengths of the shafts will be sampled. These measurements will be compared with the specifications set for the assembly operations.

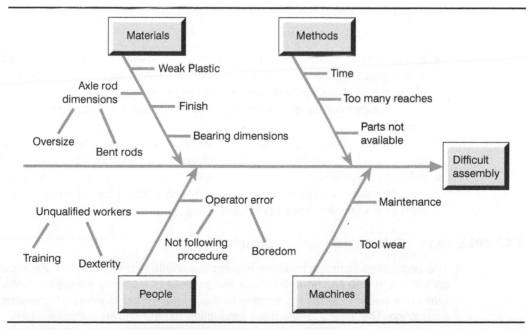

Figure 13.25   Cause-and-Effect Diagram (Example 13.10)

2. *Choose a rational subgroup size to be sampled.*   Subgroups, and the samples composing them, must be homogeneous. A **homogeneous subgroup** *will have been produced under the same conditions, by the same machine, the same operator, the same mold, and so on.* Homogeneous lots can also be designated by equal time intervals. Samples should be taken in an unbiased, random fashion. They should be representative of the entire population. The letter n is used to designate the number of samples taken within a subgroup. When constructing $\overline{X}$ and R charts, keep the subgroup sample size constant for each subgroup taken.

Decisions concerning the specific size of the subgroup (n, or the number of samples) require judgment. Sampling should occur frequently enough to detect changes in the process. Ask, "How often is the system expected to change?" Examine the process and identify the factors causing change in the process. To be effective, sampling must occur as often as the system's most frequently changing factor. Once the number and the frequency of sampling have been selected, they should not be changed unless the system itself has changed.

When gathering sample data, it is important to have the following information in order to properly analyze the data:

1. *Who* will be collecting the data?
2. *What* aspect of the process is to be measured?
3. *Where* or at what point in the process will the sample be taken?
4. *When* or how frequently will the process be sampled?
5. *Why* is this particular sample being taken?
6. *How* will the data be collected?
7. *How many* samples will be taken (subgroup size)?

Some other guidelines to be followed include:

- The larger the subgroup size, the more sensitive the chart becomes to small variations in the process average. This will provide a better picture of the process because it allows the investigator to detect changes in the process quickly.
- A larger subgroup size makes for a more sensitive chart, but it also increases inspection costs.
- Destructive testing may make large subgroup sizes unfeasible. For example, it would not make sense for a fireworks manufacturer to test each and every one of its products.
- Subgroup sizes smaller than four do not create a representative distribution of subgroup averages. Subgroup averages are nearly normal for subgroups of four or more even when sampled from a nonnormal population.

EXAMPLE 13.11   Selecting Subgroup Sample Size

The production from the machine making the shafts first examined in Example 13.10 is consistent at 150 per hour. Because the process is currently exhibiting problems, your team has decided to take a sample of five measurements every 10 minutes from the production. The values for the day's production run are shown in Figure 13.26.

3. **Collect the data.** To create a control chart, an amount of data sufficient to accurately reflect the statistical control of the process must be gathered. A minimum of 20 subgroups of sample size n = 4 is suggested. Each time a subgroup of sample size n is taken, an average is calculated for the subgroup. To do this, the individual values are recorded, summed, and then divided by the number of samples in the subgroup. This average, $\overline{X}_i$, is then plotted on the control chart.

DEPT.	Roller		PART NAME	Shaft	
PART NO.	1		MACHINE	1	
GROUP	1		VARIABLE	length	

Subgroup	1	2	3	4	5
Time	07:30	07:40	07:50	08:00	08:10
Date	07/02/95	07/02/95	07/02/95	07/02/95	07/02/95
1	11.95	12.03	12.01	11.97	12.00
2	12.00	12.02	12.00	11.98	12.01
3	12.03 ①	11.96	11.97	12.00	12.02
4	11.98	12.00	11.98	12.03	12.03
5	12.01	11.98	12.00	11.99	12.02
$\overline{X}$	11.99 ②	12.00	11.99	11.99	12.02
Range	0.08 ③	0.07	0.04	0.06	0.03

Subgroup	6	7	8	9	10
Time	08:20	08:30	08:40	08:50	09:00
Date	07/02/95	07/02/95	07/02/95	07/02/95	07/02/95
1	11.98	12.00	12.00	12.00	12.02
2	11.98	12.01	12.01	12.02	12.00
3	12.00	12.03	12.04	11.96	11.97
4	12.01	12.00	12.00	12.00	12.05
5	11.99	11.98	12.02	11.98	12.00
$\overline{X}$	11.99	12.00	12.01	11.99	12.01
Range	0.03	0.05	0.04	0.06	0.08

Subgroup	11	12	13	14	15
Time	09:10	09:20	09:30	09:40	09:50
Date	07/02/95	07/02/95	07/02/95	07/02/95	07/02/95
1	11.98	11.92	11.93	11.99	12.00
2	11.97	11.95	11.95	11.93	11.98
3	11.96	11.92	11.98	11.94	11.99
4	11.95	11.94	11.94	11.95	11.95
5	12.00	11.96	11.96	11.96	11.93
$\overline{X}$	11.97	11.94	11.95	11.95	11.97
Range	0.05	0.04	0.05	0.06	0.07

Figure 13.26   Values for a Day's Production

Subgroup	16	17	18	19	20
Time	10:00	10:10	10:20	10:30	10:40
Date	07/02/95	07/02/95	07/02/95	07/02/95	07/02/95
1	12.00	12.02	12.00	11.97	11.99
2	11.98	11.98	12.01	12.03	12.01
3	11.99	11.97	12.02	12.00	12.02
4	11.96	11.98	12.01	12.01	12.00
5	11.97	11.99	11.99	11.99	12.01
$\overline{X}$	11.98	11.99	12.01	12.00	12.01
Range	0.04	0.05	0.03	0.06	0.03

$R = 0.05 = 12.02 - 11.97$

Subgroup	21
Time	10:50
Date	07/02/95

$\dfrac{12.00 + 11.98 + 11.99 + 11.96 + 11.97}{5} = 11.98$

1	12.00
2	11.98
3	11.99
4	11.99
5	12.02
$\overline{X}$	12.00
Range	0.04

Figure 13.26   Values for a Day's Production (*continued*)

EXAMPLE 13.12   Collecting Data

A sample of size n = 5 is taken at 10-minute intervals from the process making shafts. As shown in Figure 13.26, a total of 21 subgroups of sample size n = 5 have been taken. Each time a subgroup sample is taken, the individual values are recorded [Figure 13.26, (1)]; summed; and then divided by the number of samples taken to obtain the average for the subgroup [Figure 13.26, (2)]. This subgroup average is then plotted on the control chart [Figure 13.27, (1)].

---

4. *Determine the trial centerline for the $\overline{X}$ chart.*   The centerline of the control chart is the process average. It would be the mean, μ, if the average of the population measurements for the entire process were known. Since the value of the population mean μ cannot be determined unless all of the parts being produced are measured, in its place the grand average of the subgroup averages, $\overline{\overline{X}}$ (X double bar), is used. The grand average, or $\overline{\overline{X}}$, is calculated by summing all the subgroup averages and then dividing by the number of subgroups. This value is plotted as the centerline of the $\overline{X}$ chart:

$$\overline{\overline{X}} = \frac{\sum\limits_{i=1}^{m} \overline{X}_i}{m}$$

where

$$\overline{\overline{X}} = \text{average of the subgroup averages}$$

$$\overline{X}_i = \text{average of the ith subgroup}$$

$$m = \text{number of subgroups}$$

5. *Determine the trial control limits for the $\overline{X}$ chart.* Under a normal curve 99.73 percent of the data falls within ± 3 sigma. Figure 13.24 showed how a control chart is a time-dependent pictorial representation of a normal curve displaying the distribution of the averages of the samples taken from the process. Therefore, control limits are established at ± 3 standard deviations from the centerline for the process using the following formulas:

$$UCL_{\overline{X}} = \overline{\overline{X}} + 3\sigma_{\overline{x}}$$

$$LCL_{\overline{X}} = \overline{\overline{X}} - 3\sigma_{\overline{x}}$$

where

$$UCL_{\overline{X}} = \text{upper control limit of the } \overline{X} \text{ chart}$$

$$LCL_{\overline{X}} = \text{lower control limit of the } \overline{X} \text{ chart}$$

$$\sigma_{\overline{x}} = \text{population standard deviation of the subgroup averages}$$

The population standard deviation σ is needed to calculate the upper and lower control limits. Because control charts are based on sample data, Dr. Shewhart developed a good approximation of $3\sigma_{\overline{x}}$ using the product of an A_2 factor multiplied by $\overline{R}$, the average of the ranges. The $A_2\overline{R}$ combination uses the sample data for its calculation. $\overline{R}$ is calculated by summing the values of the individual subgroup ranges and dividing by the number of subgroups m:

$$\overline{R} = \frac{\sum\limits_{i=1}^{m} R_i}{m}$$

where

$$\overline{R} = \text{average of the ranges}$$

$$R_i = \text{individual range values for the sample}$$

$$m = \text{number of subgroups}$$

A_2, the factor that allows the approximation $A_2\overline{R} \approx 3\sigma_{\overline{x}}$ to be true, is selected based on the subgroup sample size n. See Appendix 2 for the A_2 factors.

Upon replacement, the formulas for the upper and lower control limits become

$$UCL_{\overline{X}} = \overline{\overline{X}} + A_2\overline{R}$$

$$LCL_{\overline{X}} = \overline{\overline{X}} - A_2\overline{R}$$

After calculating the control limits, we place the centerline ($\overline{\overline{X}}$) and the upper and lower control limits (UCL and LCL, respectively) on the chart. The upper and lower control limits are shown by dashed lines. The grand average, or $\overline{\overline{X}}$, is shown by a solid line. The control limits on the $\overline{X}$ chart will be symmetrical about the central line. These control limits are used to evaluate the variation in quality from subgroup to subgroup.

EXAMPLE 13.13   Calculating the X̄ Chart Centerline and Control Limits

Construction of an $\overline{X}$ chart begins with the calculation of the centerline, $\overline{\overline{X}}$. Using the 21 subgroups of sample size n = 5 provided in Figure 13.26, we calculate $\overline{\overline{X}}$ by summing all of the subgroup averages based on the individual samples taken and then dividing by the number of subgroups, m:

$$\overline{\overline{X}} = \frac{11.99 + 12.00 + 11.99 + \cdots + 12.00}{21}$$

$$= \frac{251.77}{21} = 11.99$$

This value is plotted as the centerline of the $\overline{X}$ chart [Figure 13.27, (2)].

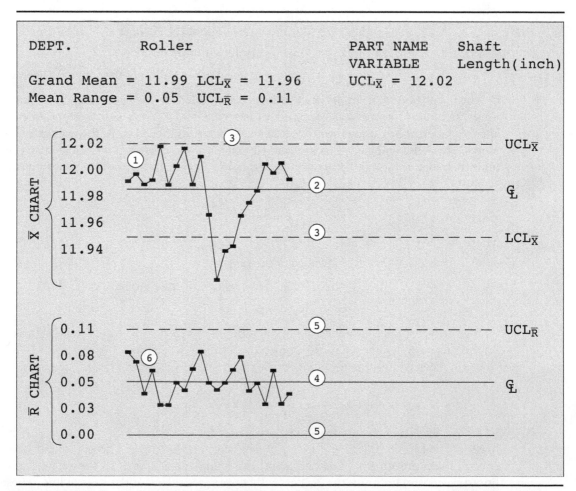

Figure 13.27   X̄ and R Control Charts for Roller Shaft Length (Example 13.13)

$\overline{R}$ is calculated by summing the values of the individual subgroup ranges and dividing by the number of subgroups, m:

$$\overline{R} = \frac{0.08 + 0.07 + 0.04 + \cdots + 0.04}{21}$$

$$= \frac{1.06}{21} = 0.05$$

The A_2 factor for a sample size of five is selected from the table in Appendix 2. The values for the upper and lower control limits of the $\overline{X}$ chart are calculated as follows:

$$UCL_{\overline{X}} = \overline{\overline{X}} + A_2\overline{R}$$
$$= 11.99 + 0.577(0.05) = 12.02$$
$$LCL_{\overline{X}} = \overline{\overline{X}} - A_2\overline{R}$$
$$= 11.99 - 0.577(0.05) = 11.96$$

Once calculated, the upper and lower control limits (UCL and LCL, respectively) are placed on the chart [Figure 13.27, (3)].

6. *Determine the trial control limits for the R chart.*   When an X chart is used to evaluate the variation in quality from subgroup to subgroup, the range chart is a method of determining the amount of variation in the individual samples. The importance of the range chart is often overlooked. Without the range chart, or the standard deviation chart to be discussed later, it would not be possible to fully understand process capability. Where the $\overline{X}$ chart shows the average of the individual subgroups, giving the viewer an understanding of where the process is centered, the range chart shows the spread or dispersion of the individual samples within the subgroup. The calculation of the spread of the measurements is necessary to determine whether the parts being produced are similar to one another. If the product displays a wide spread or a large range, then the individuals being produced are not similar to each other.

Individual ranges are calculated for each of the subgroups by subtracting the highest value in the subgroup from the lowest value. These individual ranges are then summed and divided by the total number of subgroups to calculate $\overline{R}$, the centerline of the R chart. The range chart upper and lower control limits are calculated in a manner similar to that used for the $\overline{X}$ chart limits.

$$\overline{R} = \frac{\sum\limits_{i=1}^{m} R_i}{m}$$
$$UCL_R = \overline{R} + 3\sigma_R$$
$$LCL_R = \overline{R} - 3\sigma_R$$

where

$$UCL_R = \text{upper control limit of the R chart}$$
$$LCL_R = \text{lower control limit of the R chart}$$
$$\sigma_R = \text{population standard deviation of the subgroup ranges}$$

To estimate the standard deviation of the range σ_R for the R chart, the average of the subgroup ranges ($\overline{R}$) multiplied by the D_3 and D_4 factors is used:

$$UCL_R = D_4\overline{R}$$
$$LCL_R = D_3\overline{R}$$

Along with the value of A_2, the values of D_3 and D_4 are found in the table in Appendix 2. These values are selected on the basis of the subgroup sample size n.

When displayed on the R chart, the control limits should theoretically be symmetrical about the centerline ($\overline{R}$). However, because range values cannot be negative, a value of zero is given for the lower control limit with sample sizes of six or less. This results in an R chart that is asymmetrical. As with the $\overline{X}$ chart, control limits for the R chart are shown with a dashed line. The centerline is shown with a solid line.

EXAMPLE 13.14   Calculating the R-Chart Centerline and Control Limits

Constructing an R chart is similar to creating an $\overline{X}$ chart. To begin the process, individual range values are calculated for each of the subgroups by subtracting the highest value in the subgroup from the lowest value [Figure 13.26, (3)]. Once calculated, these individual range values (R_i) are plotted on the R chart [Figure 13.27, (6)].

To determine the centerline of the R chart, individual range values (R_i) are summed and divided by the total number of subgroups to give $\overline{R}$ [Figure 13.27, (4)].

$$\overline{R} = \frac{0.08 + 0.07 + 0.04 + \cdots + 0.04}{21}$$

$$= \frac{1.06}{21} = 0.05$$

With n = 5, the values of D_3 and D_4 are found in the table in Appendix 2. The control limits for the R chart are calculated as follows:

$$UCL_R = D_4\overline{R}$$
$$= 2.114(0.05) = 0.11$$
$$LCL_R = D_3\overline{R}$$
$$= 0(0.05) = 0$$

The control limits are placed on the R chart [Figure 13.27, (5)].

7. *Examine the process: control-chart interpretation.*   Correct interpretation of control charts is essential to managing a process. Understanding the sources and potential causes of variation is critical to good management decisions. Managers must be able to determine whether the variation present in a process is indicating a trend that must be dealt with or is merely random variation natural to the process. Misinterpretation can lead to a variety of losses, including:

■ Blaming people for problems that they cannot control
■ Spending time and money looking for problems that do not exist

- Spending time and money on process adjustments or on new equipment, neither of which is necessary
- Taking action where no action is warranted
- Asking for worker-related improvements where process or equipment improvements need to be made first

If a process is understood and adjustments have been made to stabilize the process, then the benefits are many. Once the performance of a process is predictable, there is a sound basis for making plans and decisions concerning the process, the system, and the output. Costs to manufacture the product or provide the service become predictable. Quality levels can be determined and then compared with expectations. The effects of changes made to the process can be measured and evaluated with greater accuracy and reliability.

Process Variation

Because variation is present in all aspects of our lives, we already have developed an understanding of what is usual or unusual variation. For instance, on the basis of a six-month history of commuting, we may expect our commute to work to take 25 minutes, give or take a minute or two. We would be surprised if the commute took only 15 minutes. We would look for an assignable cause: Perhaps traffic was lighter because we left earlier. By the same token, we would be upset if the commute took 40 minutes, and we would want to know why. A traffic accident could be the assignable cause that meant such an increase in commuting time.

State of Process Control

A process is considered to be in a state of control, or **under control,** *when the performance of the process falls within the statistically calculated control limits and exhibits only chance, or common, causes.* When a process is under control, it is considered stable, and the amount of future variation is predictable. A stable process does not necessarily meet the specifications set by the designer or exhibit minimal variation; a stable process merely has a predictable amount of variation.

There are several benefits to a stable process with predictable variation. When the process performance is predictable, there is a rational basis for planning. It is fairly straightforward to determine costs associated with a stable process. Quality levels from time period to time period are predictable. When changes, additions, or improvements are made to a stable process, the effects of the change can be determined quickly and reliably.

When an assignable cause is present, the process is considered unstable, out of control, or beyond the expected normal variation. In an unstable process the variation is unpredictable; that is, the magnitude of the variation could change from one time period to another. Quality assurance analysts need to determine whether the variation that exists in a process is common or assignable. To treat an assignable cause as a chance cause could result in a disruption to a system or a process that is operating correctly except for the assignable cause. To treat chance causes as assignable causes is an ineffective use of resources because the variation is inherent in the process.

When a system is subject to only chance causes of variation, 99.73% of the parts produced will fall within $\pm 3\sigma$. Therefore, if 1,000 subgroups are sampled, 997 of the subgroups will

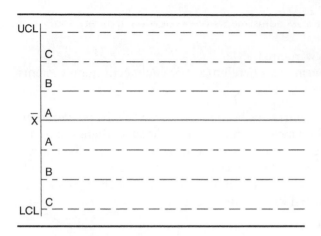

Figure 13.28   Zones on a Control Chart

have values within the upper and lower control limits. Based on the normal curve, a control chart can be divided into three zones (Figure 13.28). Zone A is ± 1 standard deviation from the centerline and should contain approximately 68.3% of the calculated sample averages or ranges. Zone B is ± 2 standard deviations from the centerline and should contain 27.2% (95.5% − 68.3%) of the points. Zone C is ± 3 standard deviations from the centerline and should contain only approximately 4.2% of the points (99.7% − 95.5%). With these zones as a guide, a control chart exhibits a state of control when:

1. Two thirds of the points are near the center value.
2. A few of the points are on or near the center value.
3. The points appear to float back and forth across the centerline.
4. The points are balanced (in roughly equal numbers) on both sides of the centerline.
5. There are no points beyond the control limits.
6. There are no patterns or trends on the chart.

While analyzing $\overline{X}$ and R charts, take a moment to study the scale of the range chart. The spread of the upper and lower control limits will reveal whether a significant amount of variation is present in the process. This clue to the amount of variation present may be overlooked if the R chart is checked only for patterns or out-of-control points.

Identifying Patterns

A process that is not under control or is unstable displays patterns of variation. Patterns signal the need to investigate the process and determine whether an assignable cause can be found for the variation. Figures 13.29 through 13.34 display a variety of out-of-control conditions. The patterns in these figures have been exaggerated to make them obvious.

Trends or Steady Changes in Level   A trend is a steady, progressive change in where the data are centered on the chart. Figure 13.29 displays a downward trend on the R chart. Note that the points were found primarily in the upper half of the control chart at the beginning of the process and on the lower half of the chart at the end. The key to identifying

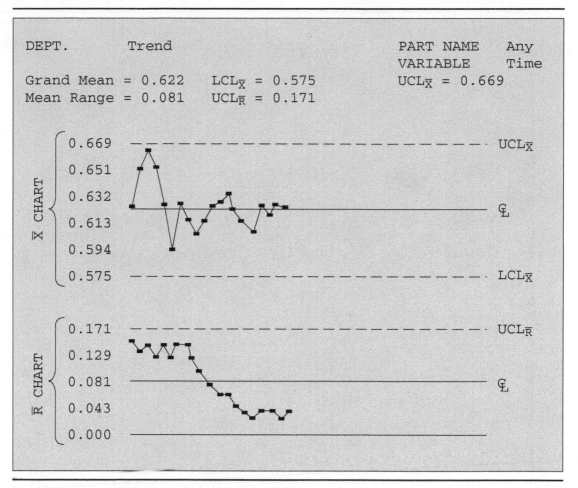

Figure 13.29   Trend Showing a Decrease in Variation

a trend or steady change in level is to recognize that the points are slowly and steadily working their way from one level of the chart to another.

A trend may appear on the $\overline{X}$ chart because of tool or die wear, a gradual deterioration of the equipment, a buildup of chips, a slowly loosening work-holding device, a breakdown of the chemicals used in the process, or some other gradual change.

R-chart trends could be due to changes in worker skills, shifting of work-holding devices, or wearing of equipment. Improvements would lead to less variation; increases in variation would reflect a decrease in skill or a change in the quality of the incoming material.

An oscillating trend would also need to be investigated (Figure 13.30). In this type of trend the points oscillate up and down for approximately 14 points or more. This could be due to a lack of homogeneity, perhaps a mixing of the output from two machines making the same product.

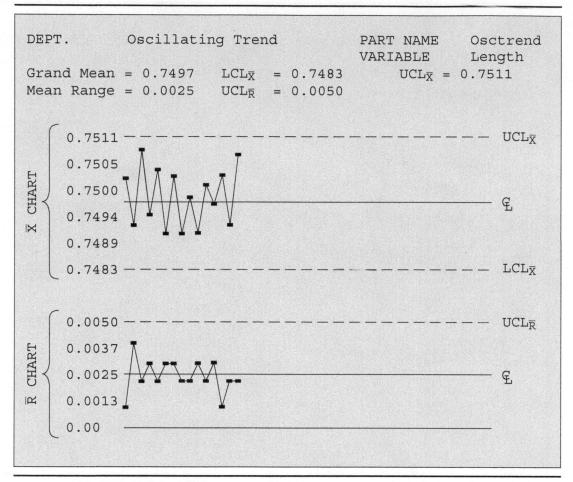

Figure 13.30   An Oscillating Trend

Change, Jump, or Shift in Level   Figure 13.31 displays what is meant by a change, jump, or shift in level. Note that the process begins at one level (Figure 13.31a) and jumps quickly to another level (Figure 13.31b) as the process continues to operate. Causes for sudden shifts in level tend to reflect some new, fairly significant difference in the process. For the $\overline{X}$ chart, causes include new machines, dies, or tooling; minor failure of a machine part; new or inexperienced workers; new batches of raw material; new production methods; or changes to the process settings. For the R chart, potential sources of jumps or shifts in level causing a change in the process variability or spread include a new or inexperienced operator, a sudden increase in the play associated with gears or work-holding devices, or greater variation in incoming material.

Runs   A process can be considered out of control when there are unnatural runs present in the process. Imagine tossing a coin. If two heads occur in a row, the onlooker would probably agree that this occurred by chance. Although the probability of the coin landing with

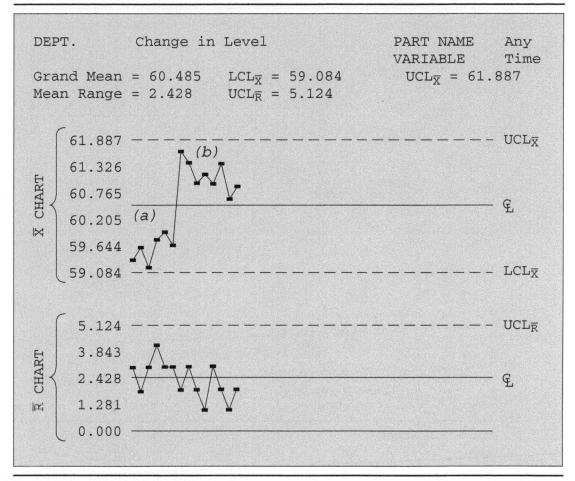

Figure 13.31   Change in Level

heads showing is 50-50, no one expects coin tosses to alternate between heads and tails. If, however, an onlooker saw someone toss six heads in a row, that onlooker would probably be suspicious that this set of events is not due to chance. The same principle applies to control charts. While the points on a control chart do not necessarily alternate above and below the centerline in a chart that is under control, the points are normally balanced above and below the centerline. A cluster of seven points in a row above or below the centerline would be improbable and would likely have an assignable cause. The same could be said for situations where 10 out of 11 points or 12 out of 14 points are located on one side or the other of the centerline (Figure 13.32a, b, c). A run may also be considered a trend if it displays increasing or decreasing values.

Runs on the $\overline{X}$ chart can be caused by temperature changes; tool or die wear; gradual deterioration of the process; or deterioration of the chemicals, oils, or cooling fluids used in the process. Runs on the R chart signal a change in the process variation. Causes for these

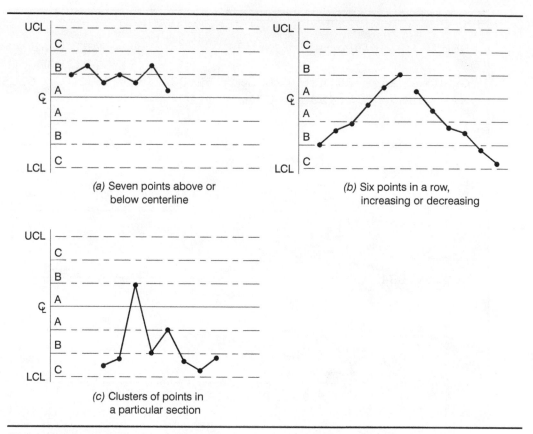

(a) Seven points above or
below centerline

(b) Six points in a row,
increasing or decreasing

(c) Clusters of points in
a particular section

Figure 13.32   Runs

R-chart runs could be a change in operator skill, either an improvement or a decrement, or a gradual improvement in the homogeneity of the process because of changes in the incoming material or changes to the process itself.

Recurring Cycles   Recurring cycles are caused by systematic changes related to the process. When investigating what appears to be cycles (Figure 13.31) on the chart, it is important to look for causes that will change, vary, or cycle over time. For the $\overline{X}$ chart, potential causes are tool or machine wear conditions, an accumulation and then removal of chips or other waste material around the tooling, maintenance schedules, periodic rotation of operators, worker fatigue, periodic replacement of cooling fluid or cutting oil, or changes in the process environment such as temperature or humidity. Cycles on an R chart are not as common; an R chart displays the variation or spread of the process, which usually does not cycle. Potential causes are related to lubrication cycles and operator fatigue.

Cycles can be difficult to locate because the entire cycle may not be present on a single chart. The frequency of inspection could potentially cause a cycle to be overlooked. For example, if the cycle occurs every 15 minutes and samples are taken only every 30 minutes, then it is possible for the cycle to be overlooked.

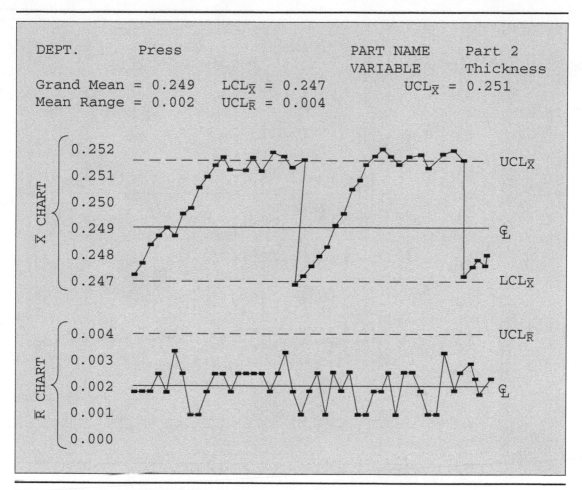

Figure 13.33   Cycle in Part Thickness

Two Populations   When a control chart is under control, approximately 68% of the sample averages will fall within $\pm 1\sigma$ of the centerline. When a large number of the sample averages appear near or outside the control limits, two populations of samples might exist. "Two populations" refers to the existence of two (or more) sources of data (Figure 13.34).

On an $\overline{X}$ chart the different sources of production might be due to the output of two or more machines being combined before sampling takes place. It might also occur because the work of two different operators is combined or two different sources of raw materials are brought together in the process. A two-population situation means that the items being sampled are not homogeneous.

This type of pattern on an R chart signals that different workers are using the same chart or that the variation is due to the fact that raw materials are coming from different suppliers.

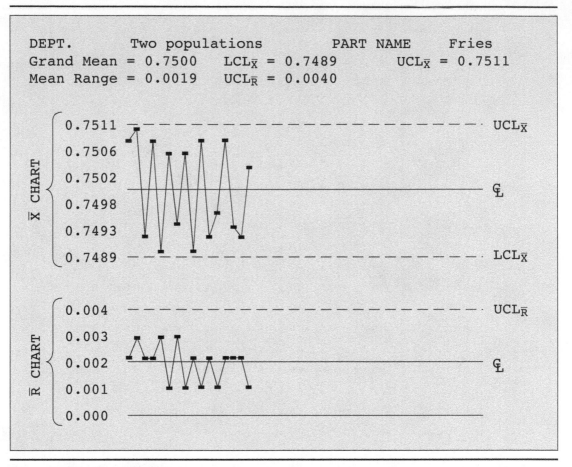

Figure 13.34   Two Populations

EXAMPLE 13.15   Examining the Control Charts for Example 13.10

Returning to the computer printer roller shaft example (Example 13.10), we begin an examination of the $\overline{X}$ and R charts by investigating the R chart, which displays the variation present in the process. Evidence of excessive variation would indicate that the process is not producing consistent product. The R chart (Figure 13.27) exhibits good control. The points are evenly spaced on both sides of the centerline, and there are no points beyond the control limits. There are no unusual patterns of trends in the data. Given these observations, it can be said that the process is producing parts of similar dimensions.

Next the $\overline{X}$ chart is examined. An inspection of the $\overline{X}$ chart reveals an unusual pattern occurring at points 12, 13, and 14. These measurements are all below the lower control limit. When compared with other samples throughout the day's production, the parts produced during the time when samples 12, 13, and 14 were taken were much shorter than parts produced during other times in the production run. A glance at the R chart reveals that the range of the individual measurements taken in the samples is small; that is, the parts produced during samples 12, 13, and 14 are all similar in size. An investigation into the cause of the production of undersized parts needs to take place.

Process Capability

Process capability *refers to the ability of a process to produce products or provide services capable of meeting the specifications set by the customer or designer.* As discussed in the last chapter, variation affects a process and may prevent the process from producing products or services that meet customer specifications. Reducing process variability and creating consistent quality increase the viability of predictions of future process performance (Figure 13.35). Knowing process capability gives insight into whether the process will be able to meet future demands placed on it. Determining the process capability aids industries in meeting their customer demands. Manufacturers of products and providers of services who know the process capability

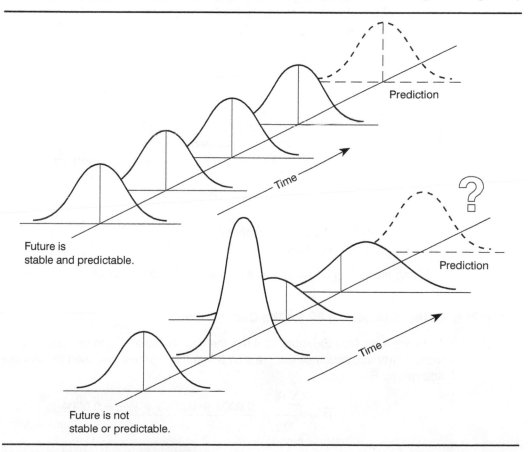

Figure 13.35   Future Predictions

can pass this information on to the customer. It can then be used to assist in decisions concerning product or process specifications, appropriate production methods, equipment to be used, and time commitments.

Process capability indices are mathematical ratios that quantify the ability of a process to produce products within specifications. The capability indices compare the spread of the individuals created by the process with the specification limits set by the customer or designer. The 6 sigma spread of the individuals can be calculated for a new process that has not produced a significant number of parts or for a process currently in operation. In either case, a true 6 sigma value cannot be determined until the process has achieved stability, as described by the $\overline{X}$ and R charts. If the process is not stable, the calculated values may or may not be representative of the true process capability.

Calculating 6$\hat{\sigma}$

Assuming that the process is under statistical control, we use the following method to calculate 6$\hat{\sigma}$ of a new process:

1. Take the past 20 subgroups, using a sample size of 4 or more.
2. Calculate the range, R, for each subgroup.
3. Calculate the average range, $\overline{R}$:

$$\overline{R} = \frac{\sum_{i=1}^{m} R_i}{m}$$

where

R_i = individual range values for the subgroups
m = number of subgroups

4. Calculate the estimate of the population standard deviation, $\hat{\sigma}$:

$$\hat{\sigma} = \frac{\overline{R}}{d_2}$$

where d_2 is obtained from Appendix 2.
5. Multiply the population standard deviation by 6.

Using more than 20 subgroups will improve the accuracy of the calculations.

EXAMPLE 13.16   Calculating 6$\hat{\sigma}$ for the Clutch Plate

The engineers from Example 13.8 use the data in Table 13.2 to calculate 6$\hat{\sigma}$ for the clutch plate. Thirty subgroups of sample size 5 and their ranges are used to calculate the average range, $\overline{R}$:

$$\overline{R} = \frac{\sum_{i=1}^{m} R_i}{m} = \frac{0.0003 + 0.0003 + \cdots + 0.0004}{21}$$

$$= 0.0003$$

Next, the engineers calculate the estimate of the population standard deviation, $\hat{\sigma}$:

$$\hat{\sigma} = \frac{\overline{R}}{d_2} = \frac{0.0003}{2.326} = 0.0001$$

Using a sample size of 5, they take the value for d_2 from Appendix 2.
   To determine $6\hat{\sigma}$, they multiply the population standard deviation by 6:

$$6\hat{\sigma} = 6(0.0001) = 0.0006$$

They now compare this value with the specification limits to determine how well the process is performing.

The Capability Index

Once calculated, the σ values can be used to determine several indices related to process capability. *The* **capability index** C_P *is the ratio of tolerance (USL − LSL) and* $6\hat{\sigma}$:

$$C_p = \frac{USL - LSL}{6\hat{\sigma}}$$

where

C_p = capability index

USL − LSL = upper specification limit − lower specification limit, or tolerance

The capability index is interpreted as follows: If the capability index is larger than 1.00, a Case I situation exists (Figure 13.36). This is desirable. The greater this value, the better.

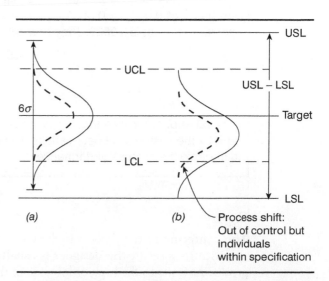

**Figure 13.36   Case I: 6σ, <
USL − LSL**

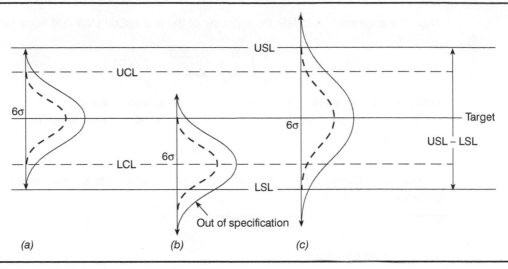

Figure 13.37 Case II: $6\sigma = $ USL $-$ LSL

If the capability index is equal to 1.00, then a Case II situation exists (Figure 13.37a). This is not optimal, but it is feasible. If the capability index is less than 1.00, then a Case III situation exists (Figure 13.38a). Values less than 1 are undesirable and reflect the inability of the process to meet the specifications.

EXAMPLE 13.17 Finding the Capability Index

The engineers working with the clutch plate in Example 13.8 are working with specification limits of 0.0625 ± 0.0003. The upper specification limit is 0.0628, and the lower specification limit is 0.0622. They now calculate, C_p:

$$C_p = \frac{\text{USL} - \text{LSL}}{6\hat{\sigma}} = \frac{0.0628 - 0.0622}{0.0006}$$

$$= 1.0$$

A value of 1.0 means that the process is just capable of meeting the demands placed on it by the customer's specifications. To be on the safe side, changes will need to occur to improve the process performance.

C_{pk}

The centering of the process is shown by C_{pk}. A process operating in the center of the specifications set by the designer is usually more desirable than one that is consistently producing parts to the high or low side of the specification limits. In Figure 13.39 all three

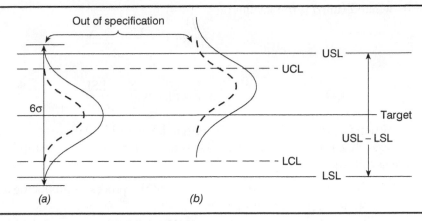

Figure 13.38   Case III: $6\sigma >$ USL − LSL

distributions have the same C_p index value of 1.3. Although each of these processes has the same capability index, they represent three different scenarios. In the first situation the process is centered as well as capable. In the second, a further upward shift in the process would result in an out-of-specification situation. The reverse holds true in the third situation. C_p does not take into account the centering of the process. *The ratio that reflects how the process is performing in terms of a nominal, center, or target value is C_{pk}, which can be calculated using the following formula:*

$$C_{pk} = \frac{Z(\min)}{3}$$

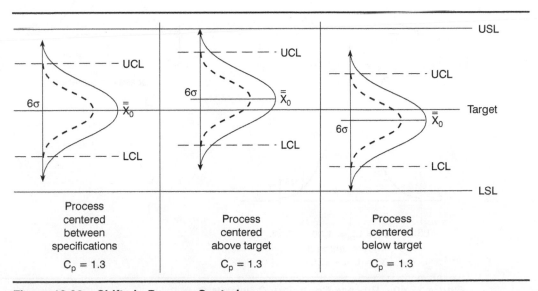

Figure 13.39   Shifts in Process Centering

where Z(min) is the smaller of

$$Z(USL) = \frac{USL - \overline{X}}{\hat{\sigma}}$$

$$\text{or } Z(LSL) = \frac{\overline{X} - LSL}{\hat{\sigma}}$$

When $C_{pk} = C_p$, the process is centered. Figure 13.40 illustrates C_p and C_{pk} values for a process that is centered and one that is off center. The relationships between C_p and C_{pk} are as follows:

1. When C_p has a value of 1.0 or greater, the process is producing a product capable of meeting specifications.
2. The C_p value does not reflect process centering.
3. When the process is centered, $C_p = C_{pk}$.
4. C_{pk} is always less than or equal to C_p.
5. When C_p is greater than or equal to 1.0 and C_{pk} has a value of 1.00 or more, it indicates that the process is producing a product that conforms to specifications.
6. When C_{pk} has a value less than 1.00, it indicates that the process is producing a product that does not conform to specifications.

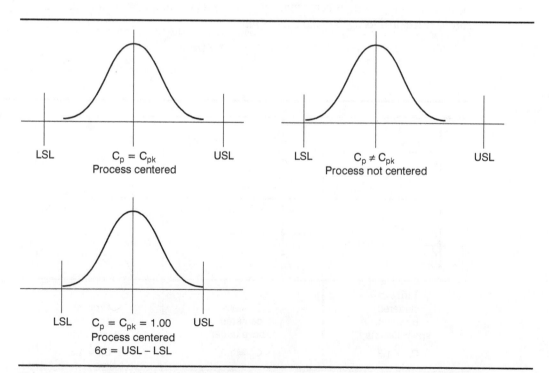

Figure 13.40   Process Centering: C_p versus C_{pk}

7. A C_p value of less than 1.00 indicates that the process is not capable.
8. A C_{pk} value of zero indicates that the process average is equal to one of the specification limits.
9. A negative C_{pk} value indicates that the average is outside the specification limits.

EXAMPLE 13.18   Finding C_{pk}

Determine C_{pk} for the roller shaft values in Example 13.16. The average, $\overline{X}$, is equal to:

$$C_{pk} = \frac{Z(min)}{3}$$

where

$$Z(min) = \text{smaller of } \frac{(USL - \overline{X})}{\hat{\sigma}} \text{ or } \frac{(\overline{X} - LSL)}{\hat{\sigma}}$$

$$Z(USL) = \frac{(0.0628 - 0.0627)}{0.0001} = 1$$

$$Z(LSL) = \frac{(0.0627 - 0.0622)}{0.0001} = 5$$

$$C_{pk} = \frac{1}{3} = 0.333$$

A C_{pk} value of less than 1 means that the process is not capable. Because the C_p value (1.0 from Example 13.17) and the C_{pk} value (0.333) are not equal, the process is not centered between the specification limits.

WHAT HAPPENS DURING THE "DO" PHASE?

Step 7. Do: Selecting and Implementing the Solution

We have been applying problem-solving techniques to find the root cause of a problem. Once the cause has been identified, it is time to propose potential solutions. This begins the Do section of the PDSA cycle, the portion of the cycle that attracts everyone's attention. So great is the desire to do something that many problem solvers are tempted to reduce the amount of time spent on planning to virtually nothing. The temptation to immediately propose solutions must be ignored. The best solutions are those that solve the true problem. They are found only after the root cause of that problem has been identified. The most significant portion of the problem-solving effort must be concentrated in the Plan phase.

It is important to recognize that applying these techniques does not mean that taking care of the immediate problem should be ignored. Immediate action should be taken to rectify any situation that does not meet the customer's reasonable needs, requirements, and expectations. However, these quick fixes are just that—a quick fix of a problem for the short term; they simply allow time for a long-term solution to be found. In no situation should a quick fix be considered the end of a problem. Problems are solved only when recurrences do not happen.

Selecting and implementing the solution is a matter of the project team's choosing the best solution for the problem under examination. The solution should be judged against four general criteria:

1. The solution should be chosen on the basis of its potential to prevent a recurrence of the problem. A quick or short-term fix to a problem will only mean that time will be wasted in solving this problem again when it recurs in the future.
2. The solution should address the root cause of the problem. A quick or short-term fix that focuses on correcting the symptoms of a problem will waste time because the problem will recur in the future.
3. The solution should be cost effective. The most expensive solution is not necessarily the best solution for the company's interests. Solutions may necessitate determining the company's future plans for a particular process or product. Major changes to the process, system, or equipment may not be an appropriate solution for a process or product that will be discontinued in the near future. Technological advances will need to be investigated to determine if they are the most cost-effective solutions.
4. The solution should be capable of being implemented within a reasonable amount of time. A timely solution to the problem is necessary to relieve the company of the burden of monitoring the current problem and its associated quick fixes.

EXAMPLE 13.19   Glove Box Solutions

Because the team was able to identify the root causes of the glove box latch problem as improper alignment, improper positioning, and low material strength, they decided to make the following changes part of their solution:

1. Redesign the glove box latch. This solution was chosen to counteract low material strength.
2. Reposition the glove box door, striker, and hinge. This solution was chosen to counteract improper positioning and alignment. They also hoped that this change would eliminate potential squeaks and rattles.
3. Reinforce the glove box latch. This solution was chosen to counteract breakage. By increasing the material at the latch position on the glove box door, they hoped to eliminate breakage. They also decided to use a stronger adhesive to reinforce the rivets securing the latch to the door.

Members of the problem-solving team often implement the solution. Critical to ensuring the success of the solution implementation is assigning responsibilities to specific individuals and holding them accountable for accomplishing the task. Knowing who will be doing what and when will help ensure that the project stays on track.

WHAT HAPPENS DURING THE "STUDY" PHASE?

Step 8. Study: Evaluating the Solution—the Follow-Up

Once implemented and given time to operate, problem-solving actions are checked to see whether the problem has truly been solved. During the Study stage we study the results and ask, "Is the solution we've chosen working? What did we learn?" To determine whether the solution has worked, the measures of performance created in step 4 should be applied. Prior data collected during the analysis phase of the project should be compared with present data taken from the process. Control charts, histograms, and run charts can be used to monitor the process, both before and after. If these formats were used in the original problem analysis, a direct comparison can be made to determine how well the solution is performing. If the solution is not correcting the problem, then the PDSA process should begin again to determine a better solution.

EXAMPLE 13.20   Evaluating the Solution

The instrument panel warranty team implemented their solutions and used the measures of performance developed in Example 13.5 to study the solutions in order to determine whether the changes were working. The Pareto chart in Figure 13.41 provides information about warranty claims made after the changes were implemented. When this figure is compared to Figure 13.7, showing warranty claims before the problem-solving team went

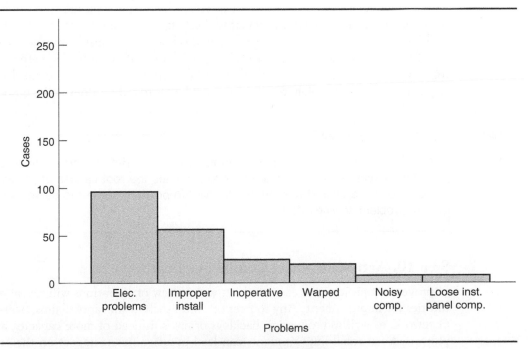

Figure 13.41   Pareto Chart of Instrument Panel Problems

into action, the improvement is obvious. Warranty costs declined in proportion to the decreased number of claims, to just under $25,000. A process measure also tracked the length of time to implement the changes—a very speedy five days.

WHAT HAPPENS DURING THE "ACT" PHASE?

Step 9. Act: Ensuring Permanence

The final stage, Act, involves making the decision to adopt the change, abandon it, or repeat the problem-solving cycle. If the change is adopted, then efforts must be made to ensure that the new methods have been established so that the new level of quality performance can be maintained. Now that a follow-up investigation has revealed that the problem has been solved, it is important that improved performance continue. From Figure 13.4 it can be seen that "ensuring permanence" is part of the action phase. This phase of the quality improvement process exists to ensure that the new controls and procedures stay in place. It is easy to believe that the "new and better" method should be utilized without fail; however, in any situation where a change has taken place, there is a tendency to return to old methods, controls, and procedures when stress is increased. It is a bit like switching to an automatic shift after having driven a manual shift car for a number of years. Under normal driving conditions, drivers will not utilize their left legs to operate a clutch pedal that does not exist in an automatic gear shift car. But place those same individuals in an emergency stop situation, and chances are they will attempt to activate a nonexistent clutch pedal at the same time as they brake. Under stress, people have a tendency to revert to their original training.

To avoid a lapse into old routines and methods, controls must be in place to remind people of the new method. Extensive training and short follow-up training are very helpful in ingraining the new method. Methods must be instituted and follow-up checks must be put in place to prevent problem recurrences from lapses to old routines and methods.

EXAMPLE 13.21   Standardize Improvements

Plastics and Dashes asks each team to write up a brief but formal discussion of the problem-solving steps their team took to eliminate the root causes of problems. These discussions are shared with others involved in problem solving to serve as a guide for future problem-solving efforts.

Step 10. Act: Maintaining Continuous Improvement

Improvement projects are easy to identify. A review of operations will reveal many opportunities for improvement. Any sources of waste, such as warranty claims, over-time, scrap, or rework, as well as production backlogs or areas in need of more capacity, are potential projects. Even small improvements can lead to a significant impact on the organization's financial statement.

How Do We Know It's Working?

Why are orders late? That's the question that leaders at JQOS wanted an answer to. Every week, a customer phone call would activate a frenzied search for the late parts, a rescheduling of other orders, and inevitable overtime. Once the order was expedited, everyone would calm down until the next phone call. This firefighting mode of operating costs the company money, but no one searched for a solution until leadership declared no more firefighting. JQOS leadership appointed a team of fire marshals to utilize Dr. Deming's PDSA technique to isolate and analyze the problem, determine the root cause, and develop and implement a permanent solution.

PLAN:

Recognize a problem exists: Orders are late.

Form team: Supervisor, representatives from receiving, production, and shipping, scheduler.

Clearly define the problem: In order to clearly define the problem, the team studied orders for the past six months. Using a check sheet, they categorized the problems with the orders (Figure 13.42). From this, they created a Pareto diagram (Figure 13.43). Analysis of the Pareto diagram pointed to an unexpected source of the late orders: the material had not been ordered.

Develop performance measures: Now that the problem has been clearly defined, appropriate performance measures can be created. The team chose:
    Number of late orders
    Number of incorrect material orders
    Number of material orders not received

Analyze the problem and processes: Since the problem clearly relates to material not being ordered, the order processing clerk joined the team. Together, the team created a cause-and-effect diagram to study the reasons behind late orders (Figure 13.44).

Determine possible causes: The team focused on the portions of the cause-and-effect diagram related to order processing and material ordering. A lack of procedures for processing

Job routing incorrect	////
Lead time too short	///// //
Material late	///// ///// ///
Material not received	////// /
Material not ordered	///// ///// /////
Incorrect material ordered	//////
Parts damaged during production	////
Parts failed inspection	///
Customer order not processed	//

Figure 13.42   Check Sheet

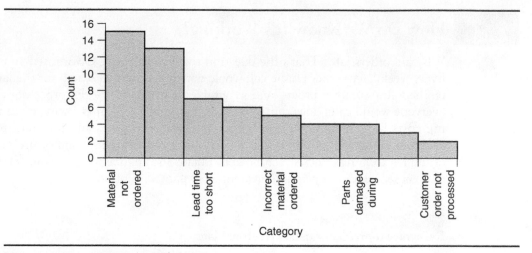

Figure 13.43   JQOS Pareto Diagram

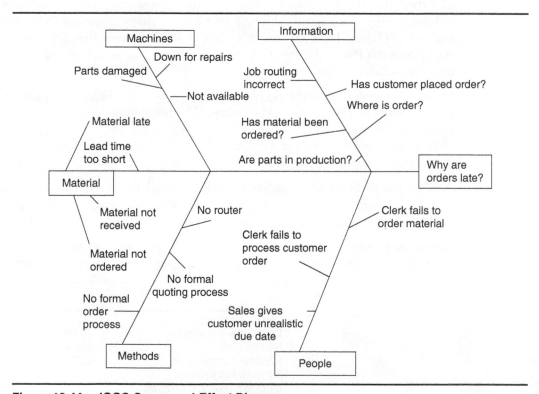

Figure 13.44   JQOS Cause-and-Effect Diagram

paperwork hampered placing orders correctly. Material often didn't get ordered because the paperwork sat lost in a pile on someone's desk. Other problems included not knowing what steps the paperwork had been through or what had or had not been ordered. No marks or stamps or notations told anyone about the work-in-process. Everyone relied on their own memory or someone else's. In the end, it was hard to determine who had the order and what had been done to it.

DO:

Select and implement a solution: At this point, the team brought in JQOS leadership to help devise an appropriate solution to this problem. They did so because the solutions the team contemplated all required significant investment in technology, both software and hardware.

STUDY:

Evaluate the solution: Because the cross-functional team worked together to develop a solution that worked for all departments, they were eager to see that it worked. Using their performance measures, they evaluated the new software during its first two months of use. Twice the vendor came back in to make minor modifications to the reports and screens. Twice the vendor returned to train additional people on the use of the software. The final result: the number of late orders dramatically decreased. The number of incorrect material orders and material not ordered decreased to nearly zero.

ACT:

Ensure permanence: The system rapidly became so popular that all computers in the plant were linked to the system. Every employee received training on the new system. The new system increased order processing efficiency since automated spreadsheets allows quotes to be adjusted quickly. When accepted by the customer, quotes can be turned into orders, complete with shop traveler, invoice, packing list, and material purchase order. When the material purchase order is sent via computer to the steel vendor, an e-mail reply notifying receipt of the order is requested. Confirmation of the order to the customer is sent via e-mail. Job progress through the plant can be tracked as the parts move from operation to operation. Barcoded shipping labels allow tracking right to the customer's facility. Taken together, the new order processing system supports the entire supply chain by tracking and confirming the order process from customer quote to final part delivery.

Continuous improvement: When the number of late orders did not decrease to zero, the team was surprised. They used the most recent data to create another Pareto diagram of reasons for late orders (Figure 13.45). Their next effort will focus on enhancing communication between sales and the shop floor. They discovered that the sales department had been providing unrealistic order due dates to please the customer.

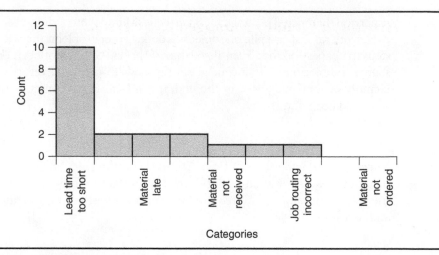

Figure 13.45 JQOS Pareto Diagram 2

LESSONS LEARNED

Teaching the tools of quality improvement and problem solving is actually the easiest part of the quality improvement process. Helping individuals and groups apply those techniques in a problem-solving format is critical and difficult. Upper-management involvement in selecting issues to be investigated is important to the success of a quality improvement program. Brainstorming and Pareto analyses also help identify where problem-solving efforts should be concentrated. Teach people the techniques and then use brainstorming to encourage them to uncover problems in their own area and begin to solve them. Upper management should be involved in the entire process, from education to implementation. They should be the ones providing their people with a push in the right direction.

1. Problem solving is the isolation and analysis of a problem and the development of a permanent solution.
2. A structured problem-solving process prevents problem reoccurrence by finding the root cause of the problem.
3. The Planning stage of problem solving includes recognizing a problem exists, forming a quality improvement team, clearly defining the problem, developing performance measures, analyzing the problem or process, and determining possible causes.
4. The Do stage of problem solving focuses on selecting and implementing a solution.
5. The Study stage of problem solving evaluates the solution to determine whether it solved the problem.
6. The Act stage of problem solving includes activities that ensure the permanence of the solution and encourages more continuous improvement.

Are You a Quality Management Person?

Are you a firefighter or a fire marshal? Check out the quality management person questions and find out whether you are a quality management person. How do you answer them on a scale of 1 to 10?

Do you see problems as opportunities?

1 2 3 4 5 6 7 8 9 10

Do you strive to improve your situation through the use of knowledge and skills?

1 2 3 4 5 6 7 8 9 10

Do you take an organized approach to a problem?

1 2 3 4 5 6 7 8 9 10

When you tackle a problem, do you just start doing something, anything, and see what happens?

1 2 3 4 5 6 7 8 9 10

When you tackle a problem, do you make a plan first?

1 2 3 4 5 6 7 8 9 10

Before making changes, do you study a situation or act on gut feeling?

1 2 3 4 5 6 7 8 9 10

When you tackle a problem, do you establish a baseline?

1 2 3 4 5 6 7 8 9 10

When you tackle a problem, do you use measures to study your progress?

1 2 3 4 5 6 7 8 9 10

When you tackle a problem, do you try to find the real cause of the problem or just patch things up the best you can?

1 2 3 4 5 6 7 8 9 10

Do problems, once you think you have them solved, come back to haunt you?

| 1 | 2 | 3 | 4 | 5 | 6 | 7 | 8 | 9 | 10 |

When appropriate, do you involve other people in finding a solution to a problem?

| 1 | 2 | 3 | 4 | 5 | 6 | 7 | 8 | 9 | 10 |

Do you slide back into old routines after making improvements?

| 1 | 2 | 3 | 4 | 5 | 6 | 7 | 8 | 9 | 10 |

Chapter Questions

Problem Solving

1. Good root cause identification and problem-solving efforts begin with a clear problem statement. Why is a well-written problem statement necessary?

2. Bicycles are being stolen at a local campus. Campus security is considering changes in bike rack design, bike parking restrictions, and bike registration to try to reduce thefts. Thieves have been using hacksaws and bolt cutters to remove locks from the bikes. Create a problem statement for this situation. How will an improvement team use the problem statement?

Pareto Charts

3. During the past month, a customer-satisfaction survey was given to 200 customers at a local fast-food restaurant. The following complaints were lodged:

Complaint	Number of Complaints
Cold Food	105
Flimsy Utensils	20
Food Tastes Bad	10
Salad Not Fresh	94
Poor Service	15
Food Greasy	9
Lack of Courtesy	5
Lack of Cleanliness	25

Create a Pareto chart with this information.

4. A local bank is keeping track of the different reasons people phone the bank. Those answering the phones place a mark on their check sheet in the rows most representative of the customers' questions. Give the following check sheet, make a Pareto diagram:

Credit Card Payment Questions	254
Transfer Call to Another Department	145
Balance Questions	377
Payment Receipt Questions	57
Finance Charges Questions	30
Other	341

Comment on what you would do about the high number of calls in the "Other" column.

5. PT Tool Inc. manufactures aircraft landing gear. The completed gear must perform to rigid specifications. Because of the expensive nature of the product, the landing gear must also meet customer expectations for fit and finish. To gather information about nonconformities that are occurring in their shop, a problem-solving team has utilized check sheets to record the nonconformities that they find on the parts during final inspection. When they encounter a problem, the inspectors check the appropriate category on the check sheet. Create a Pareto diagram from the check sheet. Based on the diagrams, where should PT Tool be concentrating their improvement efforts?

Check Sheet of Finish and Operational Flaws

Finish Flaws

Scratches	////////
Dents	//
Surface finish disfigurations in paint	////
Damage to casing	/
Wrong color	/

Operational Flaws

Mounting plate location off-center	////////////
Nonfunctional electrical system	//
Activation switch malfunction	/
Motor failure	////

6. Create a Pareto diagram using the check sheet provided in Problem 5 and the following information about the individual costs associated with correcting each type of nonconformity. Based on your Pareto diagram showing the total costs associated with each type of nonconformity, where should PT Tool be concentrating their improvement efforts? How is this focus different from that in Problem 5?

Scratches	$145
Dents	$200
Surface finish disfigurations in paint	$954
Damage to casing	$6,500
Wrong color	$200
Mounting plate location off-center	$75
Nonfunctional electrical system	$5,000
Activation switch malfunction	$300
Motor failure	$420

Brainstorming

7. Brainstorm 10 reasons why the university computer might malfunction.

8. Brainstorm 10 reasons why a customer may not feel the service was adequate at a department store.

WHY-WHY Diagrams

9. A mail-order company has a goal of reducing the amount of time a customer has to wait in order to place an order. Create a WHY-WHY diagram about waiting on the telephone. Now that you have created the diagram, how would you use it?

10. Create a WHY-WHY diagram for this problem statement: Customers leave the store without making a purchase.

Cause-and-Effect Diagrams

11. What role does a cause-and-effect diagram play in finding a root cause of a problem?

12. A customer placed a call to a mail-order catalog firm. Several times the customer dialed the phone and received a busy signal. Finally, the phone was answered electronically, and the customer was told to wait for the next available operator. Although it was a 1–800 number, he found it annoying to wait on the phone until his ear hurt. Yet he did not want to hang up for fear he would not be able to get through to the firm again. Using the problem statement "What makes a customer wait?" as your base, brainstorm to create a cause-and-effect diagram. Once you have created the diagram, how would you use it?

13. Create cause-and-effect diagrams for (a) a car that won't start, (b) an upset stomach, and (c) a long line at the supermarket.

Histograms

14. NB Plastics uses injection molds to produce plastic parts that range in size from a marble to a book. Parts are pulled off the press by one operator and passed on to

another member of the team to be finished or cleaned up. This often involves trimming loose material, drilling holes, and painting. After a batch of parts has completed its cycle through the finishing process, a sample of five parts is chosen at random and certain dimensions are measured to ensure that each part is within certain tolerances. This information (in mm) is recorded for each of the five pieces and evaluated. Create a histogram. Are the two operators trimming off the same amount of material? How do you know?

Part Name: Mount
Critical Dimension: 0.654 ± 0.005
Tolerance: ±0.001
Method of Checking: Caliper

Date	Time	Press	Oper	Samp 1	Samp 2	Samp 3	Samp 4	Samp 5
9/20/92	0100	#1	Jack	0.6550	0.6545	0.6540	0.6540	0.6545
9/20/92	0300	#1	Jack	0.6540	0.6540	0.6545	0.6545	0.6545
9/20/92	0500	#1	Jack	0.6540	0.6540	0.6540	0.6540	0.6535
9/20/92	0700	#1	Jack	0.6540	0.6540	0.6540	0.6540	0.6540
9/21/92	1100	#1	Mary	0.6595	0.6580	0.6580	0.6595	0.6595
9/21/92	1300	#1	Mary	0.6580	0.6580	0.6585	0.6590	0.6575
9/21/92	1500	#1	Mary	0.6580	0.6580	0.6580	0.6585	0.6590
9/21/92	0900	#1	Mary	0.6575	0.6570	0.6580	0.6585	0.6580

15. PL Industries machines shafts for rocker-arm assemblies. The existing machining cell is currently unable to meet the specified tolerances consistently. PL Industries has decided to replace the existing machines with CNC turning centers. The following data are from a runoff held at the CNC vendor's facility. The indicated portion of the shaft was chosen for inspection. The diameter specifications for the round shaft are 7.650 + 0.02 mm. In order to prevent scrap for this particular operation, there is no lower specification. The data are scaled from 7.650; that is, a value of 0.021 is actually 7.671 mm. Create a histogram with the data.

0.011	0.013	0.018	0.007	0.002	0.020	0.014	0.006	0.002
0.006	0.004	0.003	0.010	0.015	0.011	0.020	0.020	0.012
0.015	0.004	0.009	0.020	0.012	0.011	0.012	0.004	0.017
0.010	0.011	0.018	0.015	0.010				

16. Gold, measured in grams, is used to create circuit boards at MPL Industries. The following measurements of gold usage per batch of circuit boards have been recorded in a tally sheet. Create a histogram with the following information. Describe the shape, location, and spread of the histogram.

125   /
126   /
127
128   ///
129   ////
130
131
132   ////
133
134   ////
135   ////   /
136   ////
137   ////
138   ////   /
139
140   ////
141   ////   /
142   ////   /
143   ////   /
144   ////
145   ////   ///
146   ////   ////
147   ////   //
148   ////   /
149   ////
150   ////

$\overline{X}$ and R Charts

17. A large bank establishes $\overline{X}$ and R charts for the time required to process applications for its charge cards. A sample of five applications is taken each day. The first four weeks (20 days) of data give

$$\overline{\overline{X}} = 16 \text{ min} \qquad \overline{s} = 3 \text{ min} \qquad \overline{R} = 7 \text{ min}$$

Based on the values given, calculate the centerline and control limits for the $\overline{X}$ and R charts.

18. The data below are $\overline{X}$ and R values for 25 subgroups of size n = 4 taken from a process filling bags of fertilizer. The measurements are made on the fill weight of the bags in pounds.

Set up an $\overline{X}$ and R chart on this process. Interpret the chart. Does the process seem to be in control? If necessary, assume assignable causes and revise the trial

Subgroup Number	$\overline{X}$	Range
1	50.3	0.73
2	49.6	0.75
3	50.8	0.79
4	50.9	0.74
5	49.8	0.72
6	50.5	0.73
7	50.2	0.71
8	49.9	0.70
9	50.0	0.65
10	50.1	0.67
11	50.2	0.65
12	50.5	0.67
13	50.4	0.68
14	50.8	0.70
15	50.0	0.65
16	49.9	0.66
17	50.4	0.67
18	50.5	0.68
19	50.7	0.70
20	50.2	0.65
21	49.9	0.60
22	50.1	0.64
23	49.5	0.60
24	50.0	0.62
25	50.3	0.60

control limits. If the average fill of the bags is to be 50.0 pounds, how does this process compare?

19. RM Manufacturing makes thermometers for use in the medical field. These thermometers, which read in degrees Celsius, are able to measure temperatures

Subgroup	Average Temperature	Range
1	3.06	0.10
2	3.03	0.09
3	3.10	0.12
4	3.05	0.07
5	2.98	0.08
6	3.00	0.10
7	3.01	0.15

8	3.04	0.09
9	3.00	0.09
10	3.03	0.14
11	2.96	0.07
12	2.99	0.11
13	3.01	0.09
14	2.98	0.13
15	3.02	0.08

to a level of precision of two decimal places. Each hour, RM Manufacturing tests eight randomly selected thermometers in a solution that is known to be at a temperature of 3°C. Use the following data to create and interpret an $\overline{X}$ and R chart. Based on the desired thermometer reading of 3°C, interpret the results of your plotted averages and ranges.

20. Interpret the $\overline{X}$ and R charts in Figure 1.

21. Describe how both an $\overline{X}$ and R chart would look if they were under normal statistical control.

22. $\overline{X}$ charts describe the accuracy of a process, and R charts describe the precision. How would accuracy be recognized on an $\overline{X}$ chart? How would precision be recognized on an R chart?

23. Why are the use and interpretation of an R chart so critical when examining an $\overline{X}$ chart?

24. A hospital is using $\overline{X}$ and R charts to record the time it takes to process patient account information. A sample of five applications is taken each day. The first four weeks' (20 days') data give the following values:

$$\overline{\overline{X}} = 16 \text{ min} \qquad \overline{R} = 7 \text{ min}$$

If the upper and lower specifications are 21 minutes and 13 minutes, respectively, calculate $6\hat{\sigma}$, C_p, and C_{pk}. Interpret the indices.

25. For the data in Problem 18, calculate $6\hat{\sigma}$, C_p, and C_{pk}. Interpret the indices. The specification limits are 50 ± 0.5.

26. From the information in Problem 19, calculate $6\hat{\sigma}$, C_p, and C_{pk}. Interpret the indices. The specification limits are 3 ± 0.05.

27. A quality analyst is checking the process capability associated with the production of struts, specifically, the amount of torque used to tighten a fastener. Twenty-five samples of size 4 have been taken and were used to create $\overline{X}$ and R charts. The values for these charts are as follows: The upper and lower control limits for the $\overline{X}$ chart are 74.80 Nm and 72.37 Nm, respectively. $\overline{X}$ is 73.58 Nm. $\overline{R}$ is 1.66. The specification limits are 80 Nm $\pm$ 10. Calculate $6\hat{\sigma}$, C_p, and C_{pk}. Interpret the values.

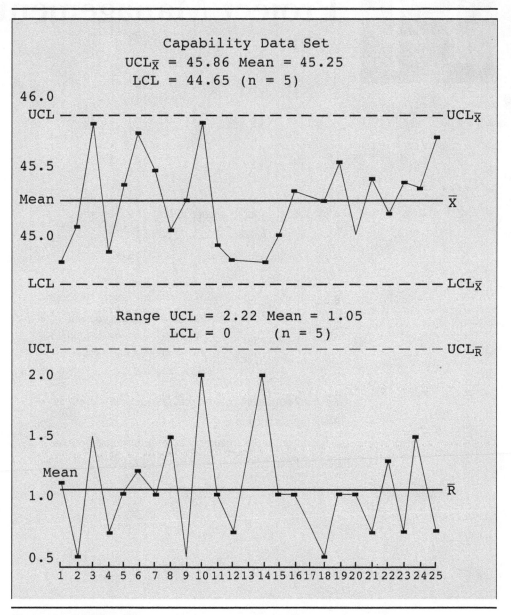

Figure 1   Problem 20

14 Project Management

Why do organizations engage in project management?

What are the characteristics of a project?

How are projects selected?

What is a project proposal?

What are the components of an effective project proposal?

What are the characteristics of clear project goals and objectives?

What are the components of an effective project plan?

How are projects scheduled?

How are resources budgeted?

What are contingency plans and change control systems?

What is project control?

What must a project manager do in order to be effective?

Lessons Learned

Learning Opportunities

1. To become familiar with how to create project proposals and plans
2. To become familiar with how to establish clear goals and objectives
3. To become familiar with how projects are scheduled and budgeted
4. To understand what it takes to be an effective project manager

How do you eat an elephant?
One bite at a time.

Unknown

WHY DO ORGANIZATIONS ENGAGE IN PROJECT MANAGEMENT?

At first glance, many projects may seem to be as daunting as eating an elephant. Where does one begin? Where should efforts be focused? Will there be enough time? Projects can seem overwhelming, yet they are essential for an effective organization to set their strategic plans in motion. By identifying projects that support the goals and objectives established in the plan, an organization can move toward its ultimate vision (Figure 14.1). Without projects, people continue to work on their daily activities without the benefit of a longer-term focus. Projects consume time and effort outside of these day-to-day activities, and employees must utilize good project management skills in order to integrate their project work with regular activities. Knowledge and skills in the area of project management are necessary anytime an employee can answer yes to the following questions:

Do assignments have to be complete by a specified deadline?
Do several tasks have to be accomplished during the day?
Is there a limited set of resources with which to complete these tasks?
Is the involvement of other people necessary in order to complete the work?
Do supervisors, colleagues, or customers ever change their mind about what they want?
Do the team members have a clear idea of what they are trying to accomplish for the ultimate user?
Does the team understand the perspectives of the people affected by the project?
Are the constraints and directives that govern the project known?
Is the project broken down into manageable chunks?
When disagreements arise, does the team have the skills to build effective agreements?

WHAT ARE THE CHARACTERISTICS OF A PROJECT?

Projects have three basic characteristics: performance, cost, and time. Performance refers to what the project seeks to accomplish. Unlike day-to-day activities, projects are unique, one-time occurrences created to fulfill specific goals for the organization. *Cost* refers to the resources needed to complete a project. Most projects must be completed with a limited set of resources. These resources may be related to skills, time, money, equipment, facilities, or knowledge. Projects are complex and typically involve a variety of people from a number of areas within an organization. Projects normally have a specific time frame for completion. Project beginnings and endings are clearly defined. To complete a project, tasks must be sequenced, with one phase or activity being completed before another is begun.

HOW ARE PROJECTS SELECTED?

Effective organizations select projects based on a project's ability to contribute to one or all three of the following: customer-perceived value and satisfaction, the organization's financial strength, or the organization's operational necessities. For customer-focused organizations, many projects will be selected based on their ability to increase customer value, satisfaction, and retention. Projects may be chosen in order to enable the organization to maintain its competitive edge. These projects may involve the development of a

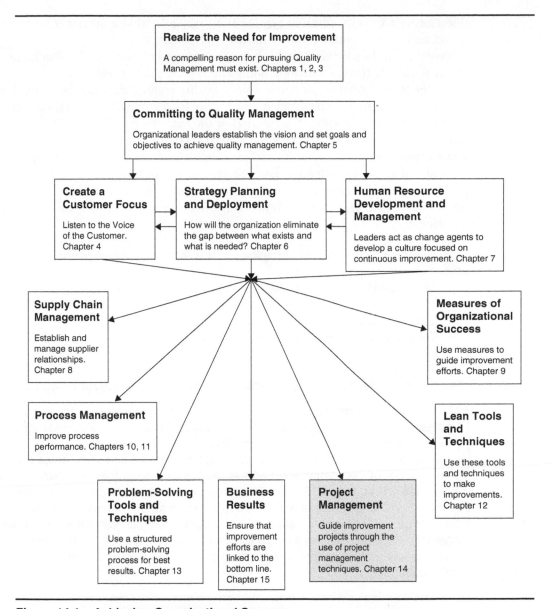

Figure 14.1   Achieving Organizational Success

new product or service or an extension to a product line or the development of a product-
or service-enhancing feature. These projects will ultimately enhance an organization's
financial success. Some projects are operational necessities, such as meeting government
regulations or repairing or replacing aging equipment. Regardless of the reasons behind

selection of a project, effective organizations recognize that a project must be financially sound and must provide a payback for the investment.

Projects can be identified through the use of external customer feedback, internal customer feedback (from the people doing the work), competitive benchmarking, existing measures of performance, or costs-of-poor-quality studies. Any gap between desired performance and actual performance is a candidate for a project. Projects should concentrate on a problem area for either the internal or the external customer. A focused project will have boundaries and will not be so large as to become unmanageable. Projects have problem statements that are specific and measurable and that identify the gap between the desired and actual performance. Each project idea must be evaluated on feasibility, benefits, costs, implementation time, people and material requirements, and any other factors important to the organization. When selecting a project, consider asking the following questions:

> What is the investment to implement this suggested improvement?
> What is the projected return?
> What are the projected benefits?
> When will the benefits be accrued?
> How long will the project take to complete?
> How will the return on investment be measured?

EXAMPLE 14.1   Computer Technology Transition and Upgrade Project Selection

Max's Munchies recently invited vendors to submit proposals for a technological upgrade of all of its desktop and laptop computers for all of its nationwide locations. The project has been titled Computer Technology Transition and Upgrade.

Max's Munchies maintains manufacturing facilities in 4 cities and sales offices in 12. The majority of the computers, 1,600 in all, are located at the manufacturing facilities. The sales offices have 400 more. The company uses two computer platforms: desktops and laptops. Although these are limited to a narrow range of models, one of the goals of the Computer Technology Transition and Upgrade project is to create more uniformity company-wide by selecting a single desktop model and a single laptop model. The company has selected the desired operating system and software applications that these computers must run.

WHAT IS A PROJECT PROPOSAL?

A **project proposal** is a document that provides clear information concerning the goals and objectives that a particular project hopes to achieve. Along with this information, the project proposal discusses how the project supports the overall mission, goals, and objectives of the organization (Figure 14.2). Project proposals seek to answer the questions proposed during the project selection process.

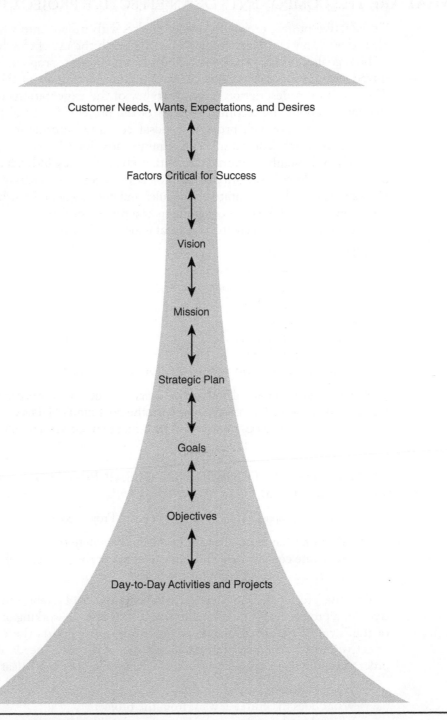

Customer Needs, Wants, Expectations, and Desires

Factors Critical for Success

Vision

Mission

Strategic Plan

Goals

Objectives

Day-to-Day Activities and Projects

Figure 14.2   Creating Alignment

WHAT ARE THE COMPONENTS OF AN EFFECTIVE PROJECT PROPOSAL?

An effective project proposal provides readers with insight into what needs to be accomplished and how it will get accomplished. Through the use of clearly stated mission, deliverables, goals, and objectives associated with the project, proposals sell the project. As an introduction, proposals provide background information about the need for the project. They contain a description or an overview of the expectations of the project, including details about the technical aspects of the project. Essential tasks are outlined and delineated. A thorough project proposal contains information concerning financial requirements, time constraints, and administrative and logistical support for the project. The proposal usually contains information about the key individuals associated with the project, including the identity of the project manager. Basic areas of performance responsibility are assigned, and tentative schedules and budgets are established. Figure 14.3 gives a brief summary of the typical components of a project proposal.

The project proposal creates a general understanding of:

- What is needed
- What is going to be done
- Why it is going to be done
- Who is going to do it
- When it will be done
- Where it will be done
- How it will be done
- How much it will cost

A good project proposal is effective. It clearly describes the project that can be implemented. Its proposed solution aligns with the goals and objectives of the organization and eliminates the root causes of problems that affect customers and employees.

WHAT ARE THE CHARACTERISTICS OF CLEAR PROJECT GOALS AND OBJECTIVES?

Effective project proposals state clear project goals. Project goals are established for three reasons:

1. To state what must be accomplished to complete the mission
2. To create commitment and agreement about the project goals among participants
3. To create clarity of focus for the project

Effective project managers recognize the importance of establishing clear goals and objectives for a project. They are careful to ensure that anyone working on the project, regardless of the level of her involvement, understands and supports these goals and objectives. Effective project goals are often stated in terms of the users' needs. Key questions to ask in order to ensure that a project proposal has a customer focus and clear project goals include:

Who are the end users?
What does the end user want from the project?
What does the end user say the project should do?

The Project Proposal

General Project Description

Provide a detailed description of the project that includes a statement of the project goals. Describe the major project subsystems or deliverables. Include a preliminary layout design if applicable. Address any special client requirements, including how they will be met.

- What is the purpose of the project?
- What is the scope of the project?
- Why should the project be selected?
- What is the life cycle (time from beginning to end) of the project?
- What is the complexity level of the project?
- Is there anything that makes the project unique?
- What measures will be used to judge the project's performance?

Implementation Plan

The implementation plan section contains a brief description listing the major components of the project, with time estimates for each component. This section also includes preliminary cost estimates and a preliminary schedule for the major project components.

- What are the project deliverables?
- How will these deliverables be met?
- What level of risk is associated with the project?
- What are the costs associated with the project?
- What are the time estimates for the components of the project?

Logistic Support and Administration

This section describes the facilities, equipment, and skills that are needed for the entire project.

- What difficulties may be encountered during the construction or implementation phase of the project?
- What contingency plans exist?
- Who will be the project manager?
- Who else will be necessary to help with this project?

Figure 14.3   Project Proposal Guidelines

An effective goal statement is specific, measurable, and realistic; members involved in the project agree with the statement; and time frames have been established to meet the goals. When goals are specific, they are so well defined that anyone with a basic knowledge of the project can understand them and recognize what the project is trying to accomplish. Measurable goals allow those involved in the project to judge how the project is progressing toward its mission. Agreement among project participants about the overall goals of a project is critical because without it, the project has little chance of achieving success.

Participants include but are not limited to customers, leadership, and affected departments and individuals. Given that a project is to be accepted based on its ability to meet the mission, goals, and objectives of an organization, insisting that the project be realistic seems unnecessary. Here the word *realistic* refers to the need to be aware of the time frame established for the project, as well as the manpower and financing available to the project. Limited resources, whether time or money, or unrealistic expectations given the time, money, and talent available are detrimental to a project. When a project is realistic, it is achievable given the resources, knowledge, and time available. Projects are selected to support the overall objectives of an organization. If a project has not been completed within established time periods, chances are that the organization has missed an opportunity for success. For this reason, it is critical that the project goals have clearly stated time frames for accomplishment.

Project objectives are the specific tasks required to accomplish the project goals; they clearly align with and support the project goals. In some instances, several project objectives will be necessary to ensure that the organization accomplishes a specific goal. They define who is responsible for accomplishing the goal, what resources are necessary, and what inputs will be needed. As with project goals, project objectives must also be specific, measurable, and realistic; participants must agree with them; and time frames must be established.

Establishing clear goals with supporting objectives helps effective project managers keep projects on track. Effective project managers use these goals and objectives as a way to reinforce the commitment of individuals to the project and the team. Well-written goals and objectives enhance communication, keeping everyone associated with a project aware of his role and what he needs to do in order to keep the project on track. Goals and objectives also make it easy to see how far the project has progressed and what still needs to be done. Project proposals are submitted to organizational leadership, who will judge each project based on its ability to help the organization meet its mission, goals, and objectives.

EXAMPLE 14.2   Computer Technology Transition and Upgrade Project Goals and Objectives

Max's Munchies has been working with vendors to clearly define the mission, goals, and objectives of the Computer Technology Transition and Upgrade project. The request for proposals sent to vendors states the mission as:

> *to upgrade all desktop and laptop computers to new models so the company can process more advanced applications in accounting, computer-aided design, and new business systems.*

This mission is supported by the following goals and objectives:

- Goal: Upgrade existing computers
  Supporting objectives:
    - Specify a single desktop and laptop model that runs the appropriate software
    - Install appropriate software on the new computers

> - Goal: Remove existing computers
>   *Supporting objectives:*
>   - Delete software on the old computers
>   - Dispose of existing computers in an environmentally friendly manner
> - Goal: Exchange all computers in 15 weeks
>   *Supporting objective:*
>   - Handle the logistics of the exchange with the users

WHAT ARE THE COMPONENTS OF AN EFFECTIVE PROJECT PLAN?

Once a proposal has been accepted, it becomes the framework or foundation of a project plan. Project plans are significantly more detailed than project proposals. Projects have three interrelated objectives: meeting the budget, finishing on schedule, and meeting the performance specifications set by the client. A good project plan enables an organization to accomplish all three. Although a project plan may be modified several times during the project, effective project plans remain key to organizational success because no project ever goes exactly according to plan, but well-planned projects are less likely to go astray. Project plans provide information about:

- Mission and deliverables
- Specific goals and objectives supporting the mission and its deliverables
- Tasks required to meet the goals and objectives
- Technicalities of who, what, where, when, why, and how
- Schedules—the time needed to support each aspect of the plan
- Resources—what is needed to support each aspect of the plan
- Cost analysis
- Value analysis
- Personnel—who is needed to support each aspect of the plan
- Personnel—responsibilities and assignments
- Evaluation measures for keeping the project on track
- Risk analysis—what could go wrong and how it will be handled
- Project change management process

Figure 14.4 gives a brief summary of the typical components of a project plan.

EXAMPLE 14.3 Computer Technology Transition and Upgrade Project Plan

The contractor whose proposal was accepted for the Computer Technology Transition and Upgrade project submitted the following project plan.

Project Mission

The contractor is responsible for developing and deploying a process for the replacement of all 2,000 computers—laptop and desktop—in a manner that minimizes user and

---

The Project Plan

The Project Plan

In the project plan the mission and deliverables are clarified. The plan also identifies the who, what, where, when, why, and how aspects of the project. The plan details how the project will be accomplished.

Project Plan Elements

- Overview (the mission and the deliverables; what will the final outcome be?)
- Objectives (specific objectives supporting the mission)
- General Approach (technicalities of who, what, where, when, why, how)
- Contractual Aspects (specifics of who is required to do what)
- Schedules (what time is needed to support each aspect of the plan?)
- Resources (what is needed to support each aspect of the plan?)
- Personnel (who is needed to support each aspect of the plan?)
- Evaluation Measures (performance, effectiveness, cost; how will the project be kept on track?)
- Potential Problems (what could go wrong? how will it be dealt with?)

Figure 14.4   Project Plan Guidelines

productivity disruption. This process includes retrieving the old equipment from the user, setting up the new equipment for the user by loading saved files and new software, and disposing of the old equipment. This service must be accomplished 15 weeks from the start date.

Project Goals and Supporting Objectives

The specific duties of the contractor are as follows:

Goal: Preparation Services

Objectives:

1. Assist Max's Munchies with the process of receiving and warehousing new computers. Max's Munchies will provide computer hardware and software and network cable connections.
2. Assist Max's Munchies in receiving old computers from users.
3. Perform the work on-site, including setup and connection of server cabling and table-top equipment placement. Space where the contractor can perform the upgrade and exchange must be provided at each Max's Munchies location. Contractor's equipment must be utilized as much as possible during the changeover procedure.

Goal: Deployment Services

Objectives:

1. Burn a CD of the user's data from the old PC for backup; provide the CD to the user.
2. Transfer data from the user's old computer to the new computer.

3. Load hardware and software onto new computers.
4. Maintain records of the asset exchange with the user.
5. Provide assistance to users following setup.
6. Provide a Help Desk while the new computers are being deployed.

Goal: Remediation Services

Objectives:

1. Remove all data/computer programs from old computers before disposal.
2. Dispose of all old computers in an environmentally friendly manner.

Goal: Project Control

Information about project control is provided in Example 14.7.

Goal: Schedule

The mission is to be accomplished in 15 weeks. The schedule is provided in Example 14.4.

Measures of Performance

- Price per computer
- Number of user difficulties experienced during changeover
- Amount of user downtime hours
- Time to effect changeover
- Changeovers per shift

Contingency Plans

- Five extra computers have been set aside in case the count of computers exceeds 2,000.
- Extra staffing will be available in case the rates of changeover are not high enough.

HOW ARE PROJECTS SCHEDULED?

Project planners need to know how much time is available to complete the project. It is also helpful to know whether any flexibility exists with this deadline. For this reason, planners create project schedules. *Schedules convert a project plan into an operating timetable.* This timetable is used to monitor and control project activity by showing the relationships among dates, times, activities/tasks, and people. For some people, combining longer-term project expectations with pressing day-to-day activities is often detrimental to the completion of a project. There is always the tendency to put the project off just a little longer because of the perception that more time will be available in the future. However, rarely does this time exist, resulting in a rush to complete the project on time or the abandonment of the project due to time constraints. Project schedules help remind people of the importance of working on the project on a regular basis.

How Do We Know It's Working?

Grinding is a key operation at JQOS. One of their product selling points is their ability to O.D. grind to very tight tolerances. This saves customers money because the parts do not have to go to secondary operations. To support this key operation, JQOS leadership selected a project focused on enhancing grinder performance. Here is a portion of the project plan the project team created.

Mission: Enhance grinder performance

Goals and Objectives:

Goal 1: Upgrade 10 existing grinders
  Objective: Perform preventive maintenance on each grinder

Goal 2: Improve operator comfort and reduce possibility for injury
  Objective: reduce lifting, bending, and twisting by operator as he or she loads grinder
  Objective: study lifting requirements and determine if lift assists are needed
  Objective: study work height and determine if adjustments are needed

Goal 3: Improve setup procedures
  Objective: reduce non–value-added activities
  Objective: create standardized procedures for setup
  Objective: create standardized procedures for operation

Goal 4: Increase part throughput
  Objective: reduce non–value-added activities
  Objective: reduce error potential
  Objective: pair each grinder with a lathe under one operator

Measures of Performance:

1. Throughput
2. Injury rate
3. Grinder ability to hold tolerance
4. First-pass quality

Contingency Plans

It may be necessary to split grinder production into three sizes of shafts: small, medium, and large. If this is necessary, grinders and lathes will be moved and grouped according to size. For larger production runs, it may not be possible to combine grinders with lathes. If this is necessary, a set of grinders and a lathe will be set aside from the rest.

Schedule:

Week 1:
  Goal 1: Maintenance and upgrades for first grinder
  Goals 2, 3, and 4: Film and study setups

Week 2:
  Goal 1: Maintenance and upgrades for second grinder
  Goals 2, 3, and 4: Propose changes to setup procedures

Week 3:
  Goal 1: Maintenance and upgrades for third grinder
  Goals 2, 3, and 4: Propose changes to operating procedures and determine changes needed to improve human factors and safety

Week 4:
  Goal 1: Maintenance and upgrades for fourth grinder
  Goals 2, 3, and 4: Contact vendors for human factors changes

Week 5:
  Goal 1: Maintenance and upgrades for fifth grinder
  Goals 2, 3, and 4: Select and order needed components for human factors and safety changes and get approval of new setup and operational procedures
  Goals 2, 3, and 4: Begin pairing grinders with lathes

Week 6:
  Goal 1: Maintenance and upgrades for sixth grinder
  Goals 2, 3, and 4: Continue pairing grinders with lathes, install new components as they arrive from vendor, and begin training new setup and operational procedures

Week 7:
  Goal 1: Maintenance and upgrades for seventh grinder
  Goals 2, 3, and 4: Continue pairing grinders with lathes, install new components as they arrive from vendor, and continue training

Week 8:
  Goal 1: Maintenance and upgrades for eighth grinder
  Goals 2, 3, and 4: Wrap up modifications and machine movements

Week 9:
  Goal 1: Maintenance and upgrades for ninth grinder
  Goals 2, 3, and 4: Study how modifications are performing

Week 10:
  Goal 1: Maintenance and upgrades for tenth grinder
  Goals 2, 3, and 4: Make any necessary modifications to new methods and procedures

Week 11:
  Complete project

In order to create a schedule, project planners must know the following: the tasks, the order in which they must be completed, when they must be completed, and the rate at which they can be completed. To schedule a project, planners lay out the tasks and activities associated with a project plan according to the time it will take to complete them.

Once a list of activities has been created, time estimates for each activity are derived. Starting with the project due date, the activities are stepped backwards in time, eventually reaching a start date as all the activities are accounted for. Realistic time estimates need to be created in order to determine how much total time it will take to complete the activities and tasks associated with a project. As the project progresses, it is not unusual for technical difficulties to arise. Solving these difficulties often takes longer than originally planned. In other cases, materials and manpower are unavailable or are late, resulting in changes to schedules and task sequencing that alter time estimates.

A project schedule is often monitored through the use of checkpoints and milestones. Milestones represent long-term or major events that have been or need to be completed for a project. To reach a milestone, a series of smaller activities or tasks need to have taken place (Figure 14.5). Projects unfold as a logical sequence of activities take place or as tasks are completed. Therefore, the relationship between these activities is critical to any project. When taken out of order, these activities or tasks waste time and effort. Checkpoints are smaller points throughout a project that are used to judge how far the project is from completion.

Gantt charts, the program evaluation and review technique (PERT), and the critical path method are excellent tools for monitoring the complex links of activities associated with a project. Available on Microsoft Project®, these charts are invaluable when scheduling a project. A Gantt chart, shown in Figure 14.6 for the Computer Technology Transition and Upgrade Project, enables the user to keep track of the flow and completion of various tasks associated with a project. The chart promotes the identification and assignment of clear-cut tasks while enabling users to visualize the passing of time. Divisions on the chart represent both an amount

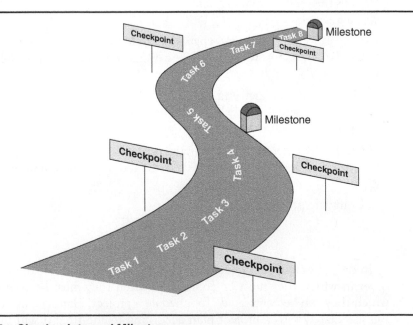

Figure 14.5   Checkpoints and Milestones

EXAMPLE 14.4 Computer Technology Transition and Upgrade Project Schedule

The contractor whose project proposal was accepted for the Computer Technology Transition and Upgrade project submitted the following schedule with the project plan.

Schedule

Major milestones and deployment rates are as follows. This schedule outlines the time frame and the pace of activity.

Phase 1: Project Preparation and Pilot Rollout

Week 1   Contractor's Project Manager begins.
Week 2   Pilot Rollout begins: 10 PCs per day exchanged. Goal: 50 for the week.
Week 3   Pilot Rollout continues: 12 PCs per day exchanged. Goal: 60 for the week.
Week 4   Pause for review of rollout. Some rollout will occur as necessary for priority needs (estimate 20 for the week).
          Phase 1 Checkpoint: 130
          Phase 1 Milestone: 130

Phase 2: Rollout

Week 5   Main Rollout begins: 20/day, 100 for the week.
Week 6   Main Rollout: 25/day, 125 for the week.
Week 7   Main Rollout: 30/day, 150 for the week.
Week 8   Main Rollout: 30/day, 150 for the week.
          Phase 2 Checkpoint: 525
          Phase 2 Milestone: 655

Phase 3: Rollout

Week 9   Main Rollout: 40/day, 200 for the week.
Week 10   Main Rollout: 40/day, 200 for the week.
Week 11   Main Rollout: 40/day, 200 for the week.
Week 12   Main Rollout: 40/day, 200 for the week
          Phase 3 Checkpoint: 800
          Phase 3 Milestone: 1,455

Phase 4: Wind-Down

Week 13   Main Rollout: 40/day, 200 for the week.
Week 14   Main Rollout: 40/day, 200 for the week.
Week 15   Finishing activities: 30/day, 150 for the week.
          Phase 2 Total: 550
          Phase 2 Milestone: 2,005

of time and a task to be done. A line drawn horizontally through a space shows the amount of work actually done compared to the amount of work scheduled to be done.

A PERT chart improves upon a Gantt chart by showing the relationships between tasks (Figure 14.7). Unlike the Gantt chart, which is a list of tasks, the PERT chart enables the

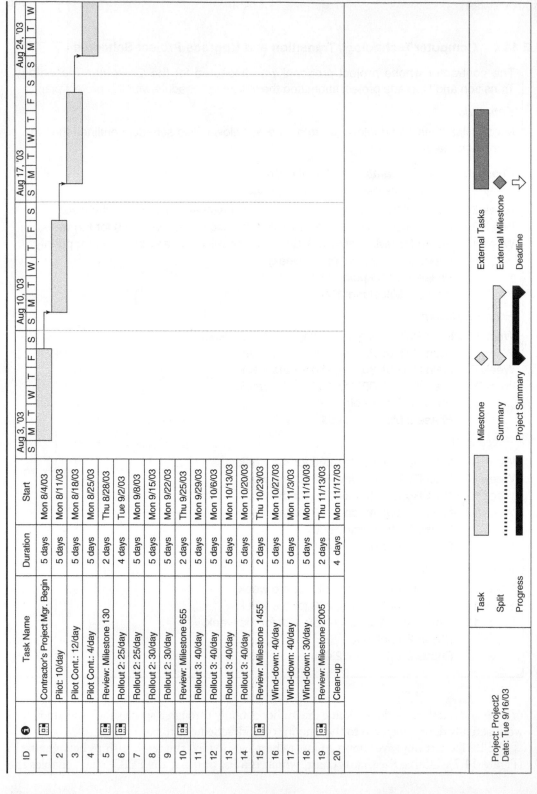

ID	⊕	Task Name	Duration	Start	Aug 3, '03	Aug 10, '03	Aug 17, '03	Aug 24, '03
1	▣	Contractor's Project Mgr. Begin	5 days	Mon 8/4/03				
2		Pilot: 10/day	5 days	Mon 8/11/03				
3		Pilot Cont.: 12/day	5 days	Mon 8/18/03				
4		Pilot Cont.: 4/day	5 days	Mon 8/25/03				
5	▣	Review: Milestone 130	2 days	Thu 8/28/03				
6	▣	Rollout 2: 25/day	4 days	Tue 9/2/03				
7		Rollout 2: 25/day	5 days	Mon 9/8/03				
8		Rollout 2: 30/day	5 days	Mon 9/15/03				
9		Rollout 2: 30/day	5 days	Mon 9/22/03				
10	▣	Review: Milestone 655	2 days	Thu 9/25/03				
11		Rollout 3: 40/day	5 days	Mon 9/29/03				
12		Rollout 3: 40/day	5 days	Mon 10/6/03				
13		Rollout 3: 40/day	5 days	Mon 10/13/03				
14		Rollout 3: 40/day	5 days	Mon 10/20/03				
15	▣	Review: Milestone 1455	2 days	Thu 10/23/03				
16		Wind-down: 40/day	5 days	Mon 10/27/03				
17		Wind-down: 40/day	5 days	Mon 11/3/03				
18		Wind-down: 30/day	5 days	Mon 11/10/03				
19	▣	Review: Milestone 2005	2 days	Thu 11/13/03				
20		Clean-up	4 days	Mon 11/17/03				

Project: Project2
Date: Tue 9/16/03

Task		Milestone	◇	External Tasks	
Split		Summary		External Milestone	◆
Progress		Project Summary		Deadline	⇩

Figure 14.6   Gantt Chart

470

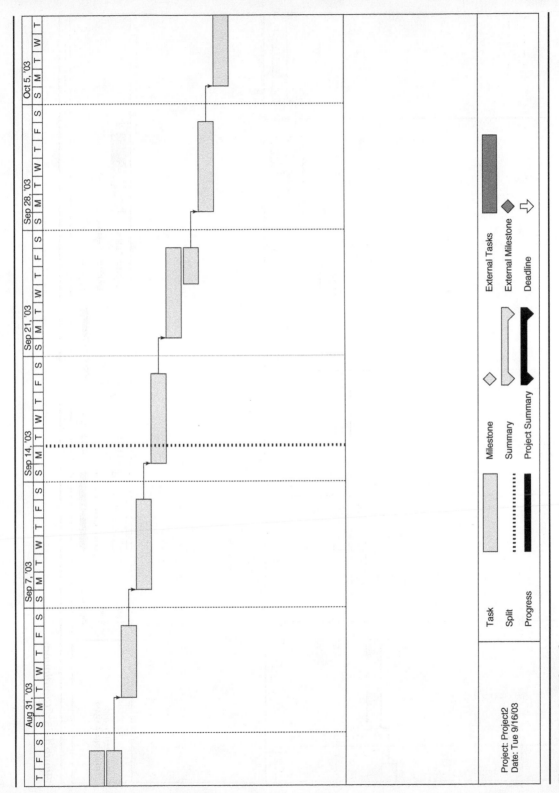

Figure 14.6 *(continued)*

471

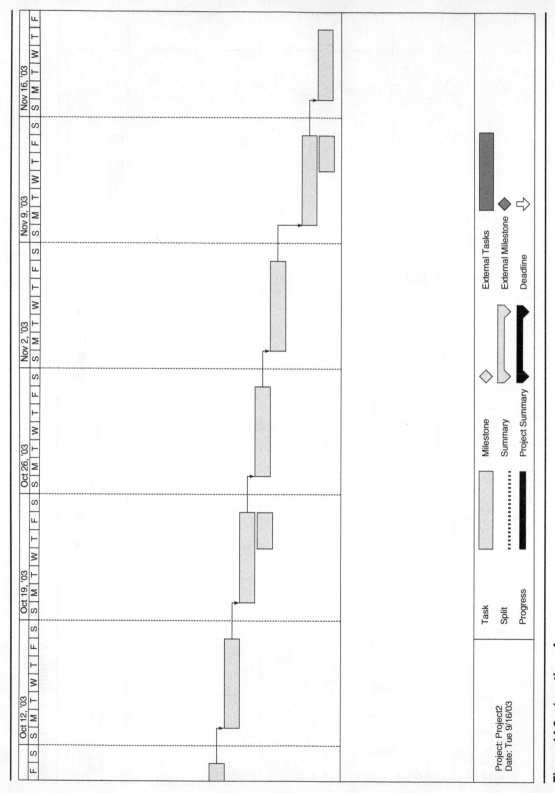

Figure 14.6 *(continued)*

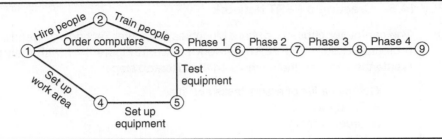

Figure 14.7   PERT Network (Example 14.5)

project to be viewed as an integrated whole. Because it coordinates and synchronizes many tasks, it is well designed to handle complex projects. To create a PERT network:

1. Compile a list of events/tasks/activities.
2. Determine the relationships between the activities (predecessors, successors).
3. Begin constructing the diagram from the end, working back to the beginning. Place the key events/tasks/activities identified in step 1 on the diagram between the nodes. Related nodes, those with predecessors and successors, are linked as shown in Figure 14.8.

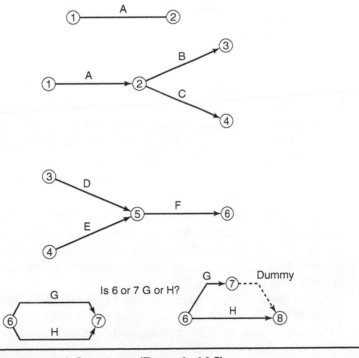

Figure 14.8   Common Network Structures (Example 14.5)

EXAMPLE 14.5   Creating a PERT Network

The contractor chosen for Max's Munchies Computer Technology Transition and Upgrade project submitted the PERT network shown in Figure 14.7 with their project plan. To create the network, they completed the following steps:

1. Compile a list of events/tasks/activities.
   - Set up work area
   - Order computers
   - Hire people to perform exchange/upgrade
   - Train people to perform exchange/upgrade
   - Phase 1
   - Phase 2
   - Phase 3
   - Phase 4
   - Set up equipment
   - Test equipment
2. Determine the relationships between the activities (predecessors and successors).

Task	Predecessor
Set up work area (A)	—
Order computers (B)	—
Hire people to perform exchange/upgrade (C)	—
Train people to perform exchange/upgrade (D)	C
Phase 1 (E)	A, B, C, D, I
Phase 2 (F)	E
Phase 3 (G)	E, F
Phase 4 (H)	E, F, G
Set up equipment (I)	A
Test equipment (J)	A, I

3. Begin constructing the diagram from the end, working back to the beginning. The key events/tasks/activities identified in step 1 are placed on the diagram between the nodes as shown in Figure 14.8.

The critical path method (CPM) builds on PERT by adding the concept of cost per unit time. CPM gets its name from its ability to determine the longest series of interrelated events that must be completed in a project—the critical path. To use this chart, one must estimate both the times and the costs associated with the activities. The following example describes the creation of a CPM.

If the speed of the project needs to be increased for any reason, then the path can be "crashed" or shortened by determining which activities can be done more quickly. The CPM also reveals the cost of doing each of these activities more quickly. For instance, if Max's Munchies wishes to complete the project in less than 28 weeks, they must decide whether to speed up the ordering of the computers or one of the four phases. Each week of

Task	Predecessor	Weeks		Cost	
		Normal	Crash	Normal	Crash
Set up work area (A)	—	2	2	$25	0
Order computers (B)	—	13	7	$20	$5/wk
Hire people to perform exchange/upgrade (C)	—	6	3	$12	$6/wk
Train people to perform exchange/upgrade (D)	C	5	4	$25	$10/wk
Phase 1 (E)	A, B, C, D, I	4	3	$100	$20/wk
Phase 2 (F)	E	4	3	$100	$20/wk
Phase 3 (G)	E, F	4	3	$100	$30/wk
Phase 4 (H)	E, F, G	3	2	$100	$40/wk
Set up equipment (I)	A	5	4	$13	$2/wk
Test equipment (J)	A, I	2	1	$10	$3/wk

Figure 14.9   Precedents, Timing, and Costs (1,000) (Example 14.6)

increased speed in order processing will cost $5,000 compared to increasing the speed of the actual transition at a cost of $20,000 to $40,000.

EXAMPLE 14.6   Creating a CPM

The steps to creating a critical path begin much the same as those of creating a PERT network. Building on Example 14.5, to finish the CPM, follow these steps:

1. Compile a list of tasks/activities (see Example 14.5).
2. Determine the relationships between the activities (predecessors and successors). See Example 14.5 and Figure 14.9.
3. Determine the costs and times associated with the activities (Figure 14.9).
4. Begin constructing the diagram from the end, working back to the beginning (Figure 14.10).
5. Add up each of the individual paths to determine the critical path. The critical path for this example is 1-3-6-7-8-9 and equals 28 weeks. The other paths are 1-4-5-6-7-8-9 (24 weeks) and 1-2-3-6-7-8-9 (26 weeks).

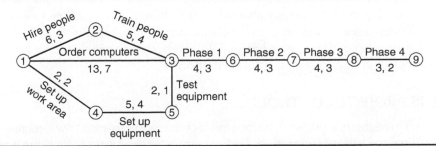

Figure 14.10   CPM (Example 14.6)

HOW ARE RESOURCES BUDGETED?

Budgets are plans for allocating and monitoring the use of scarce resources. They set the overall estimates for the costs associated with a project. For budgets to be realistic, it is critical that the people actually closest to the work take part in determining the time and money needed for a project. In order to create a budget, project managers study the key activities taking place and estimate as accurately as possible the time and money required for completion. A complete budget has information about any income or revenue expected during the life of the project, all expenses related to the project, and cash flow projections and their timing. Budgets are not arbitrary; effort should be taken to ensure that realistic cost and timing information has been found. Budgets should be monitored throughout the life of a project in order to keep project costs on track. Variances should be reported as soon as they are found. These variances can point to problems and recommended actions necessary to keep the project on target. Budgets can enable those working on a project to be aware of how their actions add to or take away from the end result of the project.

WHAT ARE CONTINGENCY PLANS AND CHANGE CONTROL SYSTEMS?

Projects, due to their very nature, are complex. For every project, there is a risk of failure. Organizing a project into manageable sections can reduce the fear of failure as well as the potential for failure. Effective project managers ask, "What if?" They work with their project team members to identify potential activities or events that may derail a project. *Contingency plans are created to ensure that the project team is ready to handle potential problems.* Although not all problems can be foreseen, a project plan that includes contingency plans keeps the team flexible and aware that they may be asked to make adjustments to their project plan some time during its lifetime. Contingency plans were included in the project plan submitted in Example 14.3.

Effective project managers recognize that clients make changes to a project as it progresses. All project proposals and plans should contain a description of how requests for changes in a project's plan, budget, schedule, or performance deliverables will be handled. An effective change control system has steps in place that review the requested changes for both content and procedure and that identify how the change will impact the project. The impact must be reflected in adjustments to the project's performance objectives and to the schedule and budget. Once accepted, change orders become part of the overall project plan. Part of the job of a change control system is to clearly communicate any changes to any person or part of the project affected by the change. The best way to ensure that this critical communication occurs is to have all changes approved in writing by all appropriate representatives of the impacted areas. Ultimately, the change should be made only if its benefits outweigh the costs of its implementation.

WHAT IS PROJECT CONTROL?

Throughout a project, effective project managers monitor the progress that the project is making toward completion. Performance, cost, and time—the three aspects of a project— all need to be monitored and controlled in order to ensure project success. *Performance* of the project refers to the end result and the steps taken to get to it. The performance of a project may be affected by unexpected technical problems, quality or reliability problems, or insurmountable technical difficulties. For instance, the completion of a building may be delayed because the foundation must be redone due to technical problems with the concrete being used. Performance of a project may also be affected by insufficient resources brought about by poor planning, logistical problems, or underestimating. Some performance problems may be brought about by the project's end user, for example, if the end user has made a variety of changes to the project's original specifications.

Project control and monitoring involves gathering and appraising information on how the project's activities compare with the project plan. Actual progress is tracked against the performance measures established in the project plan. These performance measures help a project manager assess how time, money, and other resources have been used to produce the expected outcomes. Costs need to be monitored during a project for many reasons. Technical difficulties may require more resources than originally planned. Client-related changes in specifications may have changed the scope of the work significantly, thus affecting the total costs of the project. The budget may have been inadequate in the first place due to inadequate estimates, poor projections for inflation, or additional costs due to client-related changes.

EXAMPLE 14.7 Computer Technology Transition and Upgrade Project Control

The project plan submitted by the contractor for Max's Munchies contained this section detailing project control goals and objectives.

Goal: Project Control

Objectives:

1. The contractor will provide a project manager who is responsible for the entire scope of the project and for the direction of the contractor's personnel involved in the project. This individual will be the single point of contact.
2. The contractor's project manager will develop the schedule and logistical planning for the Computer Technology Transition and Upgrade project. The schedule will include milestones based on the number of computers to be upgraded. The milestones will become the basis for the forecast on each phase of the project. The schedule will include resource planning indicating the level of staffing necessary to achieve project milestones.
3. The contractor's project manager will schedule and oversee all phases of the Computer Technology Transition and Upgrade project.
4. The contractor's project manager will supervise on-site personnel directly or will assign supervisory person(s) during periods of critical processes.
5. The contractor's project manager will participate in review and problem-solving meetings scheduled over the course of the project.
6. The contractor will create processes and procedures that meet the objectives of schedule, achieve user satisfaction in the equipment exchange, and attain inventory control.

Costs can also get out of control when the project costs are not watched closely and corrective cost control was not exercised in time. As discussed in detail in Chapter 9, measures of performance may be market driven, financial in nature, focused on internal processes, or related to organizational learning and growth.

By closely monitoring the performance measures associated with a project, an alert project manager can be prepared to respond quickly to deviations in order to keep the project on track and under control. Although very few projects have not had their goals and objectives modified in some way, from their beginning to their end, careful project control enables a project manager to minimize the effects of these changes on the overall project.

Audits, covered in Chapter 15, may also be used to study a project's progress and performance. Audits may be conducted at any time during a project. Project audits focus on project performance deliverables, costs, and time schedules. Project audits use evaluation measures to compare what was actually planned with what actually occurred. Project auditors hope to learn more about the relationship between performance deliverables, costs, and schedules. They hope to gain insight into how to improve the project's current state, how to reduce costs, and how to increase the speed of the project. Audits are often used to identify problems or mistakes. The point of an audit is not to affix blame but to find remedies for the problems and mistakes and to prevent them from happening in the future. Audits often reveal how future projects can be better organized and managed.

WHAT MUST A PROJECT MANAGER DO IN ORDER TO BE EFFECTIVE?

An effective project manager achieves the desired results within budget, on time, and according to the desired standards. Effective project managers realize that in order to accomplish what needs to be done on time and within budget, they must take time to plan their projects. Once they have created a good plan, effective project managers manage their plan.

Unlike functional managers, project managers are generalists with knowledge and experience in a wider variety of areas. Project managers are responsible for organizing, directing, planning, and controlling the events associated with a project. They deal with budgets and schedules. Responsibility for a project rests on their shoulders, and they must understand what needs to be done, when it must be done, and where the resources will come from. Throughout a project the manager must clarify misunderstandings; calm upset clients, leaders, and team members; and meet the client's demands—all the while keeping the project on time and within budget. Project managers are responsible for finding the necessary resources, motivating personnel, dealing with problems as they arise, and making project goal trade-offs. In essence, an effective project manager does whatever is necessary to keep the project on schedule, within budget, and able to meet performance expectations. Project managers must be prepared to make adjustments to schedules, budgets, and resources in order to deal with the unexpected. For this reason, they must be good at recognizing the early signs of problems and must be able to cope with stressful situations. Effective project managers utilize the checkpoints, activities, and time estimates established in a project plan to guide those working on the project. Following a clearly laid out project schedule, with clearly delineated responsibilities, enables effective project managers to keep their projects on track in terms of time, performance, and cost. Clear project plans enable the effective project manager to direct people individually as well as a team.

Project managers manage people as well as projects. To do so, effective project managers schedule frequent progress reports. These meetings allow the project manager to react quickly when he or she recognizes that a difficulty has arisen. Effective meetings are essential when working on a project. As discussed in Chapter 7, meeting minutes—when created on a timely basis—enable everyone involved to understand at a glance what action items exist and who is responsible for getting them done.

Due to its very nature, the people associated with a project are temporary. The project work is often assigned in addition to employees' regular jobs. If this is the case, how does a project manager maintain the commitment and involvement of these individuals in order to get the project done? A project manager must motivate these individuals. Motivation can be achieved in a variety of ways, including increasing an employee's visibility in the organization. In other words, ensure that people working on a project are recognized for their work and their accomplishments. Project managers also have it within their power to create interesting and challenging possibilities for their team members. As discussed in Chapter 7, people are much more motivated to perform when their assigned tasks enable them to use and stretch the talents, skills, and knowledge they already possess. Another powerful tool a project manager possesses is praise. People like being recognized both publicly and privately for a job well done.

LESSONS LEARNED

Quality management systems, Six Sigma programs, and the Malcolm Baldrige National Quality Award criteria recognize that projects managed by people with good project management skills are crucial to the success of an organization. When chosen because they support the overall goals and objectives of an organization, projects move an organization toward its vision and mission. Without projects, organizations lack a long-term focus and instead focus solely on day-to-day activities.

Conversely, poorly managed projects are costly for an organization. A project that does not achieve its goals and objectives or one that is not aligned with the goals and objectives of the organization wastes time, money, and other resources. Misdirected projects, poorly planned projects, or poorly guided projects are demoralizing for an organization's members. Care must be taken to ensure that projects are appropriately selected to support an organization's mission and goals. Once selected, a project must be served well by its project manager and team in order to ensure its successful completion.

1. Projects have three characteristics: performance, cost, and time.
2. Projects are selected based on their ability to contribute to one or all three of the following: customer-perceived value and satisfaction, the organization's financial strength, or the organization's operational necessities.

3. A project proposal provides clear information concerning the goals and objectives that a particular project hopes to achieve.

4. A good project proposal answers the questions of who, what, where, when, why, how, and how much.

5. Project goals and objectives create clarity.

6. A project plan provides the details of how the project will be accomplished.

7. Schedules convert a project plan into an operating timetable.

8. Gantt and PERT charts are used to schedule and track projects.

9. Budgets are plans for allocating and monitoring the use of scarce resources.

10. Contingency plans are created to ensure that the project team is ready to handle potential problems.

11. An effective project manager achieves the desired results within budget, on time, and according to the desired standards.

Are You a Quality Management Person?

On a day-to-day basis, we are all faced with projects to do. How do you manage your projects? Check out the quality management person questions and find out whether you are a quality management person. How do you answer them on a scale of 1 to 10?

Do you select your projects based on how they align with your personal goals and objectives?

1	2	3	4	5	6	7	8	9	10

When selecting a project, do you consider a project's performance, cost, and time objectives?

1	2	3	4	5	6	7	8	9	10

When faced with choosing among multiple projects, do you compare and contrast the benefits or costs associated with projects that will require the same resources, like time and money?

1	2	3	4	5	6	7	8	9	10

Do your projects have clear goals and objectives?

| 1 | 2 | 3 | 4 | 5 | 6 | 7 | 8 | 9 | 10 |

Do you plan out your projects in advance?

| 1 | 2 | 3 | 4 | 5 | 6 | 7 | 8 | 9 | 10 |

Do you schedule your project work into your day-to-day activities?

| 1 | 2 | 3 | 4 | 5 | 6 | 7 | 8 | 9 | 10 |

Do you have projects just waiting to be finished?

| 1 | 2 | 3 | 4 | 5 | 6 | 7 | 8 | 9 | 10 |

Do you start something and fail to finish it?

| 1 | 2 | 3 | 4 | 5 | 6 | 7 | 8 | 9 | 10 |

Are you getting things done on time?

| 1 | 2 | 3 | 4 | 5 | 6 | 7 | 8 | 9 | 10 |

Do you budget for your projects?

| 1 | 2 | 3 | 4 | 5 | 6 | 7 | 8 | 9 | 10 |

Do you have contingency plans in place for problems that might arise?

| 1 | 2 | 3 | 4 | 5 | 6 | 7 | 8 | 9 | 10 |

Chapter Questions

1. Why is it important that project management skills be applied?
2. Describe the three basic project characteristics in terms of a project you have worked on.

3. Why is completing a project like eating an elephant?

4. How is a project selected?

5. What are the components of an effective project proposal?

6. Consider a project you are working on either at work or at school. Using the guidelines presented in this chapter, write a project proposal. Why should your project proposal be accepted? Are the reasons for selecting your project apparent in your proposal?

7. What are the components of an effective project plan?

8. Consider a project you are working on either at work or at school. Using the guidelines presented in this chapter, write a project plan. Are the project's goals and objectives clearly stated? How do you know?

9. Create a Gantt chart for a project you are working on either at work or at school.

10. What differentiates a Gantt chart from a PERT chart from a CPM?

11. A not-for-profit organization is interested in buying caramels and selling them to raise money. Create a PERT chart for the following information:

Task	Predecessor	Weeks		Cost	
		Normal	Crash	Normal	Crash
Design ads (A)	—	2	2	$250	0
Order stock (B)	—	12	8	$200	$35/wk
Organize salespeople (C)	—	6	3	$120	$60/wk
Place ads (D)	A	3	2	$25	$10/wk
Select distribution sites (E)	C	4	3	$100	$20/wk
Assign distribution sites (F)	C, E	4	3	$100	$10/wk
Distribute stock to salespeople (G)	C, E, F	2	1	$100	$25/wk
Sell caramels (H)	E, F, G	5	3	$100	$40/wk

12. Complete a CPM for the information in Question 11.

13. Suppose that the leaders of the project completed in Question 12 wanted to speed up their project by three weeks. What would be the most cost-effective way of accomplishing that?

14. What does it mean to keep a project under control? How is a project controlled?

15. What are contingency plans? Why is it important to have contingency plans?

16. What is a change control system? How are change control systems structured? What are they used for?

17. What does it take to be an effective project manager?

15 Business Results

Why are business results key to effective organizations?

How does auditing allow an effective organization to see its progress?

What types of audits exist?

How is an audit designed?

What do effective organizations audit?

What happens when the audit is over?

How do the Malcolm Baldrige National Quality Award criteria encourage a focus on business results?

Lessons Learned

Learning Opportunities

1. To understand why business results are key to effective organizations
2. To understand what an audit is, how it is performed, and what audits are used for

Create a constancy of purpose toward improvement of product and service, with the aim to become competitive and to stay in business and to provide jobs.

Dr. W. Edwards Deming

WHY ARE BUSINESS RESULTS KEY TO EFFECTIVE ORGANIZATIONS?

The first of Dr. Deming's 14 points begins with "create a constancy of purpose," encouraging organizations to focus on their vision. Dr. Deming goes on to say that the path toward the vision should "aim to become competitive and to stay in business and to provide jobs." One of the key drawbacks of the original quality movement was its lack of focus on business results. Quality management systems, the Malcolm Baldrige National Quality Award, and Six Sigma programs, on the other hand, have business results as their main focus.

Effective organizations compete based on value creation, delivery, and market performance. As discussions in this text have shown, high-quality products and services come from good practices. Good people must be hired and trained. Good leaders must provide direction and focus. Good communication must exist. Good designs must exist for products and services. Good materials and equipment must flow into well-designed processes. Good tools and techniques must be implemented to improve processes to meet market needs. As markets evolve, good continuous training and education must support the changing environment. Good supplier and customer relationships must exist. Quality management provides the framework needed for pursuing organizational excellence (Figure 15.1).

Effective organizations focus on results. They are always measuring their progress toward their goals and objectives. They are constantly evaluating their customers, products and services, overall financial and market performance, employee satisfaction and performance, leadership, and culture. The knowledge they gain by measuring and monitoring their performance is put to good use to focus on key quality and customer requirements, identify marketplace product and service differentiators, and determine appropriate process improvements. Their improvement activities are linked to improving the overall financial health and market performance of the organization. Results, or the outcomes provided, drive the choice of project selected (Figure 15.2). Strong performance satisfies customers and develops loyalty, repeat business, and referrals.

How Do We Know It's Working?

Leaders at JQOS have linked their vision, mission, goals, and objectives with the day-to-day activities of the firm (Figure 15.3). Their performance measures have helped them stay on track and on target during the past two years of significant change. Weekly and monthly, they review these measures and make the necessary adjustments to stay the course that they have set for themselves. Their results show their success. Sales have risen from $4 million to $8 million a year. The number of jobs handled weekly by the plant has risen from 30 to 60 per week. Lead time has fallen from four to six weeks to nearly two weeks. First-pass quality has increased from 85% to 95%. Profit has risen from 2% to 8%, a figure that includes money spent on improvements. The number of employees has risen from 26 to 36, primarily due to adding a second shift to handle the increase in customer orders. Even with the additional employees, the ratio between JQOS's sales and the number of people they employ is strong, $4/26 to $8/38. By making improvements to their

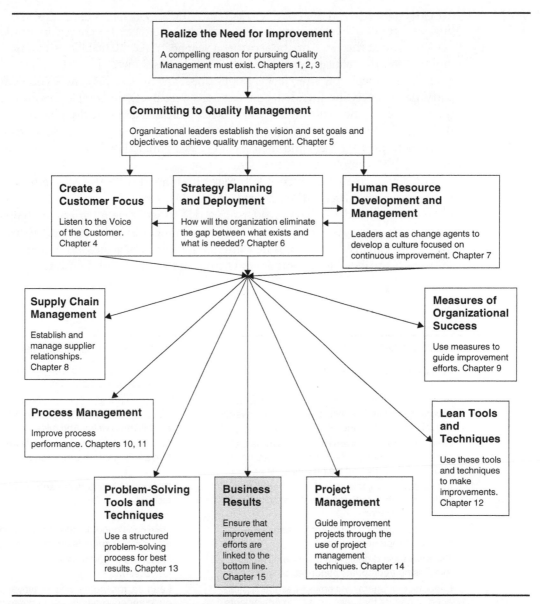

Figure 15.1 Achieving Organizational Success

billing and account receivable processes, despite the significant increase in sales, the dollar amount of outstanding receivables has not increased.

As an interesting side effect of all these changes, employee attitudes have changed dramatically. When working, they show a sense of camaraderie and community. Their dress and personal hygiene have risen noticeably. Never was this change in attitude more evident than at the two Christmas parties a year apart. At the first Christmas party, three months after the management changeover, employees showed their lack of pride in their workplace through low attendance, sloppy dress, and attitude. At the second Christmas party, nearly everyone attended, conversations were lively, and employees dressed up for the occasion. Participation at other events has increased dramatically. Having completed their first "million-dollar" quarter, employees were rewarded with free passes to a nearby amusement park for them and their families. Two hundred and twelve people attended, or nearly 100% employee participation.

The approach that enabled JQOS to set itself apart from the competition was not trying to reinvent itself with new products but rather making significant improvements to key systems and processes. Leadership at JQOS will face a variety of challenges in the coming years. As they continue to improve processes, their strategic focus is moving toward expansion of the business into light assemblies, gears, and pinions. A second shift has been added to support growing customer demand and their future plans.

Results-Driven Improvements	Activity-Based Improvements
Goals and objectives are specifically linked to the voice of the customer.	Goals and objectives are generally linked to the voice of the customer.
Improvement efforts have goals and objectives established based on their alignment with overall organization vision and mission.	Improvement efforts have goals and objectives that may or may not be aligned with the overall organization vision and mission.
Goals and objectives are refined based on their ability to contribute to the overall financial health and market position of the organization.	Improvement efforts have goals and objectives that are larger in scale or long-term focused.
Improvement projects are selected based on their ability to contribute to the overall financial health of the organization.	Improvement efforts are selected in a random fashion.
Training is specific to immediate needs for a project or improvement effort.	Training is general, often focused on vocabulary rather than tools and techniques.
Progress toward goals and objectives is monitored using appropriate performance measures.	Measures of performance are broad and unfocused.
Leaders, managers, and employees are encouraged to ensure that changes and improvements will actually result in achieving organizational goals and objectives.	Leaders, managers, and employees are encouraged to participate in improvement teams.
Leaders and managers are actively involved in the improvement process as process owners, providing funding and support.	Leaders and managers serve as cheerleaders, not as process owners.
Positive finanacial results are expected, not hoped for.	Activities are measured (counted) rather than end results.

Figure 15.2   Results-Driven Improvement Efforts versus Activity-Based Improvement Efforts

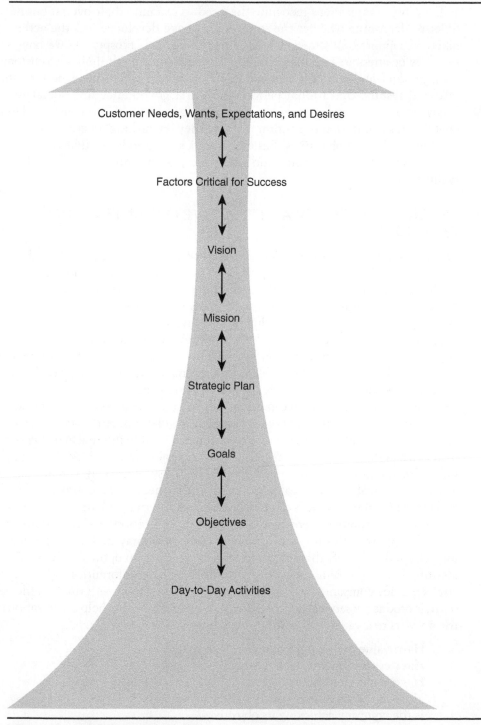

Customer Needs, Wants, Expectations, and Desires

Factors Critical for Success

Vision

Mission

Strategic Plan

Goals

Objectives

Day-to-Day Activities

Figure 15.3   Creating Alignment

489

Effective organizations recognize the need to examine their overall business results in order to determine whether the strategy they have developed and the actions they have taken to support that strategy are helping the business prosper. As we have seen in this text, key business areas that are examined include but are not limited to customer satisfaction, product and service performance, financial and marketplace performance, human resource results, operational performance, and governance and social responsibility. Studying business results, using measures of performance like those discussed in Chapter 9, enables an organization to confirm whether they are making progress toward their vision, mission, goals, and objectives. Business results should indicate that the goals and objectives described in the organization's strategic plan are not pipe dreams but achievable realities.

HOW DOES AUDITING ALLOW AN EFFECTIVE ORGANIZATION TO SEE ITS PROGRESS?

Studying current performance levels, trends over time, and comparison data for the key measures and indicators allows an organization to see what is working and what is not. Audits are designed to appraise the activities, practices, records, or policies of an organization; they determine whether a company has the ability to meet or exceed a standard. Audits provide many benefits, including allowing an organization to see what is really taking place. This insight can enable the organization to take action before potential problems become serious issues. Information provided by the audit can point to where action is necessary in order to contain the problem. Audits enable organizations to see which processes, policies, procedures, and practices are effective and which are not.

A variety of circumstances can initiate an audit. Audit programs may be part of meeting customer contract requirements or government regulations. Audits do not have to have an outside instigator; it is not unusual to see a company utilize internal auditing systems to verify its own performance to key metrics. Audits may also examine aspects of equipment, software, documentation, and procedures. Audits of systems such as material handling or accounts receivable can reveal poor practices that need to be improved. Supplier quality and record-keeping practices can also be checked. Internal and supplier audits allow a company the opportunity to verify conformance to specifications and procedures. Since audits identify opportunities for improvement, companies may perform a product or service integrity audit to verify that the process is performing in an optimal fashion. Using audits to identify process problems reduces opportunities for nonconformities. Whatever the reason, audits provide companies with information concerning their performance, the performance of their product or service, and areas for improvement. Audits help organizations determine the answers to a variety of questions including:

Have value-driven improvements been made?
Have costs of quality been reduced?
Have lead times been reduced?
How do the measures of performance measure up?
Is the company achieving its objectives?

Are the correct procedures being followed?

Are new and more efficient methods of performance documented and used where applicable?

Are records being properly retained and used to solve production problems?

Are preventive maintenance schedules being followed?

The frequency of audits varies according to need. Areas having a significant effect directly on product creation, service provision, and product or service safety or quality are targeted for more frequent audits. When audits are not based on customer or government requirements, their frequency is usually related to the need to balance audit effectiveness with economics. To decrease the impact and disruption caused by a major company-wide audit, an organization may choose to audit one or two at a time—areas, systems, or processes—until all key areas have been evaluated.

Audits should be a positive experience used to improve a system. When deficiencies are uncovered, they should be seen as opportunities to look for solutions, not to affix blame. An audit involves comparisons, checks of compliance, and discoveries of discrepancies. Because the news audits provide is not always positive, those conducting an audit may not be well received by the area being audited. To be successful, good auditors should be polite, objective, and professional. Audits should be conducted in an objective and factual manner. Audits are not subjective assessments against personal standards. The auditor's opinion should be unbiased, unfiltered, and undistorted. The objectives, criteria, and measures against which an area is to be compared should be well defined before the audit is begun. In some instances a great deal of perseverance is necessary to find the information desired. Those involved in an audit should be notified as early as possible about the scope and breadth of the audit.

WHAT TYPES OF AUDITS EXIST?

Audits are designed to determine whether deficiencies exist between actual performance and desired standards. They can cover the entire company, a division of a company, or any portion of the processes that provide a product or service. Audits may be focused on product development and design, material procurement, billing, order taking, or production. A customer may request an audit prior to awarding a contract to a supplier. Other types of audits include product design audits, preproduction audits, compliance audits, production audits, and supplier-quality system audits.

HOW IS AN AUDIT DESIGNED?

Auditing is a process. It requires step-by-step consistent planning, implementation, study, and action on the results in order to be effective. Applying the Shewhart Plan-Do-Study-Act cycle, made popular by Dr. Deming, can help a company create an organized system for conducting an audit and using the information to make improvements. Typical auditing programs include a planning phase, the actual audit, reports recommending improvement, and follow-up action plans (Figure 15.4).

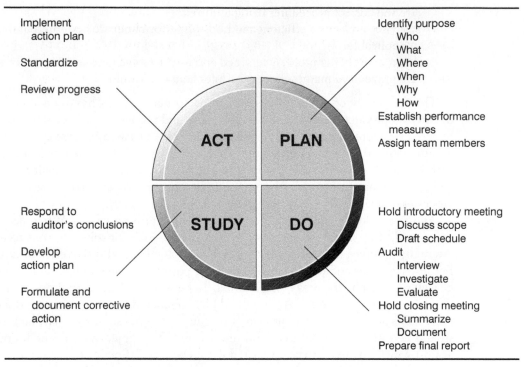

Figure 15.4   The Auditing Process

1. **Plan:** To begin, those planning an audit need to identify its purpose or objective. A statement of purpose clarifies the focus of the audit. Following this, planners will need to identify the who, what, where, when, why, and how as related to the audit:

- Who is to be audited?
- Who is to perform the audit?
- What does this audit hope to accomplish?
- What are the performance measures?
- Where the audit will be conducted?
- When will the audit be conducted?
- Why is the audit being conducted?
    - Will the audit judge conformance to standards?
    - What are the critical standards?
- How will the audit be conducted?

Those about to be audited should be informed by an individual in a position of authority. Also, clear statements of the reasons behind the audit (why), the performance measures (what), and the procedures (how) should be given to those being audited. An audit is a valuable tool for improvement. It is important to determine how the results will be used and who will have access to the results before the auditing process begins.

2. **Do:** Using the information clarified in the planning phase, the audit is conducted. Audits go smoother if everyone affected by the audit is informed in advance. This removes the feeling that a witch hunt is about to occur. Often an introductory meeting is held by the participants to discuss the scope, objectives, schedule, and paperwork considerations. After the opening meeting, examiners begin the process of reviewing the process, product, or system under study. Auditors may require access to information concerning quality systems, equipment operation procedures, preventive maintenance records, inspection histories, or planning documents. Auditors may conduct interviews with those involved in the process of providing a product or service. How the auditor phrases questions can have an effect on the information gathered. Those answering the questions may provide operational answers instead of verbatim procedural answers. Report whether the person is able to do what is reasonable and normally expected. Any and all information related to the area under study is critical for the success of the audit. During the review process, auditors document their findings. Auditors are expected to investigate and report the facts. Avoid blindsiding a person, and tell them about what the report will include. Be unbiased when reporting results. Avoid conflicts of interest. Their findings are presented in a general summary at a closing meeting of the participants. Within a short period of time, perhaps 10 to 20 days, the auditors will prepare a written report that documents their findings, conclusions, and recommendations. Complete and clearly written audit reports are the most useful.

3. **Study:** Audits provide information about the participants' strengths, weaknesses, and areas for improvement. Upon receipt, the auditor's report is read by the participants in the audit. During this phase of the audit cycle, they respond to the report and develop an action plan based on the recommendations of the auditors. This action plan should specify the actions, assign responsibilities, and provide the time frame required to deal with the issues raised by the audit.

4. **Act:** Once adopted, the action plan becomes the focus of the improvement activities related to the audited area. Auditors and company administrators should follow up at predetermined intervals to evaluate the status of the continuous improvement action plan. This follow-up ensures that the recommendations and conclusions reached by the auditors and supported by an action plan assist the company in reaching its continuous improvement goals.

WHAT DO EFFECTIVE ORGANIZATIONS AUDIT?

Effective organizations audit their systems to ensure that leadership and the strategic plan are focused on improving and enhancing their customer and market focus, product and service performance, financial performance, human resource development and management, process, and operational performance.

Customer-focused results should be gauged by key measures or indicators of customer satisfaction and dissatisfaction. This information may be gathered and sorted by customer groups or market segments. Comparative data—for example, past versus present performance, customer retention, or competitive assessments—is valuable when determining how an organization is doing from the customer's point of view. Effective organizations monitor and study the levels and trends of these key measures or indicators of customer-perceived value. They are careful to include data concerning customer loyalty and retention, customer-generated referrals, and other aspects that reveal how they build relationships with their customers. The organization may choose to gather objective information and data themselves from their customers, or they may employ an independent organization to gather the information for them.

Effective organizations carefully study results related to the products and services they offer. This information concerning key product and service performance results may be broken down by product groups, customer groups, or market segments. Comparative data, including levels and trends of key measures or indicators of product and service performance, is gathered and audited as it relates to what is important to the organization's customers. Competitors' performance results also provide organizations with insight into how they are doing.

Financial and market results often receive the greatest amount of focus during audits. In this arena the levels and trends of key measures or indicators of financial performance may include such measures of financial return and economic value as revenues, profits, growth rates, earnings, and return on investments. Levels and trends of key measures or indicators of marketplace performance may include market share or position, business growth, and new market performance. Other indicators and measures were presented in Chapter 9.

Human resource development and management often receives less emphasis than the other areas. However, it is critical to determine whether the human resource policies and practices of the organization have resulted in a strong organization. Organizations are only as good as the people comprising them. For this reason, audits of human resource development and management focus on work system performance, employee learning and development, and employee well-being and satisfaction. Appropriate measures for the human resource area include the number and kind of job rotation systems, work layout improvement efforts, employee retention data, internal promotion rates, supervisor/employee ratios, and job classifications. Employee learning and development might be judged by the number of employee innovations or suggestions. The number of courses completed by employees or on-the-job performance improvements and cross-training rates are also good indicators.

As discussed in Chapters 9 and 11, process and operational performance measures or indicators should focus on key value creation processes. These measures that may be audited may include productivity comparisons, cycle time reductions, supplier or partner performance, and other appropriate indicators of organizational effectiveness and

EXAMPLE 15.1   Processes and Procedures Auditing

A supplier of oxygen valves for the oxygen system in a military aircraft has been the focus of an audit. The auditors are trying to determine whether proper processes and procedures for the design, manufacture, and testing of the oxygen valves have been followed.

The military has contracted with PL Inc. to supply 120 oxygen valves for a military airplane. These valves are supplied in lot sizes of 10 and are produced on an as-needed basis. Production and use of the valves has been completed up to number 96 of the 120 requested. The valves are inspected using the Approved Test Plan (ATP) developed at the awarding of the contract.

The audit was triggered by an oxygen leak in one of the valves. While troubleshooting the failure, the investigators found liquid mercury in the valve, a dangerous occurrence for the personnel using the valves.

When first questioned during the audit about the procedure used, PL Inc. stated that they had changed the Approved Test Plan (ATP) to use a liquid mercury manometer to check for leaks at part serial number 096. Their records showed that only four valves were tested in this manner. The mercury manometer was found to be the contamination source of mercury in the valves. PL Inc. also stated that the manometer was not used to test any other products.

The auditor asked two questions:

1. Why did they change the ATP 24 units before the end of the production run?
2. Why did they buy the mercury manometer, a special piece of testing equipment, to test just 24 remaining valves?

These questions caused PL Inc. to change their story. The new story was that the ATP originally called for testing with a mercury manometer, so the ATP was never changed in the first place. What happened was that the inspector who normally tested the valves retired. Unlike the previous inspector, the new inspector followed the ATP and performed the inspection with the manometer rather than with air pressure gages. All units tested by the retired inspector used the air pressure gages and therefore were not contaminated by the mercury manometer.

The auditor asked:

1. Why didn't the retired inspector know how the units were to be tested according to the ATP?
2. Why didn't the supervisor ensure that the ATP was being followed?
3. Why were pressure gages used in the first place? If the ATP was not changed, then the mercury manometer should have always been used.

In response, PL Inc. stated that the previous inspector took Polaroid pictures of the original setup when the first piece was inspected and used these pictures when testing the valves. The previous inspector did not refer to the ATP, which would have instructed her to use the mercury manometer. All units tested through serial number 095 were tested using air pressure, although the approved test plan called for the use of a liquid mercury manometer, which provides a more accurate test.

The auditors asked to watch the inspector perform the test. During this activity, the auditors had to remind everyone present that the inspector was to perform the test, not the supervisor (who was in the process of doing the test for the inspector). The supervisor claimed that he was helping because portions of the setups were cumbersome and could not be performed by the inspector alone—but, no, he was not always available to help. Once allowed to perform the test herself, the inspector performed the tests out of sequence. When the tests called for placing the valve in an explosion-proof container for a high-pressure test, the inspector merely placed the valve in a pan of water!

While discussing the ATP procedure violations, the auditors also noticed some other discrepancies as they watched the inspector test the valves. Although the test was being held in a designated clean room, with everyone present attired in hair nets, smocks, and booties, upon entering the room, the auditors noticed they walked over a dirty, tacky mat. They also noticed that when they slid their hands along the front edge of the workbenches in the room, their hands turned black and gritty with dirt!

If you were the auditor in this situation, what would your findings be? What do you see as the potential consequences of failing this audit?

efficiency. Key support processes should also be audited using productivity rates, cycle times, supplier or partner performance, and other measures of effectiveness and efficiency.

Organizations have governance and social responsibilities. Audits in this area look for evidence of fiscal accountability, ethical behavior, regulatory and legal compliance, and organizational citizenship supporting the community/communities where the organization is located.

WHAT HAPPENS WHEN THE AUDIT IS OVER?

Whether the audit is internal or external, the recommendations of the auditors are critical to an organization seeking to improve. It is leadership's response to these results that matters. When discrepancies between the actual and desired performance are uncovered, it is up to leadership to examine the gap and create and implement a plan of action to deal with the gap.

EXAMPLE 15.2   After the Audit

The oxygen valve supplier of Example 15.1 has just received the results of the audit. The auditors studied the processes and procedures for the design, manufacture, and testing of the oxygen valves. The audit report stated that systemic shortcomings with processes and procedures for the design, manufacture, and testing of the oxygen valves exist. If the errors are left uncorrected, the company will be considered noncompliant with regulations, resulting in the loss of the contract.

Specifically the auditors found:

- Inadequate inspections
- Compromised quality control due to shortcomings in parts inspection
- No up-to-date company-approved testing procedures
- General lack of awareness about approved testing procedures among both the inspector/testers and the process engineers
- Failure on the part of inspectors to learn correct procedures, instead adopting a "Do as I do" imitation of procedures performed (incorrectly) by their predecessors
- Inadequate work instructions
- Incomplete or overly complex processes

The supplier faces serious consequences if it fails to make improvements based on the information provided in the audit report. Among these consequences are loss of business, incurrence of fines, product liability suits (imagine not being able to breathe at high altitudes because your oxygen regulator failed!), and loss of customer goodwill.

Leaders at the supplier are taking the recommendations of the auditors seriously, knowing that improvements are critical to the organization's future success. Where the audit has revealed discrepancies between the actual and desired performance, leadership has assigned teams to examine the gap and then determine and implement a plan of action to correct it. Members of the organization's leadership are serving on each team. They recognize that in the short term, they will be incurring many costs to correct the current situation. These costs include the cost of testing all previous produced units and replacing existing faulty units and the cost of replacing testing equipment. As the teams make improvements, other costs will also be incurred.

HOW DO THE MALCOLM BALDRIGE NATIONAL QUALITY AWARD CRITERIA ENCOURAGE A FOCUS ON BUSINESS RESULTS?

The Results category (Section 7.0) of the Malcolm Baldrige National Quality Award (MBNQA) criteria recognizes that an organization's performance levels and improvement efforts in key business areas are vital to its success. The criteria are interested in whether alignment between the customer's needs, requirements, and expectations and the day-to-day activities of the organization exists. The criteria investigate the key business areas of customer satisfaction, product and service performance, financial and marketplace performance, human resource results, operational performance, and organizational governance and social responsibility. Current levels and trends for key performance measures are used as indicators of an organization's ability to achieve solid business results in these critical areas. Overall, this section of the award criteria is interested in whether an organization has been able to create value for its customers while achieving business success. When evaluating how effective an organization is in the area of business results, ask these questions based on the MBNQA criteria:

Leadership

- Has leadership focused the organization's efforts on converting policies into actions that have in turn translated to improved customer satisfaction and customer-perceived value?
- Has leadership focused the organization's efforts on converting policies into actions that have translated into improved financial performance?
- Have business results reflected that leadership has communicated priorities and held to them?

Strategic Planning

- Has a strategic plan been created and implemented that has translated to improved customer satisfaction and customer-perceived value?
- Has a strategic plan been created and implemented that has translated to positive results for financial performance?

Customer and Market Focus

- Has the organization's improved ability to determine the needs, wants, and requirements of the customers translated into improved customer satisfaction and customer-perceived value?
- Has the organization's improved ability to determine the needs, wants, and requirements of customers translated into positive results for financial performance?

Measurement, Analysis, and Knowledge Management

- Has the information being gathered, analyzed, disseminated, and used enabled the organization to improve customer satisfaction and customer-perceived value?
- Has the information being gathered, analyzed, disseminated, and used enabled the organization to improve performance as exhibited by positive financial performance?

Workforce Focus

- Have personnel policies, practices, and reward systems enabled the organization to enhance customer satisfaction and customer-perceived value?
- Have personnel policies and reward systems enhanced the organization's performance as exhibited by positive financial performance?

Process Management

- Do key processes and critical projects support enhanced customer satisfaction and customer-perceived value?
- Do key processes and critical projects support improvements to financial performance?
- Have improvements made to key processes enhanced financial performance?
- Have improvements made to key processes increased customer satisfaction and customer-perceived value?

Results

- Do the organization's business results reflect improvement efforts?
- Are the organization's measures measuring the right things?
- Have business results been used to measure the gap between what the organization proposed to do versus what it actually did?
- What are the organization's product and service performance results?
- How does the organization's performance compare with other best-in-field companies?
- What are the organization's customer-focused performance results?
- How does the organization measure customer-perceived value?
- What are the organization's financial and marketplace performance results?
- What are the organization's workforce-focused performance results?
- What are the trends in the organization's workforce environment, including health, safety, and security?
- What are the organization's process effectiveness results?
- What are the organization's leadership results?

LESSONS LEARNED

Twelve key areas necessary for organizational effectiveness have been discussed in this text, including:

- Organizational Philosophy
- Leadership
- Strategic Planning
- Customer- and Market-Focused Organization
- Human Resource Development and Management
- Supply Chain Management
- Measurement, Analysis, and Knowledge Management
- Process Management
- Systematic Problem-Solving
- Quality Systems

- Project Management
- Business Results

These areas are supported by having a process-oriented organization with a systems improvement approach to management. Regardless of whether an organization decides to follow quality management principles, the MBNQA criteria, the Six Sigma methodology, Dr. Deming's teaching, or another approach to quality management, these principal activities can be found in each approach.

Having the knowledge to create an effective organization is crucial, but organizational effectiveness depends on the effective performance of the individuals within the organization. Dr. Deming said it best when he concluded his 14 points with, "Put everybody in the company to work to accomplish the transformation." Everyone must be involved in the effort of creating and sustaining organizational effectiveness. Leadership helps create an effective organization by guiding, directing, and motivating their workforce toward good performance. Effective organizations align customer requirements with clearly defined operational processes, procedures, and systems that are continually monitored and improved. This focus will ultimately increase market share, sales, number of customers, and other vital statistics, while decreasing customer complaints, employee turnover, costs of quality, and other waste.

1. Effective organizations focus on the voice of the customer.
2. Effective organizations are results focused.
3. Effective organizations align customer requirements with clearly defined operational processes, procedures, and systems that are continuously monitored and improved.
4. Effective organizations create effective activities by analysis, innovation, information sharing, process improvement, and measurement of progress toward key strategic and operational goals.
5. Effective organizations monitor their process toward goals and objectives through the use of audits.
6. Effective organizations use the MBNQA criteria to ensure that their processes are repeatable and regularly evaluated for improvement opportunities.
7. An effective organization requires an integrated approach on their journey to performance excellence.

Are You a Quality Management Person?

It's the little day-to-day activities that make a quality management person. We'd all like to think that we are, but how do you know? Based on what you have learned in the 15 preceding chapters, the following questions should not be unexpected. They focus on results, achieving the goals and objectives you have set for yourself in order to get to your life vision. Check them out and find out whether you are a quality management person. How do you answer them on a scale of 1 to 10?

Do your goals and objectives align with your overall vision for your life?

| 1 | 2 | 3 | 4 | 5 | 6 | 7 | 8 | 9 | 10 |

Have you achieved the goals and objectives you set 12 months ago?

| 1 | 2 | 3 | 4 | 5 | 6 | 7 | 8 | 9 | 10 |

Are you actively engaged in achieving your established goals and objectives?

| 1 | 2 | 3 | 4 | 5 | 6 | 7 | 8 | 9 | 10 |

Are you on track to meet your future goals and objectives?

| 1 | 2 | 3 | 4 | 5 | 6 | 7 | 8 | 9 | 10 |

Do you have clearly stated measures of performance that enable you to determine whether you are on track to meet your future goals and objectives?

| 1 | 2 | 3 | 4 | 5 | 6 | 7 | 8 | 9 | 10 |

Do you focus on results or activities?

| 1 | 2 | 3 | 4 | 5 | 6 | 7 | 8 | 9 | 10 |

Do you set aside funding for the goals and objectives that you would like to achieve?

| 1 | 2 | 3 | 4 | 5 | 6 | 7 | 8 | 9 | 10 |

Do you audit your activities occasionally to see whether they align with your overall vision for your life?

| 1 | 2 | 3 | 4 | 5 | 6 | 7 | 8 | 9 | 10 |

Chapter Questions

1. How does auditing allow effective organizations to see their progress?
2. What do effective organizations audit?
3. What do effective organizations do with the information when an audit is completed?

4. Employees at a local bank have been trained to be polite and eager to serve customers. The bank has been able to identify several other features that are important to their customers: timeliness, accuracy, immediate service, knowledge, courtesy, convenience, and accessibility. The bank already conducts regular financial audits in order to comply with government regulations. However, they are considering auditing the remaining aspects of the bank in order to determine how they can improve. The following are some questions they have about auditing. Provide answers as if you were a consultant helping them with their audit.

   a. Besides the financial area, what areas would be appropriate to audit?
   b. Describe the steps the bank should take to complete an audit. What should take place during each step?
   c. After completing the audit, what should the bank do with the information?
   d. What benefits would auditing various areas within the bank provide for the bank?

5. What areas would a manufacturing company be interested in auditing? A hospital? An Internet company?

6. Describe the steps involved in auditing.

7. Why would a company be interested in auditing? What does a company hope to gain by auditing?

8. Select a company, either service or manufacturing related, and describe the steps and activities involved in auditing it. Be sure to indicate which areas of the company would be most appropriate to audit and why as well as who should be involved in the audit and what they should investigate.

Section 7.0: Results

The Results section of the MBNQA criteria addresses an organization's performance and improvement based on the results of customer-focused results, operational performance results, human resource results, and overall organizational effectiveness.

The key goals of this section are to:

1. Examine how Remodeling Designs, Inc. and Case Handyman Services measure their performance with the aim of achieving a high-performance organization in the following areas:
   a. Customer-focused results
      i. Customer satisfaction
      ii. Product and service performance results
   b. Financial and market results
   c. Human resource results
      i. Employee well-being, satisfaction, and development
      ii. Work system performance
   d. Organizational effectiveness results
      i. Citizenship, production, and other measures of effectiveness
2. Examine how Remodeling Designs and Case Handyman improve their business performance.

This section was created by Erich Eggers, Mike Cordonnier, and Donna Summers.

Results Questions

1. What measures are critical to your ability to run a company? What absolutely must succeed on the bottom line in order for your company to be successful?

CUSTOMER-FOCUSED RESULTS

2. What levels and trends do you use to measure your critical success factors when dealing with customer satisfaction/dissatisfaction?
3. How do you compare your customer satisfaction with competitors' levels of customer satisfaction?
4. What are your current levels of and trends in key measures/indicators of customer-perceived value, customer retention, positive referral, or other aspects of building rapport with customers?
5. How do you measure your service performance that is important to your customer?

6. How does meeting your customer needs integrate into your business results as shown on your bottom line?

7. Has your improved ability to determine the needs and wants of your customers translated into improvements of the bottom line?

HUMAN RESOURCE RESULTS

8. How do you give your employees empowerment—that is, the authority and responsibility to make decisions and take actions?

9. What are your current levels of and trends in key measures/indicators of employee well-being, satisfaction/dissatisfaction, and development?

10. How do you keep track of your employee-to-customer relationships?

11. Does your company have a reward system that reinforces customer-focused behavior? How does this show up on your business results?

12. Have your personnel policies and reward systems enhanced your performance as exhibited by the bottom line?

ORGANIZATIONAL EFFECTIVENESS RESULTS

13. What type of process does your company have to ensure that it is focused on a continuous improvement cycle, prevention, and business excellence?

14. How does your company practice good citizenship? Is it reflected in your business results?

15. How do you measure the accomplishments of organizational strategy? In other words, if you completed a job a certain way and it was good, do you keep track of how you did it?

16. How do you measure the operational performance of key design, production, and support processes?

17. Overall, have your business results been used to measure the gap between what you said you would do and what you actually accomplished?

7.0 Results

7.1 CUSTOMER-FOCUSED RESULTS

7.1a Customer Results

Remodeling Designs and Case Handyman believe that there are four keys to excellent customer relationships: communicating with the customer, doing what they said they would do, doing whatever it takes to get the job done, and keeping the job site as clean as possible. They perform extensive customer surveys to measure customer results. Figure 1 shows the Valued Client Feedback form they use. Repeat business and referrals are essential measures of business results. Over 70% of their business is repeat or referral business. They often find themselves in a "noncompetitive bid" situation because customers are so pleased with their approach to remodeling and repair. They also use warranty claims as a measure of customer satisfaction.

Remodeling Designs, Inc.
482 Windsor Park Dr.
Dayton, Ohio 45459
937-438-0031
Fax: 937-438-0501

VALUED CLIENT FEEDBACK

<<FirstName>> <<LastName>>
<<Address>>
<<City>>, <<State>> <<PostalCode>>   November 20, 2001

Dear <<FirstName>>,

Thank you for giving us the opportunity to work with you on your remodeling project. We take pride in our work and are constantly striving to maintain our high quality. Our company is guided by our Mission Statement:

It is the mission of Remodeling Designs, Inc. to provide the highest quality, personalized and comprehensive design, planning and remodeling services to discerning clients in the Miami Valley. Building upon our already established reputation for service above and beyond the client's expectations, we will continue to demand the best of ourselves to ensure our continued growth and success.

Would you help us by taking a minute to answer the following questions?

(5=Strongly Agree...1=Strongly Disagree)

The Salesperson I worked with was knowledgeable and competent.	5 4 3 2 1
The Designer I worked with was knowledgeable and competent.	5 4 3 2 1
The Craftsman I worked with was knowledgeable and competent.	5 4 3 2 1
I felt I received a "good value" for my remodeling dollars spent.	5 4 3 2 1
Remodeling Designs, Inc. set realistic expectations of how my project would be handled.	5 4 3 2 1
I felt I was treated according to the Golden Rule: Treat others as you want to be treated.	5 4 3 2 1
Remodeling Designs, Inc. "did what they had to do, to get the job done".	5 4 3 2 1
The overall remodeling experience, from beginning to end, was a good one.	5 4 3 2 1
I would recommend Remodeling Designs, Inc. to a friend.	5 4 3 2 1

We invite you to write additional comments.

_____

_____

_____

_____

_____

_____

THE BIG
50

NATIONAL
REMODELING
QUALITY
AWARD

May we use your name as a reference on future jobs?   YES   NO

www.remodelingdesigns.com • E-Mail: remdes@remodelingdesigns.com

We Design, Plan & Build Your Dreams

Figure 1   Client Feedback Form

7.2 PRODUCT AND SERVICE RESULTS

Remodeling Designs and Case Handyman believe that their four keys to excellent customer relations contribute significantly to their business results. They are well placed in the market. Their product and service differentiator is that their clients want and their companies provide excellent service, unique design skills, and high quality. Remodeling Designs and Case Handyman have experienced very few complaints concerning the products or services they provide. They believe this is due to their judicious use of daily logs to maintain contact with their customers during construction on the job. The most frequent negative comment is that the job took longer than expected. They are engaged in continuous improvement efforts to complete the jobs more quickly at the same level of quality.

7.3 FINANCIAL AND MARKET RESULTS

To track their financial and market results, Remodeling Designs and Case Handyman utilize two measures: fiscal performance and project performance. Fiscal performance is indicated by gross profit and net profit. The cost of each job includes a gross profit of about 15% on each job and an additional 45% for overhead. Project performance is measured long term by tracking their repeat and referral rate. Over 70% of their jobs fall into this category; over one third of all new leads are generated from repeat and referral business. The average job size for Remodeling Designs is $60,000; for Case Handyman it is $2,000. They were recently ranked in the top 10% remodeling businesses by the National Association of Remodeling Industries (NARI).

In 1999, they won the Miami Valley NARI Contractor of the Year Award for Residential Bathroom $30,000–$60,000. At the same time, they won the NARI Contractor of the Year Award for Residential Kitchens $30,000–$60,000. In 1998, they won the National Remodeling Quality Gold Award for Companies under $1 Million Volume, as well as the Miami Valley NARI Contractor of the Year Award for Residential Kitchens over $30,000. In 1997, they won three awards: the Miami Valley NARI Contractor of the Year for Residential Interior Remodeling, the Miami Valley NARI Contractor of the Year for Residential Kitchens under $30,000, and the Better Business Bureau Eclipse Award for Customer Service–Small Company. In 1996, they won three awards: the Miami Valley NARI Contractor of the Year for Residential Kitchens under $25,000, the Miami Valley NARI Contractor of the Year for Residential Kitchens over $25,000, and the Miami Valley NARI Contractor of the Year for Residential Interior Remodeling. They also won the Remodeling Excellence Award for Kitchen and Bath Remodeling in 1995 and the Miami Valley NARI Contractor of the Year for Residential Interior Remodeling in 1994.

7.4 HUMAN RESOURCE RESULTS

Remodeling Designs and Case Handyman measure their employees' satisfaction by using an Employee Review (see Figure 2 in the Chapter 7 case study). This review asks how the employees feel about their work, what they would like to see changed, how they would like to progress in their career and so on. Since they are a small organization, leaders are able to take the time to regularly communicate with individual employees to ask them how they are getting on with their work and what improvements could be made. Employee satisfaction ratings are high and their turnover is essentially zero.

Remodeling Designs and Case Handyman are interested in employee development. They pay for their employees to take courses leading to Certified Lead Carpenter designation. After an employee has passed the course, Remodeling Designs pays him a $100 bonus, as well as increasing his hourly wage. They are willing to pay for any appropriate courses at a college, adult school, or vocational school. Currently five out of seven of their carpenters are Certified Lead Carpenters. These efforts maintain a high level of skill within their workforce.

Employee performance is measured through the use of the Customer Quality Audit feedback form (which differs from the Valued Client Feedback form). These forms are considered to be the report cards for the lead carpenters, and they provide a "snapshot" of the carpenter's performance on a particular job. The carpenter's time spent on the job is also tracked as a measure of productivity.

7.5 ORGANIZATIONAL EFFECTIVENESS RESULTS

Remodeling Designs and Case Handyman have a greater focus on the customer than on their own business results. Their current effort to improve organizational effectiveness is focused on studying their processes to determine whether they can complete their jobs in less time. They hope to identify non–value-added activities and remove them from the processes. They have set a goal of 40% gross profit on a job; they are now at 37%. They are also studying their support processes to determine their effectiveness. Another process under study is the advertising process. Currently they spend 4% of their total income on advertisements in home shows, on television, on their website, through mailers, and so on. They are also investigating their suppliers to determine whether they offer the best price and timely delivery.

The companies' strategic planning process and deployment need to be reviewed. Although the companies have created an organizational strategy, they have not acted upon the strategy. Organizational strategy is critical to the life of the companies because they are still considered young companies in a business where companies fail easily.

7.6 GOVERNANCE AND SOCIAL RESPONSIBILITY RESULTS

Remodeling Designs and Case Handyman practice good citizenship by participating in a variety of community events. They sponsor local softball and baseball teams. They participate regularly in Christmas in April, a work project for people who are in need of home repair but cannot afford it. They took part in Rehabarama, a project in which eight homes were completely remodeled in an inner-city neighborhood in order to help start revitalization in the community. They are part of the National Ramp-A-Thon program, a group that builds wheelchair ramps throughout the community. Remodeling Designs believes in giving back to the community they live in.

Appendix 1

Z Tables

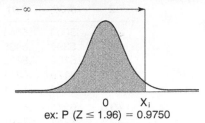

Normal Curve Areas P ($Z \leq Z_0$)

ex: P ($Z \leq 1.96$) = 0.9750

Z	0.09	0.08	0.07	0.06	0.05	0.04	0.03	0.02	0.01	0.00
−3.8	.0001	.0001	.0001	.0001	.0001	.0001	.0001	.0001	.0001	.0001
−3.7	.0001	.0001	.0001	.0001	.0001	.0001	.0001	.0001	.0001	.0001
−3.6	.0001	.0001	.0001	.0001	.0001	.0001	.0001	.0001	.0002	.0002
−3.5	.0002	.0002	.0002	.0002	.0002	.0002	.0002	.0002	.0002	.0002
−3.4	.0002	.0003	.0003	.0003	.0003	.0003	.0003	.0003	.0003	.0003
−3.3	.0003	.0004	.0004	.0004	.0004	.0004	.0004	.0005	.0005	.0005
−3.2	.0005	.0005	.0005	.0006	.0006	.0006	.0006	.0006	.0007	.0007
−3.1	.0007	.0007	.0008	.0008	.0008	.0008	.0009	.0009	.0009	.0010
−3.0	.0010	.0010	.0011	.0011	.0011	.0012	.0012	.0013	.0013	.0013
−2.9	.0014	.0014	.0015	.0015	.0016	.0016	.0017	.0018	.0018	.0019
−2.8	.0019	.0020	.0021	.0021	.0022	.0023	.0023	.0024	.0025	.0026
−2.7	.0026	.0027	.0028	.0029	.0030	.0031	.0032	.0033	.0034	.0035
−2.6	.0036	.0037	.0038	.0039	.0040	.0041	.0043	.0044	.0045	.0047
−2.5	.0048	.0049	.0051	.0052	.0054	.0055	.0057	.0059	.0060	.0062
−2.4	.0064	.0066	.0068	.0069	.0071	.0073	.0075	.0078	.0080	.0082
−2.3	.0084	.0087	.0089	.0091	.0094	.0096	.0099	.0102	.0104	.0107
−2.2	.0110	.0113	.0116	.0119	.0122	.0125	.0129	.0132	.0136	.0139
−2.1	.0143	.0146	.0150	.0154	.0158	.0162	.0166	.0170	.0174	.0179
−2.0	.0183	.0188	.0192	.0197	.0202	.0207	.0212	.0217	.0222	.0228
−1.9	.0233	.0239	.0244	.0250	.0256	.0262	.0268	.0274	.0281	.0287
−1.8	.0294	.0301	.0307	.0314	.0322	.0329	.0336	.0344	.0351	.0359
−1.7	.0367	.0375	.0384	.0392	.0401	.0409	.0418	.0427	.0436	.0446
−1.6	.0455	.0465	.0475	.0485	.0495	.0505	.0516	.0526	.0537	.0548
−1.5	.0559	.0571	.0582	.0594	.0606	.0618	.0630	.0643	.0655	.0668
−1.4	.0681	.0694	.0708	.0721	.0735	.0749	.0764	.0778	.0793	.0808
−1.3	.0823	.0838	.0853	.0869	.0885	.0901	.0918	.0934	.0951	.0968
−1.2	.0985	.1003	.1020	.1038	.1056	.1075	.1093	.1112	.1131	.1151
−1.1	.1170	.1190	.1210	.1230	.1251	.1271	.1292	.1314	.1335	.1357
−1.0	.1379	.1401	.1423	.1446	.1469	.1492	.1515	.1539	.1562	.1587
−0.9	.1611	.1635	.1660	.1685	.1711	.1736	.1762	.1788	.1814	.1841
−0.8	.1867	.1894	.1922	.1949	.1977	.2005	.2033	.2061	.2090	.2119
−0.7	.2148	.2177	.2206	.2236	.2266	.2296	.2327	.2358	.2389	.2420
−0.6	.2451	.2483	.2514	.2546	.2578	.2611	.2643	.2676	.2709	.2743
−0.5	.2776	.2810	.2843	.2877	.2912	.2946	.2981	.3015	.3050	.3085
−0.4	.3121	.3156	.3192	.3228	.3264	.3300	.3336	.3372	.3409	.3446
−0.3	.3483	.3520	.3557	.3594	.3632	.3669	.3707	.3745	.3783	.3821
−0.2	.3859	.3897	.3936	.3974	.4013	.4052	.4090	.4129	.4168	.4207
−0.1	.4247	.4286	.4325	.4364	.4404	.4443	.4483	.4522	.4562	.4602
−0.0	.4641	.4681	.4721	.4761	.4801	.4840	.4880	.4920	.4960	.5000

Z	0.00	0.01	0.02	0.03	0.04	0.05	0.06	0.07	0.08	0.09
+0.0	.5000	.5040	.5080	.5120	.5160	.5199	.5239	.5279	.5319	.5359
+0.1	.5398	.5438	.5478	.5517	.5557	.5596	.5636	.5675	.5714	.5753
+0.2	.5793	.5832	.5871	.5910	.5948	.5987	.6026	.6064	.6103	.6141
+0.3	.6179	.6217	.6255	.6293	.6331	.6368	.6406	.6443	.6480	.6517
+0.4	.6554	.6591	.6628	.6664	.6700	.6736	.6772	.6808	.6844	.6879
+0.5	.6915	.6950	.6985	.7019	.7054	.7088	.7123	.7157	.7190	.7224
+0.6	.7257	.7291	.7324	.7357	.7389	.7422	.7454	.7486	.7517	.7549
+0.7	.7580	.7611	.7642	.7673	.7704	.7734	.7764	.7794	.7823	.7852
+0.8	.7881	.7910	.7939	.7967	.7995	.8023	.8051	.8078	.8106	.8133
+0.9	.8159	.8186	.8212	.8238	.8264	.8289	.8315	.8340	.8365	.8389
+1.0	.8413	.8438	.8461	.8485	.8508	.8531	.8554	.8577	.8599	.8621
+1.1	.8643	.8665	.8686	.8708	.8729	.8749	.8770	.8790	.8810	.8830
+1.2	.8849	.8869	.8888	.8907	.8925	.8944	.8962	.8980	.8997	.9015
+1.3	.9032	.9049	.9066	.9082	.9099	.9115	.9131	.9147	.9162	.9177
+1.4	.9192	.9207	.9222	.9236	.9251	.9265	.9279	.9292	.9306	.9319
+1.5	.9332	.9345	.9357	.9370	.9382	.9394	.9406	.9418	.9429	.9441
+1.6	.9452	.9463	.9474	.9484	.9495	.9505	.9515	.9525	.9535	.9545
+1.7	.9554	.9564	.9573	.9582	.9591	.9599	.9608	.9616	.9625	.9633
+1.8	.9641	.9649	.9656	.9664	.9671	.9678	.9686	.9693	.9699	.9706
+1.9	.9713	.9719	.9726	.9732	.9738	.9744	.9750	.9756	.9761	.9767
+2.0	.9772	.9778	.9783	.9788	.9793	.9798	.9803	.9808	.9812	.9817
+2.1	.9821	.9826	.9830	.9834	.9838	.9842	.9846	.9850	.9854	.9857
+2.2	.9861	.9864	.9868	.9871	.9875	.9878	.9881	.9884	.9887	.9890
+2.3	.9893	.9896	.9898	.9901	.9904	.9906	.9909	.9911	.9913	.9916
+2.4	.9918	.9920	.9922	.9925	.9927	.9929	.9931	.9932	.9934	.9936
+2.5	.9938	.9940	.9941	.9943	.9945	.9946	.9948	.9949	.9951	.9952
+2.6	.9953	.9955	.9956	.9957	.9959	.9960	.9961	.9962	.9963	.9964
+2.7	.9965	.9966	.9967	.9968	.9969	.9970	.9971	.9972	.9973	.9974
+2.8	.9974	.9975	.9976	.9977	.9977	.9978	.9979	.9979	.9980	.9981
+2.9	.9981	.9982	.9982	.9983	.9984	.9984	.9985	.9985	.9986	.9986
+3.0	.9987	.9987	.9987	.9988	.9988	.9989	.9989	.9989	.9990	.9990
+3.1	.9990	.9991	.9991	.9991	.9992	.9992	.9992	.9992	.9993	.9993
+3.2	.9993	.9993	.9994	.9994	.9994	.9994	.9994	.9995	.9995	.9995
+3.3	.9995	.9995	.9995	.9996	.9996	.9996	.9996	.9996	.9996	.9997
+3.4	.9997	.9997	.9997	.9997	.9997	.9997	.9997	.9997	.9997	.9998
+3.5	.9998	.9998	.9998	.9998	.9998	.9998	.9998	.9998	.9998	.9998
+3.6	.9998	.9998	.9999	.9999	.9999	.9999	.9999	.9999	.9999	.9999
+3.7	.9999	.9999	.9999	.9999	.9999	.9999	.9999	.9999	.9999	.9999
+3.8	.9999	.9999	.9999	.9999	.9999	.9999	.9999	.9999	.9999	.9999

Appendix 2

$\overline{X}$ and R Chart Factors

Factors for Computing Central Lines and 3σ Control Limits for $\overline{X}$, s, and R Charts

Observations in Sample, n	Chart for Averages — Factors for Control Limits			Chart for Ranges — Factor for Central Line / Factors for Control Limits						Chart for Standard Deviations — Factor for Central Line / Factors for Control Limits				
	A	A_2	A_3	d_2	d_3	D_1	D_2	D_3	D_4	c_4	B_3	B_4	B_5	B_6
2	2.121	1.880	2.659	1.128	0.853	0	3.686	0	3.267	0.7979	0	3.267	0	2.606
3	1.732	1.023	1.954	1.693	0.888	0	4.358	0	2.574	0.8862	0	2.568	0	2.276
4	1.500	0.729	1.628	2.059	0.880	0	4.698	0	2.282	0.9213	0	2.266	0	2.088
5	1.342	0.577	1.427	2.326	0.864	0	4.918	0	2.114	0.9400	0	2.089	0	1.964
6	1.225	0.483	1.287	2.534	0.848	0	5.078	0	2.004	0.9515	0.030	1.970	0.029	1.874
7	1.134	0.419	1.182	2.704	0.833	0.204	5.204	0.076	1.924	0.9594	0.118	1.882	0.113	1.806
8	1.061	0.373	1.099	2.847	0.820	0.388	5.306	0.136	1.864	0.9650	0.185	1.815	0.179	1.751
9	1.000	0.337	1.032	2.970	0.808	0.547	5.393	0.184	1.816	0.9693	0.239	1.761	0.232	1.707
10	0.949	0.308	0.975	3.078	0.797	0.687	5.469	0.223	1.777	0.9727	0.284	1.716	0.276	1.669
11	0.905	0.285	0.927	3.173	0.787	0.811	5.535	0.256	1.744	0.9754	0.321	1.679	0.313	1.637
12	0.866	0.266	0.886	3.258	0.778	0.922	5.594	0.283	1.717	0.9776	0.354	1.646	0.346	1.610
13	0.832	0.249	0.850	3.336	0.770	1.025	5.647	0.307	1.693	0.9794	0.382	1.618	0.374	1.585
14	0.802	0.235	0.817	3.407	0.763	1.118	5.696	0.328	1.672	0.9810	0.406	1.594	0.399	1.563
15	0.775	0.223	0.789	3.472	0.756	1.203	5.741	0.347	1.653	0.9823	0.428	1.572	0.421	1.544
16	0.750	0.212	0.763	3.532	0.750	1.282	5.782	0.363	1.637	0.9835	0.448	1.552	0.440	1.526
17	0.728	0.203	0.739	3.588	0.744	1.356	5.820	0.378	1.622	0.9845	0.466	1.534	0.458	1.511
18	0.707	0.194	0.718	3.640	0.739	1.424	5.856	0.391	1.608	0.9854	0.482	1.518	0.475	1.496
19	0.688	0.187	0.698	3.689	0.734	1.487	5.891	0.403	1.597	0.9862	0.497	1.503	0.490	1.483
20	0.671	0.180	0.680	3.735	0.729	1.549	5.921	0.415	1.585	0.9869	0.510	1.490	0.504	1.470

Appendix 3

Malcolm Baldrige National Quality Award Organizational Effectiveness Evaluation

Many firms have implemented comprehensive quality improvement programs. Recent articles in trade publications, as well as nationally known magazines such as *Forbes, Business Week, Newsweek,* and *The Wall Street Journal,* have reported on what these companies have been doing to improve their organizational effectiveness.

To evaluate an organization's effectiveness utilizing the Malcolm Baldrige National Quality Award criteria, select and research an organization using the following information and question set as a guide. The remainder of the information accompanying this appendix can be used to structure an assignment and provide guidance. The Case Studies that accompanied Chapters 3 through 7, 9, 11 and 15 in this text discuss the result of students' evaluation of a local company (Remodeling Designs/Case Handyman) using this format.

ASSIGNMENT

Read one or more comprehensive articles about a company from a current business or news magazine. On the basis of your reading, assess the organizational effectiveness of the company. You may wish to contact a nearby company and work with them on this project.

Choose one or more of the seven areas found in the Malcolm Baldrige National Quality Award criteria and make your assessment along these guidelines. You may wish to strengthen your assessment by contacting the company itself. Use the following guidelines and questions to aid you while investigating the organization.

Purpose

The purpose of this evaluation is twofold:

- To become familiar with what it takes to run a world-class organization
- To provide the company being reviewed with an accurate picture of their business activities as they relate to the MBNQA

Report Format

Cover Letter

A cover letter provides a brief introduction to the report, including what it contains and who has compiled it.

Cover Sheet

1. Title of section
2. Team members
3. Description of key goal(s) of section
4. Names of those interviewed

Appendices
5. Questions posed to those interviewed (an appendix)
6. Supporting documentation (an appendix)

Report
7. Executive summary of findings
8. Executive summary of suggested areas to investigate
9. Detailed findings for each section

CREATING MBNQA REVIEWER QUESTIONS

Obtain a copy of the Malcolm Baldrige National Quality Award criteria by visiting the website www.nist.gov.

1. Study the terms in the Glossary and the requirements described in the Criteria Overview and Item Descriptions.
2. Generate a list of the factors critical to your part of the criteria.
3. Use the list you generated in number 2 to guide the development of interview questions pertinent to your area.

For example:

1. Leadership
2. Critical Factors: setting direction, building the organization, sustaining the organization, organizational learning, individual learning, innovation, stakeholders, capturing and sharing learning, translating findings, ethical behavior, legal requirements, good citizenship, public concerns, and so on.

Examples of Questions

Section 1.0: Leadership
a. What is the attitude/involvement of top management? How is this visible? Check top management's understanding of quality, their investment of time and money in quality issues, their willingness to seek help on quality management, their support of one another and subordinates, the level of importance they place on quality, and their participation in quality process.
b. What importance does the company management place on developing a quality culture? How is this visible? Check their understanding of quality, their investment of time and money in quality issues, their willingness to seek help on quality management, their support of one another and subordinates, the level of importance they place on quality, their participation in quality process, and their training of employees.
c. What is your perception of the importance of quality to this company? How is this visible? Is the company's position based on eliminating defects by inspection? Judging cost of quality by scrap and rework? Or preventing defects through design of process and product?

Section 2.0: Strategic Planning

a. What importance does the company place on quality in its strategic planning? How is this visible? Check management's understanding of quality, their investment of time and money in quality issues, how their interest is reflected in the strategic plan, the level of importance they assign the strategic plan, and whether quality control is evident throughout the plan or in just one section.

b. How does the company develop its plans and strategies for the short term? For the long term?

c. Which benchmarks does the company use to measure quality? Are the benchmarks relative to the market leader or to competitors in general? What are management's projections about the market?

d. How does the company rank the following?
- Cost of manufacturing and product provision
- Volume of output
- Meeting schedules
- Quality

e. Does the company management feel that a certain level of defects is acceptable as a cost of doing business and a way of company life?

f. Is quality first incorporated into the process at the concept development/preliminary research level? The product development level? The production/operations level? At final inspection?

g. Does the company define quality according to performance? Aesthetic issues? Serviceability? Durability? Features? Reliability? Fit and finish? Conformance to specifications?

h. Are the following measures used to evaluate overall quality?
- Zero defects
- Parts per million
- Reject or rework rate
- Cost of quality

i. The company's quality improvement program is best described as:
- There is no formal program
- The program emphasizes short-range solutions.
- The program emphasizes motivational projects and slogans.
- A formal improvement program creates widespread awareness and involvement.
- The quality process is an integral part of ongoing company operations and strategy.

j. Perceived barriers to a better company are:
- Top-management inattention
- Perception of program costs
- Inadequate organization of quality effort
- Inadequate training
- Costs of quality not computed
- Low awareness of need for quality emphasis
- Crisis management a way of life (leaves no time)

- No formal program/process for improvement
- The management system
- The workers

k. The following steps in a quality improvement plan have (have not) been taken:
- Obtained top-management commitment to establish a formal policy on quality
- Begun implementation of a formal companywide policy on quality
- Organized cross-functional improvement teams
- Established measures of quality (departmentally and in an employee evaluation system)
- Established the cost of quality
- Established and implemented a company-wide training program
- Begun identifying and correcting quality problems
- Set and begun to move toward quality goals

l. What objectives has the company set?
- Less rework
- Less scrap
- Fewer defects
- Plant utilization
- Improved yield
- Design improvements
- Fewer engineering changes (material, labor, process changes)
- Workforce training
- Improved testing
- Lower energy use
- Better material usage
- Lower labor hours per unit

Section 3.0: Customer and Market Focus

a. How does the company provide information to the customer? How easy is that information to obtain?

b. How are service standards defined?

c. How are customer concerns handled? What is the follow-up process?

d. Which statement(s) describe company efforts to provide the best customer service?
- The process has unequivocal support of top management.
- Middle management is able to make significant changes.
- Employees assume the major responsibility for ensuring customer satisfaction.
- Formal training in customer satisfaction is provided to all employees.

e. How does the company gather quality feedback from its customers?
- Through customer surveys
- Through a telephone hot line
- Through customer focus groups
- Through sales force reports
- Through service rep reports

    f. Customer complaints are received by:
- The CEO or his/her office
- The marketing or sales department
- The quality assurance staff
- Service support departments
- Customer complaint bureaus

Section 4.0: Measurement, Analysis, and Knowledge Management

a. What types of data and information does the company collect? Are these records on customer-related issues? On internal operations? On company performance? On cost and financial matters?

b. How does the company ensure the reliability of the data throughout the company? Are their records consistent? Standardized? Timely? Updated? Is there rapid access to data? What is the scope of the data?

c. How are their information and control systems used? Do they have key methods of data collection and analysis? Systematic collection on paper? Systematic collection with computers? How do they use the information collected to solve problems? Can problems be traced to their source?

Section 5.0: Workforce Focus

a. Quality is incorporated into the human resources system through:
- Job descriptions of the president, vice president, managers, and so on
- Performance appraisals
- Individual rewards for quality improvement efforts
- Hiring practices
- Education and training programs

b. How does the company encourage employees to buy in to quality?
- Through incentives
- Through stressing value to customers
- Through surveys or other forms of customer feedback
- Through the example of management

c. Has the quality message been distributed throughout the organization so that employees can use it in their day-to-day job activities? How has it been distributed?

d. If the importance of quality has been communicated throughout the corporation, how is that emphasis made visible?
- All employees are aware of the importance of quality. How is this visible?
- All employees are aware of how quality is measured. How is this visible?
- All employees are aware of the role of quality in their job. How is this visible?
- All employees are aware of how to achieve quality in their job. How is this visible?

e. Have teams been established to encourage cross-functional quality efforts?
- Yes, multifunctional teams seek continuous improvement.
- Yes, teams have been formed in nonmanufacturing and staff departments to encourage quality companywide.

    f. Which of the following topics are emphasized in employee participation groups? Is this emphasis reflected in the minutes, in attitudes, in solutions to problems?
- Product quality
- Service quality
- Quality of work life
- Productivity
- Cost
- Safety
- Energy
- Schedules
- Specifications
- Long-term versus quick-fix solutions

    g. What percentage of employees have been involved in training programs in the past three years?
- On the managerial/supervisory level?
- On the nonmanagerial level?
- Among clerical and shop workers?

    h. Which of the following topics are covered in the training programs? What is the depth of coverage?
- Process control
- Problem solving
- Data gathering and analysis
- Quality tools (Pareto, cause-and-effect charts, and so on)

Section 6.0: Process Management

    a. How are designs of products, services, and processes developed so that customer requirements are translated into design and quality requirements?

    b. How do people handle problem solving and decision making? How is this visible?
- Crisis management prevails. Quality problems arise and are fought on an ad hoc basis.
- Individuals or teams are set up to investigate major problems.
- Problem solving is institutionalized and operationalized between departments. Attempts are frequently made to blame others.
- Problems and potential problems are identified early in development. Data and history are used for problem prevention.

    c. The following quality-related costs are compiled and analyzed on a regular basis:
- Scrap
- Product liability
- Product redesign
- Repair
- Warranty claims
- Consumer contacts/concerns
- Inspection
- Specification/documentation review

- Design review
- Engineering change orders
- Service after service
- Rework
- Quality audits
- Test and acceptance
- Supplier evaluation and surveillance

d. How are costs of quality (or poor quality) calculated?
   - Costs are not computed; there is little or no awareness of total costs.
   - Direct costs are computed for rework, scrap, and returns, but the total cost of poor quality is not.
   - Total costs are computed and are related to percentage of sales or operations.
   - The costs of quality (prevention, appraisal, failure) are computed and reduced to 2% to 3% of actual sales.

e. Are periodic audits conducted to determine whether the system is meeting the goals?

f. How are supplier relationships studied?
   - Suppliers are certified by requiring evidence of statistical process control (SPC).
   - Defects in shipments are identified and suppliers must pay for them.
   - A close working relationship has been established, allowing the suppliers to participate in the design/manufacture of the products.
   - A just-in-time system has been established.
   - Suppliers are rated with a formal system based on quality levels, capacity, production facilities, and delivery on schedule.

Section 7.0: Results

a. How does the company track the key measures of product, service, and process quality?

b. How does the company benchmark itself against other companies?

c. How does the company justify expenditures on quality improvements?
   - By traditional accounting procedures (expenditures justified only if lower than cost of product failure)
   - By normal capital budgeting procedures that incorporate risk-adjusted net present value and discounted costs of capital
   - By cash flow increase
   - By the bottom line: return on investment, net profit percent, and so on.

d. The finance and accounting system has the following roles in the quality management system:
   - It calculates and tracks cost of quality.
   - It measures and reports quality trends.
   - It distributes quality data to appropriate persons.
   - It designs performance measures.
   - It compares performance to competition.
   - It develops new, innovative ways to evaluate quality investments.
   - It turns the traditional cost-accounting system into a quality management system.

e. What tangible benefits has this company seen related to its quality program? How are the benefits visible? How are they measured?
   - Increased sales
   - Increased return on investment
   - Customer satisfaction
   - Lower cost
   - Higher selling price
   - More repeat business
   - Higher market share
   - Improved cash flow

Appendix 4
Websites for Quality

American Society for Quality: www.asq.org
American Society for Quality, Statistics Division: www.asq.org/about/divisions/stats
American Society for Quality, Quality Audit Division: www.asq.org/qad
American Society for Quality, Reliability Division: www.asq-rd.org
U.S. Commerce Department's National Institute of Standards: www.nist.gov
International Organization for Standardization (ISO 9000 and 14000): www.iso.org
Automotive Industry Action Group: www.aiag.org
Quality Information: www.itl.nist.gov/div898/handbook/pmc/pmc_d.htm
Six Sigma Information: www.sixsigma.com
Quality in Healthcare: www.jcaho.org
Information on Quality: www.qualitydigest.com
Discussion Forum: www.insidequality.com
Juran: www.juran.com
Deming: http://deming.eng.clemson.edu/pub/den/deming_map/htm
European Foundation for Quality Management: www.efqm.org
Quality Standards: http://e-standards.asq.org/perl/catalog.cgi
American Productivity and Quality Center, Benchmarking Studies: www.apqc.org
Benchmarking Exchange: www.benchnet.com
Performance Measures: www.zigonperf.com
Lean Manufacturing: www.nwlean.net
Quality Function Deployment Institute: www.nauticom.net/www.qfd
American Standards Institute (ANSI): www.ansi.org
FMEA Info Centre: www.fmeainfocentre.com
Institute of Industrial Engineers: www.iienet.org
Internal Auditor: www.internal–auditor.com
Institute for Supply Management (ISM): www.ism.ws
Lean SCM: www.leanscm.net
ISO/IED Information Centre: www.standardsinfo.net
Supply-Chain Council: www.supply-chain.org

Appendix 5

Attribute Charts

ATTRIBUTES

Attributes *are characteristics associated with a product or service.* These characteristics either do or do not exist, and they can be counted. Examples of attributes include the number of leaking containers, of scratches on a surface, of on-time deliveries, or of errors on an invoice. Attribute charts are used to study the stability of processes over time, provided that a count of nonconformities can be made. Attribute charts are used when measurements may not be possible or when measurements are not made because of time or cost issues. The most difficult part of collecting attribute data lies in the need to develop precise operational definitions of what is conforming and what is not. Attribute charts do not give any indication about why the nonconformity occurred. Nor do the charts provide much detail. The charts do not provide information to answer questions such as, "Do several nonconformities exist on the same product? Is the product still usable? Can it be reworked? What is the severity or degree of nonconformance?" *When the interest is in studying the proportion of products rendered unusable by their nonconformities, a fraction nonconforming* (p) *chart, a number nonconforming* (np) *chart, or a percent nonconforming chart should be used. When the situation calls for tracking the count of nonconformities, a number of nonconformities* (c) *chart or a number of nonconformities per unit* (u) *chart is appropriate.*

CHARTS FOR NONCONFORMING UNITS

Fraction Nonconforming (p) Charts: Constant Sample Size

A *fraction nonconforming* (p) *chart is based on binomial distribution and is used to study the proportion of nonconforming products or services being provided.* This chart is also known as a *fraction defective chart* or *p chart.* A nonconforming or defective product or service is considered unacceptable because of some deviation from an expected level of performance. For a p chart, nonconformities render the product or service unusable and therefore nonconforming. Sometimes called "charts for defective or discrepant items," these charts are used to study situations where the product or service can be judged to be either good or bad, correct or incorrect, working or not working. For example, a container is either leaking or it is not, an engine starts or it does not, and an order delivered to a restaurant patron is either correct or incorrect.

Because of structure of their formulas, p charts can be constructed using either a constant or variable sample size. A p chart (Figure 1) for constant sample size is constructed using the following steps:

1. *Gather the data.* In constructing any attribute chart, careful consideration must be given to the process and what characteristics should be studied. The choice of the

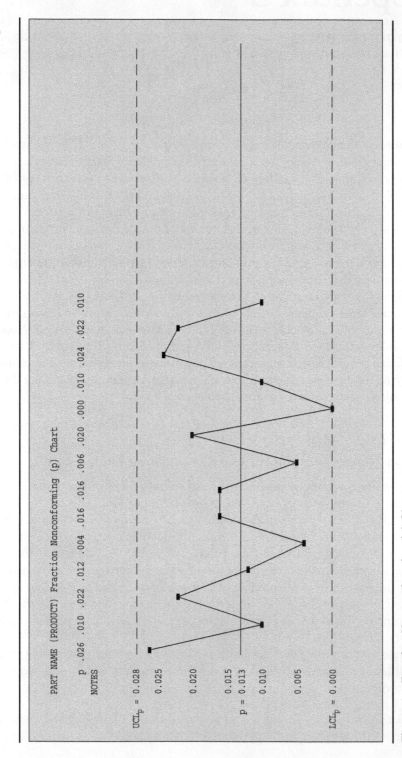

Figure 1   Fraction Nonconforming (p) Chart

attributes to monitor should center on the customer's needs and expectations as well as on current and potential problems areas. Once the characteristics have been identified, the acceptance criteria must be defined. The number nonconforming (np) is tracked. The sample sizes for attribute charts tend to be quite large (for example, n = 250). Large sample sizes are required to maintain sensitivity to detect process performance changes. The sample size should be large enough to include nonconforming items in each subgroup. When process quality is very good, large sample sizes are needed to capture information about the process. When selected, samples must be random and representative of the process.

2. *Calculate p, the fraction nonconforming.*   The fraction nonconforming (p) is plotted on a fraction nonconforming chart. As the products or services are inspected, each subgroup will yield a number nonconforming (np). The fraction nonconforming (p), plotted on the p chart, is calculated using n, the number of inspected items, and np, the number of nonconforming items found:

$$p = \frac{np}{n}$$

3. *Plot the fraction nonconforming (p) on the control chart.*   Once calculated, the values of p for each subgroup are plotted on the chart. The scale for the p chart should reflect the magnitude of the data.

4. *Calculate the centerline and control limits.*   The centerline of the control chart is the average of the subgroup fraction nonconforming. The number nonconforming values are added up and then divided by the total number of samples:

$$\text{Centerline } \bar{p} = \frac{\sum\limits_{i=1}^{n} np}{\sum\limits_{i=1}^{n} n}$$

The control limits for a p chart are found using the following formulas:

$$UCL_p = \bar{p} + 3\frac{\sqrt{\bar{p}(1 - \bar{p})}}{\sqrt{n}}$$

$$LCL_p = \bar{p} - 3\frac{\sqrt{\bar{p}(1 - \bar{p})}}{\sqrt{n}}$$

On occasion, the lower control limit of a p chart may have a negative value. When this occurs, the result of the LCL_p calculation should be rounded up to zero.

5. *Draw the centerline and control limits on the chart.*   Using a solid line to denote the centerline and dashed lines for the control limits, draw the centerline and control limits on the chart.

6. *Interpret the chart.* The interpretation of a fraction nonconforming chart is similar in many aspects to the interpretation of a variables control chart. As with variables charts, in the interpretation of attribute charts, emphasis is placed on determining whether the process is operating within its control limits and exhibiting random variation. As with a variables control chart, the data points on a p chart should flow smoothly back and forth across the centerline. The number of points on each side of the centerline should be balanced, with the majority of the points near the centerline. There should be no patterns in the data, such as trends, runs, cycles, or sudden shifts in level. All of the points should fall between the upper and lower control limits. Points beyond the control limits are immediately obvious and indicate an instability in the process. One difference between the interpretation of a variables control chart and that of a p chart is the desirability in the p chart of having points that approach the lower control limits. This makes sense because quality improvement efforts reflected on a fraction nonconforming chart should show that the fraction nonconforming is being reduced, which is the ultimate goal of improving a process. This favorable occurrence should be investigated to determine what was done right and whether there are changes or improvements that should be incorporated into the process on a permanent basis. Similarly, a trend toward zero nonconforming or a shift in level that lowers the fraction of nonconforming should be investigated.

The process capability is $\bar{p}$, the centerline of the control chart.

EXAMPLE 1   Making a p Chart

Special Plastics, Inc., has been making the blanks for credit cards for a number of years. They use p charts to keep track of the number of nonconforming cards that are created each time a batch of blank cards is run. Use the data in Table 1 to create a fraction nonconforming (p) chart.

Step 1. *Gather the data.*   The characteristics that have been designated for study include blemishes on the card's front and back surfaces; color inconsistencies; white spots or bumps caused by dirt, scratches, chips, or indentations; or other flaws. Several photographs are maintained at each operator's workstation to provide a clear understanding of what constitutes a nonconforming product.

Batches of 15,000 blank cards are run each day. Samples of size 500 are randomly selected and inspected. The number of nonconforming units (np) is recorded on data sheets (Table 1).

Step 2. *Calculate p, the fraction nonconforming.*   After each sample has been taken and the inspections are complete, the fraction nonconforming (p) is calculated using n = 500 and np, the number of nonconforming items. (Here we work p to three decimal places.) For example, for the first value,

$$p = \frac{np}{n} = \frac{20}{500} = 0.040$$

The remaining calculated p values are shown in Table 1.

Step 3. *Plot the fraction nonconforming on the control chart.*   As they are calculated, the values of p for each subgroup are plotted on the chart. The p chart in Figure 2 has been scaled to reflect the magnitude of the data.

Table 1 Data Sheet: Credit Cards

Subgroup Number	n	np	p
1	500	20	0.040
2	500	21	0.042
3	500	19	0.038
4	500	15	0.030
5	500	18	0.036
6	500	20	0.040
7	500	19	0.038
8	500	28	0.056
9	500	17	0.034
10	500	20	0.040
11	500	19	0.038
12	500	18	0.036
13	500	10	0.020
14	500	11	0.022
15	500	10	0.020
16	500	9	0.018
17	500	10	0.020
18	500	11	0.022
19	500	9	0.018
20	500	8	0.016
	10,000	312	

Step 4. *Calculate the centerline and control limits.* The centerline of the control chart is the average of the subgroup fraction nonconforming. The number of nonconforming values from Table 1 are added up and then divided by the total number of samples:

$$\text{Centerline } \bar{p} = \frac{\sum\limits_{i=1}^{n} np}{\sum\limits_{i=1}^{n} n} = \frac{312}{20(500)} = 0.031$$

The control limits for a p chart are found using the following formulas:

$$\text{UCL}_p = \bar{p} + 3\frac{\sqrt{\bar{p}(1 - \bar{p})}}{\sqrt{n}}$$

$$= 0.031 + 0.023 = 0.054$$

$$\text{LCL}_p = \bar{p} - 3\frac{\sqrt{\bar{p}(1 - \bar{p})}}{\sqrt{n}}$$

$$= 0.031 - 0.023 = 0.008$$

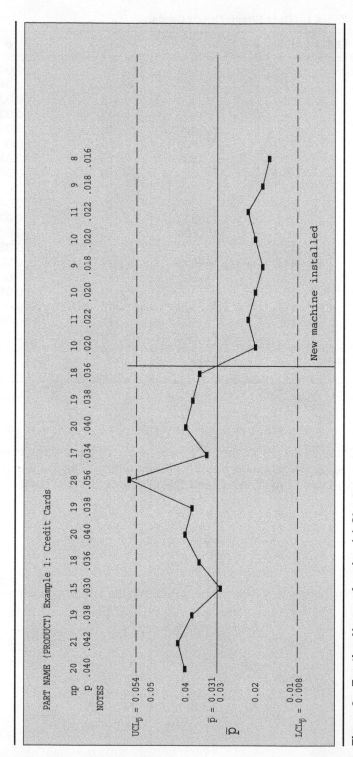

Figure 2 Fraction Nonconforming (p) Chart

Step 5. *Draw the centerline and control limits on the chart.* The centerline and control limits are then drawn on the chart (Figure 2), with a solid line denoting the centerline and dashed lines the control limits.

Step 6. *Interpret the chart.* The process capability for this chart is $\bar{p}$ 0.031, the centerline of the control chart. Point 8 in Figure 2 is above the upper control limit and should be investigated to determine whether an assignable cause exists. If one is found, steps should be taken to prevent future occurrences. When the control chart is studied for any nonrandom conditions, such as runs, trends, cycles, or points out of control, connecting the data points can help reveal any patterns. Of great interest is the significant decrease in the fraction nonconforming after point 13. This reflects the installation of a new machine.

Number Nonconforming (np) Chart

A ***number nonconforming (np) chart*** *tracks the number of nonconforming products or services produced by a process.* A number nonconforming chart eliminates the calculation of p, the fraction nonconforming.

1. *Gather the data.* We must apply the same data-gathering techniques we used in creating the p chart:
   - Designate the specific characteristics or attributes for study
   - Clearly define nonconforming product
   - Select the sample size n
   - Determine the frequency of sampling
   - Take random, representative samples

Those studying the process keep track of the number of nonconforming units (np) by comparing the inspected product with the specifications.

2. *Plot the number of nonconforming units (np) on the control chart.* Once counted, the values of np for each subgroup are plotted on the chart. The scale for the np chart should reflect the magnitude of the data.

3. *Calculate the centerline and control limits.* The centerline of the control chart is the average of the total number nonconforming. The number nonconforming (np) values are added up and then divided by the total number of samples:

$$\text{Centerline } n\bar{p} = \frac{\sum_{i=1}^{n} np}{m}$$

The control limits for an np chart are found using the following formulas:

$$\text{UCL}_{np} = n\bar{p} + 3\sqrt{n\bar{p}(1 - \bar{p})}$$

$$\text{LCL}_{np} = n\bar{p} - 3\sqrt{n\bar{p}(1 - \bar{p})}$$

On occasion, the lower control limit of an np chart may have a negative value. When this occurs, the result of the LCL_{np} calculation should be rounded up to zero.

4. *Draw the centerline and control limits on the chart.* Using a solid line to denote the centerline and dashed lines for the control limits, draw the centerline and control limits on the chart.

5. *Interpret the chart.* Number nonconforming (np) charts are interpreted in the same manner as are fraction nonconforming charts. The points on an np chart should flow smoothly back and forth across the centerline with a random pattern of variation. The number of points on each side of the centerline should be balanced, with the majority of the points near the centerline. There should be no patterns in the data, such as trends, runs, cycles, points above the upper control limit, or sudden shifts in level. Points approaching or going beyond the lower control limits are desirable. They mark an improvement in quality and should be studied to determine their cause and to ensure that it is repeated in the future. The process capability is $n\bar{p}$, the centerline of the control chart.

EXAMPLE 2   Making an np Chart

PCC Inc. receives shipments of circuit boards from its suppliers by the truckload. They keep track of the number of damaged, incomplete, or inoperative circuit boards found when the truck is unloaded. This information helps them make decisions about which suppliers to use in the future.

Step 1. *Gather the data.* The inspectors have a clear understanding of what constitutes a nonconforming circuit board by comparing the inspected circuit boards with standards. The nonconforming units are set aside to be counted. For the purposes of this example, each shipment contains the same number of circuit boards.

Circuit boards are randomly sampled from each truckload with a sample size of $n = 50$.

Step 2. *Plot the number of nonconforming units* (np) *on the control chart.* The results for the 20 most recent trucks are shown in Table 2. The values of np for each subgroup are plotted on the chart in Figure 3, which has been scaled to reflect the magnitude of the data.

Step 3. *Calculate the centerline and control limits.* The average of the total number nonconforming is found by adding up the number of nonconforming values and dividing by the total number of samples. The example is worked to two decimal places:

$$\text{Centerline } n\bar{p} = \frac{\sum_{i=1}^{n} np}{m} = \frac{73}{20} = 3.65$$

The control limits for an np chart are found using the following formulas:

$$\bar{p} = \frac{\sum_{i=1}^{n} np}{\sum_{i=1}^{n} n} = \frac{73}{1,000} = 0.073$$

Table 2   Data Sheet: PCC Inc.

Subgroup Number	n	np
1	50	4
2	50	6
3	50	5
4	50	2
5	50	3
6	50	5
7	50	4
8	50	7
9	50	2
10	50	3
11	50	1
12	50	4
13	50	3
14	50	5
15	50	2
16	50	5
17	50	6
18	50	3
19	50	1
20	50	2

$$UCL_{np} = n\bar{p} + 3\sqrt{n\bar{p}(1 - \bar{p})}$$
$$= 3.65 + 3\sqrt{3.65(1 - 0.073)} = 9.17$$
$$LCL_{np} = n\bar{p} - 3\sqrt{n\bar{p}(1 - \bar{p})}$$
$$= 3.65 - 3\sqrt{3.65(1 - 0.073)} = -1.87 = 0$$

Step 4. *Draw the centerline and control limits on the chart.*   Figure 3 uses a solid line to denote the centerline and dashed lines for the control limits.

Step 5. *Interpret the chart.*   A study of Figure 3 reveals that the chart is under control. The points flow smoothly back and forth across the centerline. The number of points on each side of the centerline are balanced, with the majority of the points near the centerline. There are no trends, runs, cycles, or sudden shifts in level apparent in the data. All of the points fall between the upper and lower control limits.

CHARTS FOR COUNTS OF NONCONFORMITIES

Charts for counts of nonconformities monitor the number of nonconformities found in a sample. Nonconformities represent problems that exist with the product or service. Nonconformities may or may not render the product or service unusable. Two types of charts recording nonconformities may be used: a count of nonconformities (c) chart or a

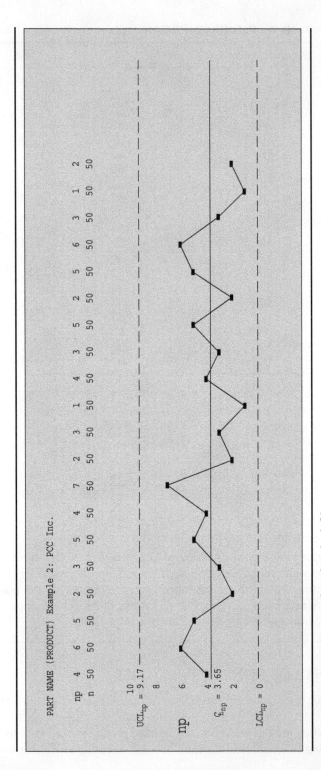

Figure 3   Number Nonconforming (np) Chart

count of nonconformities per unit (u) chart. As the names suggest, one chart is for total nonconformities in the sample, and the other chart is for nonconformities per unit.

Number of Nonconformities (c) Chart

A *number of nonconformities chart, or c chart, is used to track the count of nonconformities observed in a single unit of product or single service experience*, n = 1. Charts counting nonconformities are used when nonconformities are scattered through a continuous flow of product, such as bubbles in a sheet of glass, flaws in a bolt of fabric, discolorations in a ream of paper, or services such as mistakes on an insurance form or errors on a bill. To be charted, the nonconformities must be expressed in terms of what is being inspected, such as four bubbles in a 3-square-foot pane of glass or two specks of dirt on an 8½ × 11 = inch sheet of paper. Count-of-nonconformities charts can combine the counts of a variety of nonconformities, such as the number of mishandled, dented, missing, or unidentified suitcases to reach a particular airport carousel.

1. *Gather the data.* A clear understanding of the nonconformities to be tracked is essential for successfully applying a count-of-nonconformities chart. Gathering the data requires that the area of opportunity of occurrence for each sample taken be equal, n = 1. The rate of occurrences of nonconformities within a sample or area of opportunity (area of exposure) is plotted on the c chart. For this reason, the size of the piece of paper or fabric, the length of the steel, or the number of units must be equal for each sample taken. The number of nonconformities will be determined by comparing the inspected product or service with a standard and counting the deviations from the standard. On a c chart, all nonconformities have the same weight, regardless of the type of nonconformity. The area of opportunity for these nonconformities should be large, with a very small chance of a particular nonconformity occurring at any one location.

2. *Count and plot c, the count of the number of nonconformities, on the control chart.* As it is inspected, each item or subgroup will yield a count of the nonconformities (c). It is this value that is plotted on the control chart. The scale for the c chart should reflect the number of nonconformities discovered.

3. *Calculate the centerline and control limits.* The centerline of the control chart is the average of the subgroup nonconformities. The number of nonconformities are added up and then divided by the total number of subgroups:

$$\text{Centerline } \bar{c} = \frac{\sum_{i=1}^{n} c}{m}$$

The control limits for a c chart are found using the following formulas:

$$\text{UCL}_c = \bar{c} + 3\sqrt{\bar{c}}$$
$$\text{LCL}_c = \bar{c} - 3\sqrt{\bar{c}}$$

On occasion, the lower control limit of a c chart may have a negative value. When this occurs, the result of the LCL_c calculation should be rounded up to zero.

4. *Draw the centerline and control limits on the chart.* Using a sold line to denote the centerline and dashed lines for the control limits, draw the centerline and control limits on the chart.

5. *Interpret the chart.* Charts for counts of nonconformities are interpreted similarly to interpreting charts for the number of nonconforming occurrences. The charts are studied for changes in random patterns of variation. There should be no patterns in the data, such as trends, runs, cycles, or sudden shifts in level. All of the points should fall between the upper and lower control limits. The data points on a c chart should flow smoothly back and forth across the centerline and be balanced on either side of the centerline. The majority of the points should be near the centerline. Once again, it is desirable to have the points approach the lower control limits or zero, showing a reduction in the count of nonconformities. The process capability is $\bar{c}$, the average count of nonconformities in a sample and the centerline of the control chart.

EXAMPLE 3   Making a c Chart

Pure and White, a manufacturer of paper used in copy machines, monitors its production using a c chart. Paper is produced in large rolls, 12 feet long and 6 feet in diameter. A sample is taken from each completed roll, n = 1, and checked in the lab nonconformities. Nonconformities have been identified as discolorations, inconsistent paper thickness, flecks of dirt in the paper, moisture contents, and ability to take ink. All of these nonconformities have the same weight on the c chart. A sample may be taken from anywhere in the roll so the area of opportunity for these nonconformities is large, while the overall quality of the paper creates only a very small chance of a particular nonconformity occurring at any one location.

Step 1. *Gather the data (Table 3).*

Step 2. *Count and plot c, the count of the number of nonconformities, on the control chart.* As each roll is inspected, it yields a count of the nonconformities (c) that is recorded in Table 3. This value is then plotted on the control chart shown in Figure 4.

Step 3. *Calculate the centerline and control limits.* The centerline of the control chart is the average of the subgroup nonconformities. The number of nonconformities is added up and then divided by the total number of rolls of paper inspected:

$$\text{Centerline } \bar{c} = \frac{\sum\limits_{i=1}^{n} c}{m} = \frac{210}{20} = 10.5 \text{ rounded to } 11$$

The control limits for a c chart are found using the following formulas:

$$UCL_c = \bar{c} + 3\sqrt{\bar{c}}$$
$$= 11 + 3\sqrt{11} = 21$$
$$LCL_c = \bar{c} - 3\sqrt{\bar{c}}$$
$$= 11 - 3\sqrt{11} = 1$$

Table 3   Data Sheet: Pure and White Paper

Sample Number	c
1	10
2	11
3	12
4	10
5	9
6	22
7	8
8	10
9	11
10	9
11	12
12	7
13	10
14	11
15	10
16	12
17	9
18	10
19	8
20	9
	210

Step 4. *Draw the centerline and control limits on the chart.*   Using a solid line to denote the centerline and dashed lines for the control limits, draw the centerline and control limits on the chart (Figure 4).

Step 5. *Interpret the chart.*   Point 6 is out of control and should be investigated to determine the cause of so many nonconformities. There are no other patterns present on the chart. Except for point 6, the chart is performing in a very steady manner.

Number of Nonconformities per Unit (u) Charts

*A **number of nonconformities per unit** (u) chart, is a chart that studies the number of nonconformities in a unit* (n > 1). The u chart is very similar to the c chart; however, unlike c charts, u charts can also be used with variable sample sizes.

1. *Gather the data.*
2. *Calculate u, the number of nonconformities per unit.*   As it is inspected, each sample of size n will yield a count of nonconformities (c). The number of nonconformities per unit (u), used on the u chart, is calculated using n, the number of inspected items, and c, the count of nonconformities found:

$$u = \frac{c}{n}$$

This value is plotted on the control chart.

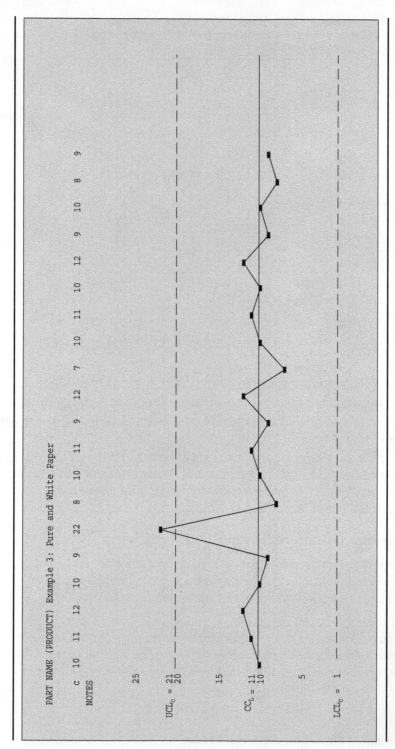

Figure 4 Chart for Nonconformities (c)

3. *Calculate the centerline and control limits.* The centerline of the u chart is the average of the subgroup nonconformities per unit. The number of nonconformities are added up and then divided by the total number of samples:

$$\text{Centerline } \bar{u} = \frac{\sum_{i=1}^{n} c}{\sum_{i=1}^{n} n}$$

The control limits for the u chart are found using the following formulas:

$$UCL_u = \bar{u} + 3\frac{\sqrt{\bar{u}}}{\sqrt{n}}$$

$$LCL_u = \bar{u} - 3\frac{\sqrt{\bar{u}}}{\sqrt{n}}$$

On occasion, the lower control limit of a u chart may have a negative value. When this occurs, the result of the LCL_u calculation should be rounded to zero.

4. *Draw the centerline and control limits on the chart.* Using a solid line to denote the centerline and dashed lines for the control limits, draw the centerline and control limits on the chart.

5. *Interpret the chart.* A u chart is interpreted in the same manner as is a p, np, or c chart. The chart should be studied for any nonrandom conditions such as runs, trends, cycles, or points out of control. The data points on a u chart should flow smoothly back and forth across the centerline. The number of points on each side of the centerline should be balanced, with the majority of the points near the centerline. Quality improvement efforts are reflected on a u chart when the count of nonconformities per unit is reduced, as shown by a trend toward the lower control limit and therefore toward zero nonconformities. The process capability is $\bar{u}$, the centerline of the control chart, the average number of nonconformities per unit.

EXAMPLE 4   Making a u Chart

At Special Plastics, Inc., small plastic parts used to connect hoses are created on a separate production line from the credit-card blanks. Special Plastics, Inc., used u charts to collect data concerning the nonconformities per unit in the process.

Step 1. *Gather the data.*   During inspection, a random sample of size 400 is taken once an hour. The hose connectors are visually inspected for a variety of nonconformities, including flashing on inner diameters, burrs on the part's exterior, incomplete threads, flashing on the ends of the connectors, incorrect plastic compound, and discolorations.

Step 2. *Calculate u, the number of nonconformities per unit.*   Table 4 shows the number of nonconformities (c) that each subgroup of sample size n = 400 yielded. The number of

Table 4   Data Sheet: Hose Connectors

Subgroup Number	n	c	u
1	400	12	0.030
2	400	7	0.018
3	400	10	0.025
4	400	11	0.028
5	400	10	0.025
6	400	12	0.030
7	400	9	0.023
8	400	10	0.025
9	400	8	0.020
10	400	9	0.023
11	400	10	0.025
12	400	11	0.028
13	400	12	0.030
14	400	10	0.025
15	400	9	0.023
16	400	22	0.055
17	400	8	0.020
18	400	10	0.025
19	400	11	0.028
20	400	9	0.023

nonconformities per unit (u) to be plotted on the chart is calculated by dividing c, the number of nonconformities found, by n, the number of inspected items. Working the example of three decimal places, for the first sample:

$$u_1 = \frac{c}{n} = \frac{10}{400} = 0.025$$

Step 3. *Calculate the centerline and control limits.*   The centerline of the u chart is the average of the subgroup nonconformities per unit. The number of nonconformities are added up and then divided by the total number of samples:

$$\text{Centerline } \bar{u} = \frac{\sum_{i=1}^{n} c}{\sum_{i=1}^{n} n} = \frac{210}{20(400)} = 0.026$$

Find the control limits for the u chart by using the following formulas:

$$UCL_u = \bar{u} + 3\frac{\sqrt{\bar{u}}}{\sqrt{n}}$$

$$= 0.026 + 3\frac{\sqrt{0.026}}{\sqrt{400}} = 0.05$$

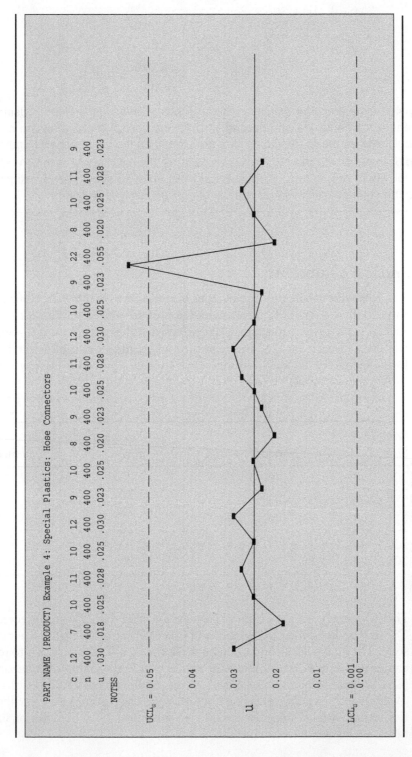

Figure 5   Chart for Nonconformities per Unit (u)

535

$$LCL_u = \bar{u} - 3\frac{\sqrt{\bar{u}}}{\sqrt{n}}$$

$$= 0.026 - 3\frac{\sqrt{0.026}}{\sqrt{400}} = 0.001$$

Step 4. *Create and draw the centerline and control limits on the chart.* Using a solid line to denote the centerline and dashed lines for the control limits, draw the centerline and control limits on the chart. Values for u are plotted on the control chart (Figure 5).

Step 5. *Interpret the chart.* Except for point 16, the chart appears to be under statistical control. There are no runs or unusual patterns. Point 16 should be investigated to determine the cause of such a large number of nonconformities per unit.

Appendix 5 Questions

1. The following table gives the number of nonconforming product found during an inspection of a series of 12 consecutive lots of galvanized washers for finish defects such as exposed steel, rough galvanizing, and discoloration. A sample size of n = 200 was used for each lot. Find the centerline and control limits for a fraction nonconforming chart. If the manufacturer wishes to have a process capability of $\bar{p} = 0.015$, is the process capable?

Sample Size	Nonconforming	Sample Size	Nonconforming
200	0	200	0
200	1	200	0
200	2	200	1
200	0	200	0
200	1	200	3
200	1	200	1

2. Thirst-Quench, Inc., has been in business for more than 50 years. Recently Thirst-Quench updated their machinery and processes, acknowledging their out-of-date style. They have decided to evaluate these changes. The engineer is to record data, evaluate the data, and implement strategy to keep quality at a maximum. The plant operates 8 hours a day, 5 days a week, and produces 25,000 bottles of Thirst-Quench each day. Problems that have arisen in the past include partially filled bottles, crooked labels, upside-down labels, and no labels. Samples of size 150 are taken each hour. Create a p chart.

Subgroup Number	Number Inspected n	Number Nonconforming np	Proportion Nonconforming p
1	150	6	0.040
2	150	3	0.020
3	150	9	0.060
4	150	7	0.047
5	150	9	0.060
6	150	2	0.013
7	150	3	0.020
8	150	5	0.033
9	150	6	0.040
10	150	8	0.053
11	150	9	0.060
12	150	7	0.047
13	150	7	0.047
14	150	2	0.013
15	150	5	0.033
16	150	7	0.047
17	150	4	0.027
18	150	3	0.020
19	150	9	0.060
20	150	8	0.053
21	150	8	0.053
22	150	6	0.040
23	150	2	0.013
24	150	9	0.060
25	150	7	0.047
26	150	3	0.020
27	150	4	0.027
28	150	6	0.040
29	150	5	0.033
30	150	4	0.027
Total	4,500	173	

3. Given the chart in Figure 1, discuss the state of control the charts are exhibiting. What percent nonconforming should the company expect in the future for process 1?

4. At Fruits and Such, a local manufacturer of frozen fruit concentrate, frozen grape juice is packed in 12-ounce cardboard cans. These cans are formed on a machine by

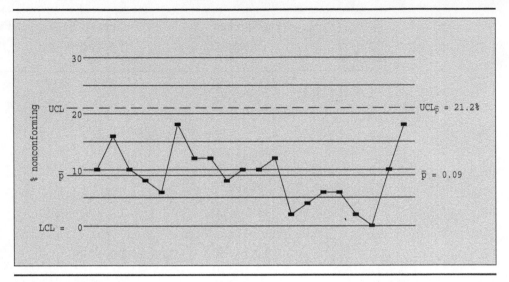

Figure 1   Question 3

spinning them from preprinted cardboard stock and attaching two metal lids, one for the top and one for the bottom. After the filling operation, samples ($n = 100$) of the production are taken hourly. Inspectors look for several things that would make the can defective: how well the strip for opening the can operates, whether there are proper crimps holding the lids to the cardboard tube, whether the lids are poorly sealed, and so on. Create and interpret an np chart for the following data.

	Defectives		*Defectives*
1	12	14	12
2	15	15	22
3	8	16	8
4	10	17	10
5	4	18	5
6	7	19	13
7	16	20	12
8	9	21	20
9	14	22	18
10	10	23	24
11	5	24	15
12	6	25	8
13	17		

5. A manufacturer of lightbulbs is keeping track of the number of nonconforming bulbs (the number that do not light). Create the centerline and control limits for a number nonconforming chart. What is the capability of the process? Assume assignable causes and revise the chart.

	n	np		n	np
1	300	10	11	300	31
2	300	8	12	300	32
3	300	9	13	300	10
4	300	12	14	300	8
5	300	10	15	300	12
6	300	11	16	300	10
7	300	9	17	300	11
8	300	10	18	300	9
9	300	12	19	300	10
10	300	11	20	300	8

6. A production line manufactures compact discs. Twice an hour, one CD is selected at random from those made during the hour. Each disc is inspected separately for imperfections (defects), such as scratches, nicks, discolorations, and dents. The resulting count of imperfections from each disc is recorded on a control chart. Use the following data to create a c chart. If the customer wants a process capability of 1, will this line be able to meet that requirement?

	Imperfections		Imperfections
1	2	14	1
2	0	15	1
3	2	16	0
4	1	17	0
5	0	18	2
6	1	19	4
7	0	20	1
8	2	21	2
9	0	22	0
10	2	23	2
11	1	24	1
12	2	25	0
13	0		

7. The Par Fore Golf Company is a producer of plastic divot fixers. The company produces the fixers in bags of 30. They consider these bags a sample size of 1. The nonconformities range from overall size to individual flaws in the production of the plastic tool. These tools are inspected four times daily. Create a chart with the following data. Assume assignable causes and revise the chart.

	Serial Number	Count of Nonconformities	Comment
1	JG100	7	
2	JG101	3	
3	JG102	5	
4	JG103	4	
5	JG104	6	
6	JG105	6	
7	JG106	17	Wrong Mold Used
8	JG107	7	
9	JG108	2	
10	JG109	0	
11	JG110	3	
12	JG111	4	
13	JG112	2	
14	JG113	0	
15	JG114	21	Not Enough Time in Mold
16	JG115	3	
17	JG116	2	
18	JG117	7	
19	JG118	5	
20	JG119	9	
		113	

8. When printing full-color glossy magazine advertising pages, editors do not like to see blemishes or pics (very small places where the color did not transfer to the paper). One person at the printer has the job of looking at samples of magazine pages (n = 20) and counting blemishes. Below are her results. The editors are interested in plotting the nonconformities per unit on a u chart. Determine the centerline and control limits. Create a chart of this process. If the editors want only two nonconformities per unit on average, how does this compare with the actual process average?

Subgroup Number	Count of Nonconformities
1	65
2	80
3	60
4	50
5	52
6	42
7	35
8	30
9	30
10	25

9. A manufacturer of holiday light strings tests 400 of the light strings each day. The light strings are plugged in, and the number of unlit bulbs are counted and recorded. Unlit bulbs are considered nonconformities, and they are replaced before the light strings are shipped to customers. These data are then used to aid process engineers in their problem-solving activities. Given the following information, create and graph the chart. How is their performance? What is the process capability?

	Sample Size	Nonconformities
1	400	50
2	400	23
3	400	27
4	400	32
5	400	26
6	400	38
7	400	57
8	400	31
9	400	48
10	400	34
11	400	37
12	400	44
13	400	34
14	400	32
15	400	50
16	400	49
17	400	54
18	400	38
19	400	29
20	400	47
Total	8,000	780

Glossary

*Accuracy In the quality field, a term that refers to how close an observed value is to a true value.

Accreditation Certification of a duly recognized body of the facilities, capability, objectivity, competence, and integrity of an agency, service, or operational group or individual to provide the specific service or operation needed. For example, The Registrar Accreditation Board accredits organizations that register companies to the ISO 9000 series standards.

*Accredited registrars Persons who perform audits for ISO 9000 standards. They are qualified organizations certified by a national body (e.g., the Registrar Accreditation Board in the U.S.).

*Activity-based costing An approach to accounting that analyzes and assigns costs to make a specific product or provide a specific service based on the resources used to make the product or provide the service.

*Alignment To create agreement between processes and activities supporting an organization's strategies, objectives, and goals.

American Society for Quality A professional, not-for-profit association that develops, promotes, and applies quality-related information and technology for the private sector, government, and academia. The Society serves more than 108,000 individuals and 1,000 corporate members in the United States and 108 other countries.

*Analysis of means (ANOM) The statistical technique used to analyze the results of experiment factors.

*Analysis of variance (ANOVA) The statistical technique used to analyze the variation present in experiment data.

*Assessment An evaluation process that includes document reviews, audits, analysis, and report of findings.

*Assignable cause The source or cause of variation in a process that can be identified and then isolated and removed from the process. It is not due to chance causes and is also termed *special cause*.

Attribute data Go/no-go information. The control charts based on attribute data include percent chart, number of affected units chart, count chart, count-per-unit chart, quality score chart, and demerit chart.

*Audit A systematic appraisal procedure that examines, evaluates, and verifies that appropriate procedures, requirements, checklists, and programs are being followed effectively.

*Autonomation Machinery designed to not only make a part or product but also to inspect each item after producing it and alerting a human if an error is detected.

Average chart A control chart in which the subgroup average, $\overline{X}$, is used to evaluate the stability of the process level.

*Balanced Scorecard Developed by Kaplan and Kaplan, a Balanced Scorecard is a method of designing and using performance measures to provide feedback in four areas: financial, customer, internal processes, and learning and growth.

---

*Definitions of asterisked terms have been provided by the author; all others come from "The Quality Glossary," *Quality Progress*, June 2007, pp. 39–59.

Benchmarking An improvement process in which a company measures its performance against that of best-in-class companies, determines how those companies achieved their performance levels, and uses the information to improve its own performance. The subjects that can be benchmarked include strategies, operations, processes, and procedures.

***Black Belt** Designation for an improvement project team leader in a Six Sigma environment.

Blemish An imperfection that is severe enough to be noticed but should not cause any real impairment with respect to intended normal or reasonably foreseeable use. (*See also* Defect, Nonconformity.)

***Bottleneck** A bottleneck occurs whenever a person or a piece of equipment does not have the capacity to fulfill the demand placed on it in the required time.

Brainstorming A technique that teams use to generate ideas on a particular subject. Each person in the team is asked to think creatively and write down as many ideas as possible. The ideas are not discussed or reviewed until after the brainstorming session.

Calibration The comparison of a measurement instrument or system of unverified accuracy to a measurement instrument or system of a known accuracy to detect any variation from the required performance specification.

***Capability** The amount of variation inherent in a stable process. Capability is determined using data from control charts and histograms from stable processes. When these indicate a stable process and a normal distribution, the indices C_p and C_{pk} can be calculated.

***Cause** The identified and isolated reason behind a defect or problem in a process or product.

Cause-and-effect diagram A tool for analyzing process dispersion. It is also referred to as the *Ishikawa diagram*, because Kaoru Ishikawa developed it, and the *fish bone diagram*, because the complete diagram resembles a fish skeleton. The diagram illustrates the main causes and subcauses leading to an effect (symptom). The cause-and-effect diagram is one of the seven tools of quality.

***Characteristic** The individual elements that define a process, function, product, or service.

c Chart or count of nonconformities chart A control chart for evaluating the stability of a process in terms of the count of events of a given classification occurring in a sample of constant size.

***Cell** An organized arrangement of people, equipment, information, and materials that enables parts or products to be made in a continuous flow with minimal material handling.

***Cellular manufacturing** A manufacturing arrangement of related cells. Work-in-progress flows from cell to cell in an organized arrangement throughout the plant.

***Changeover** A changeover occurs when a piece of equipment or a workplace is changed to allow the production of a different type of product or service from what was previously being produced or provided.

***Changeover time** The amount of time it takes to switch equipment or people from producing or providing one type of product or service to another.

Checklist A tool used to ensure that all important steps or actions in an operation have been taken. Checklists contain items that are important or relevant to an issue or situation. Checklists are often confused with check sheets and data sheets.

Check sheet A simple data-recording device. The check sheet is custom designed by the user, which allows him or her to readily interpret the results. The check sheet is one of the seven tools of quality. Check sheets are often confused with data sheets and checklists.

Common causes Causes of variation that are inherent in a process over time. They affect every outcome of the process and everyone working in the process. (*See also* Special causes.)

Company culture A system of values, beliefs, and behaviors inherent in a company. To optimize business performance, top management must define and create the necessary culture.

*Compliance Meeting established criteria, specifications, terms, standards, or regulations.

Conformance An affirmative indication or judgment that a product or service has met the requirements of a relevant specification, contract, or regulation.

*Constraint Anything that prevents a system from achieving its goal.

Continuous improvement The ongoing improvement of products, services, or processes through incremental and breakthrough improvements.

Control chart A chart with upper and lower control limits on which values of some statistical measure for a series of samples or subgroups are plotted. The chart frequently shows a central line to help detect a trend of plotted values toward either control limit.

*Control limits Statistically calculated limits placed on a control chart. The variability of the process is compared with the limits, and special and common causes of variation are identified.

Corrective action The implementation of solutions resulting in the reduction or elimination of an identified problem.

Cost of poor quality The costs associated with providing poor-quality products or services. There are four categories of costs: internal failure costs (costs associated with defects found before the customer receives the product or service), external failure costs (costs associated with defects found after the customer receives the product or service), appraisal costs (costs incurred to determine the degree of conformance to quality requirements), and prevention costs (costs incurred to keep failure and appraisal costs to a minimum).

C$_p$ A widely used process capability index. It is expressed as

$$C_p = \frac{\text{upper specification limit} - \text{lower specification limit}}{6\sigma}$$

C$_{pk}$ A process capability index showing process centering:

$$C_{pk} = \frac{|\mu - \text{nearer specification limit}|}{3\sigma}$$

*Crosby, Philip Originated the zero defects concept. He has authored many books, including *Quality Is Free*, *Quality without Tears*, *Let's Talk Quality*, and *Leading: The Art of Becoming an Executive*.

*Culture The attitudes, beliefs, values, and norms shared by a group of individuals.

Customer delight The result of delivering a product or service that exceeds customer expectations.

Customer satisfaction The result of delivering a product or service that meets customer requirements.

Customer–supplier partnership A long-term relationship between a buyer and a supplier characterized by teamwork and mutual confidence. The supplier is considered an extension of the buyer's organization. The partnership is based on several commitments. The buyer provides long-term contracts and uses fewer suppliers. The supplier implements quality assurance processes so that incoming inspection can be minimized. The supplier also helps the buyer reduce costs and improve product and process designs.

*Cycle An ordered sequence of activities that is or can be repeated.

*Cycle time The time it takes from start to finish to complete a task, create a product, or provide a service.

Defect A product's or service's nonfulfillment of an intended requirement or reasonable expectation for use, including safety considerations.

*Defective Term describing a product or service whose characteristics do not conform to the requirements placed on it.

Deming, W. Edwards (deceased) A prominent consultant, teacher, and author on the subject of quality. After Deming shared his expertise in statistical quality control to help the U.S. war effort during World War II, the War Department sent him to Japan in 1946 to help that nation recover from its wartime losses. Deming published more than 200 works, including the well-known books *Quality, Productivity and Competitive Position*, and *Out of the Crisis*. Deming, who developed the 14 points for managing, was an ASQC Honorary member.

Deming cycle *See* Plan-do-study-act cycle.

Deming Prize An award given annually to organizations that, according to the award guidelines, have successfully applied companywide quality control based on statistical quality control and that will continue the quality control in the future. Although the award is named in honor of W. Edwards Deming, its criteria are not specifically related to Deming's teachings. There are three separate divisions for the award: the Deming Application Prize, the Deming Prize for Individuals, and the Deming Prize for Overseas Companies. The award process is overseen by the Deming Prize Committee of the Union of Japanese Scientists and Engineers in Tokyo.

Documentation Written material describing the methods, procedures, or processes to be followed. Documentation may also provide supporting evidence that particular procedures are followed. Documentation may be in the form of a quality manual or process operation sheets.

*Downtime Time that is lost when equipment or people are not able to function and complete their goals.

80–20 rule A term referring to the Pareto principle, that suggests that most effects come from relatively few causes; that is, 80% of the effects come from 20% of the possible causes.

Employee involvement A practice within an organization whereby employees regularly participate in making decisions on how their work areas operate, including making suggestions for improvement, planning, goal setting, and monitoring performance.

*Error detection Determining that something has gone wrong immediately after its occurrence.

*Error proofing Designing methods or systems that prevent an error from occurring in the first place. Error proofing is also known as "mistake proofing."

External customer A person or organization who receives a product, a service, or information but is not part of the organization supplying it. (*See also* Internal customer.)

Feigenbaum, Armand V. The founder and president of General Systems Co., an international engineering company that designs and implements total quality systems. Feigenbaum originated the concept of total quality control in his book *Total Quality Control*, which was published in 1951.

*First-pass yield The percentage of items produced that meets specifications without being rerun, retested, or repaired. To calculate, divide the number of good units by the total number of units entering the process.

Fish bone diagram *See* Cause-and-effect diagram.

Fitness for use A term used to indicate that a product or service fits the customer's defined purpose for that product or service.

Flowchart A graphical representation of the steps in a process. Flowcharts are drawn to help people better understand processes. The flowchart is one of the seven tools of quality.

Funnel experiment An experiment that demonstrates the effects of tampering. Marbles are dropped through a funnel in an attempt to hit a flat-surfaced target below. The experiment shows that adjusting a stable process to compensate for an undesirable result or an extraordinarily good result will produce output that is worse than if the process had been left alone.

*Green Belt Designation for an improvement team member in a Six Sigma environment.

Histogram A graphic summary of variation in a set of data. The pictorial nature of the histogram lets people see patterns that are difficult to see in a simple table of numbers. The histogram is one of the seven tools of quality.

Inspection Measuring, examining, testing, or gauging one or more characteristics of a product or service and comparing the results with specified requirements to determine whether conformity is achieved for each characteristic.

Instant pudding A term used to illustrate an obstacle to achieving quality: the supposition that quality improvement and productivity improvement are achieved quickly through an affirmation of faith rather than through sufficient effort and education. W. Edwards Deming used this term—which was initially coined by James Bakken of the Ford Motor Co—in his book *Out of the Crisis*.

Internal customer The recipient (person or department) of another person's or department's output (product, service, or information) within an organization. (*See also* External customer.)

*Internal failure A process or product failure that is caught before the customer receives the product or before the customer receives the service.

Ishikawa, Kaoru (deceased) A pioneer in quality control activities in Japan. In 1943 he developed the cause-and-effect diagram. Ishikawa, an ASQC Honorary member, published many works, including *What Is Total Quality Control?*, *The Japanese Way*, *Quality Control Circles at Work*, and *Guide to Quality Control*. He was a member of the quality control research group of the Union of Japanese Scientists and Engineers while also working as an assistant professor at the University of Tokyo.

Ishikawa diagram *See* Cause-and-effect diagram.

ISO 9000 Series Standards A set of five individual but related international standards on quality management and quality assurance developed to help companies effectively document the quality system elements to be implemented to maintain an efficient quality system. The standards were developed by the International Organization for Standardization (ISO), a special international agency for standardization composed of the national standards bodies of 91 countries.

*ISO 14000 An environmental management standard that organizations use to monitor and manage how their organization's activities affect the environment around them.

*ISO/TS 16949 The international standard for quality management systems used in the manufacture of automobiles, subassemblies, components, and parts.

Juran, Joseph M. The chairman emeritus of the Juran Institute and an ASQC Honorary Member. Since 1924 Juran has pursued a varied career in management as an engineer, executive, government administrator, university professor, labor arbitrator, corporate director, and consultant. Specializing in managing for quality, he has authored hundreds of papers and 12 books, including *Juran's Quality Control Handbook*, *Quality Planning and Analysis* (with F. M. Gryna), and *Juran on Leadership for Quality*.

Just-in-time manufacturing An optimal material requirement planning system for a manufacturing process in which there is little or no manufacturing material inventory on hand at the manufacturing site and little or no incoming inspection.

Kaizen A Japanese term that means gradual unending improvement by doing little things better and setting and achieving increasingly higher standards. The term was made famous by Masaaki Imai in his book *Kaizen: The Key to Japan's Competitive Success*.

*** Kanban** A Japanese term for a system, usually a printed card, that enables material to flow in an orderly and efficient manner through a manufacturing process.

LCL *See* Lower control limit.

Leadership An essential part of a quality improvement effort. Organization leaders must establish a vision, communicate that vision to those in the organization, and provide the tools and knowledge necessary to accomplish the vision.

*** Lean** Optimizing the number of parts produced or services provided through the elimination of all inappropriate non–value-added activities.

*** Lean manufacturing** The improvement initiatives that focus on the elimination of waste from systems and processes.

*** Line balancing** A production or service line is considered balanced when each participant or piece of equipment in the line is performing the same amount of work in the same time. The work is evenly distributed, thus eliminating over- or underutilization of resources.

*** Logistics** The process of determining the best methods of procuring, maintaining, packaging, transporting, and storing of materials and personnel in order to satisfy customer demand.

Lower control limit Control limit for points below the central line in a control chart.

Malcolm Baldrige National Quality Award An award established by the U.S. Congress in 1987 to raise awareness of quality management and to recognize U.S. companies that have implemented successful quality management systems. Two awards may be given annually in each of three categories: manufacturing company, service company, and small business. The award is named after the late Secretary of Commerce Malcolm Baldrige, a proponent of quality management. The U.S. Commerce Department's National Institute of Standards and Technology manages the award, and ASQC administers it.

*** Mistake proofing** *See* Error proofing.

*** Nonconformance** The state that exists when a product, service, or material does not conform to the customer requirements or specifications.

Nonconformities-per-unit chart A control chart for evaluating the stability of a process in terms of the average count of events of a given classification per unit occurring in a sample; a count per unit chart.

Nonconformity The nonfulfillment of a specified requirement. (*See also* Blemish, Defect.)

*** Non–value-added** Term describing any activity or action that does not directly affect the production of a product or the provision of a service.

*** One piece flow** Manufacturing a single item at a time rather than a batch or group of items. Also known as single piece flow.

Operating characteristic curve A graph used to determine the probability of accepting lots as a function of the lots' or processes' quality level when using various sampling plans. There are three types: Type A curves, which give the probability of acceptance for an individual lot coming from finite production (will not continue in the future); Type B curves, which give the probability of acceptance for lots coming from a continuous process; and Type C curves, which for a continuous sampling plan give the long-run percentage of product accepted during the sampling phase.

*Overall equipment effectiveness** The amount of time a piece of equipment is up and running and available for use. This may also include performance efficiency and first-pass yield.

Pareto chart A graphical tool for ranking causes from most significant to least significant. It is based on the Pareto principle, which was first defined by J. M. Juran in 1950. The principle, named after 19th-century economist Vilfredo Pareto, suggests that most effects come from relatively few causes; that is, 80% of effects come from 20% of the possible causes. The Pareto chart is one of the seven tools of quality.

*Parts per million (PPM)** A number that describes the performance of a process in terms of either actual or projected defective material.

p Chart The fraction nonconforming chart, or p chart, is a control chart used to monitor the proportion nonconforming in a lot of goods.

PDSA cycle *See* Plan-do-study-act cycle.

*Plan-do-study-act cycle** A four-step process for quality improvement. In the first step (Plan), a plan to effect improvement is developed. In the second step (Do), the plan is carried out, preferably on a small scale. In the third step (Study), the effects of the plan are observed. In the last step (Act), the results are studied to determine what was learned and what can be predicted. The plan-do-study-act cycle was originally called the plan-do-check-act cycle, but in the early 1990s it was standardized as plan-do-study-act. It is still sometimes referred to as the Shewhart cycle, because Walter A. Shewhart discussed the concept in his book *Statistical Method from the Viewpoint of Quality Control*, or as the Deming cycle, because W. Edwards Deming introduced the concept in Japan.

*Precision** The ability of a system, process, or activity to repeat its actions consistently.

*Prevention costs** The costs that occur when actions are taken to prevent nonconformities in a system, process, service, or product.

Prevention vs. detection A term used to contrast two types of quality activities. Prevention refers to those activities designed to prevent nonconformances in products and services. Detection refers to those activities designed to detect nonconformances already in products and services. Another term used to describe this distinction is "designing in quality vs. inspecting in quality."

*Process** The action of taking inputs and transforming them into outputs through the performance of value-added activities.

Process capability A statistical measure of the inherent process variability for a given characteristic. The most widely accepted formula for process capability is 6σ.

Process capability index The value of the tolerance specified for the characteristic divided by the process capability. There are several types of process capability indexes, including the widely used C_{pk} and C_p.

*Process control** Using statistical process control to measure and regulate a process.

*Process map** A diagram showing the steps or activities that take place in a process.

*Process owner** The individual ultimately responsible for ensuring that the appropriate activities take place in a process.

*Production part approval process (PPAP)** The process of obtaining approval to produce parts.

Product or service liability The obligation of a company to make restitution for loss related to personal injury, property damage, or other harm caused by its product or service.

*Productivity** The amount of output as compared with the amount of input.

*Pull system An item is not produced until an item has been "pulled" by the customer. The newly produced item replenishes the system, replacing the item that was pulled.

QFD *See* Quality function deployment.

QS 9000 A quality standard utilized by the automotive industry to ensure the quality of its and its suppliers' components, subsystems, and finished products.

Quality A subjective term for which each person has his or her own definition. In technical usage, *quality* can have two meanings: (1) the characteristics of a product or service that bear on its ability to satisfy stated or implied needs and (2) a product or service free of deficiencies.

Quality assurance/quality control Two terms that have many interpretations because of the multiple definitions for the words *assurance* and *control*. For example, *assurance* can mean the act of giving confidence, the state of being certain, or the act of making certain; *control* can mean an evaluation to indicate needed corrective responses, the act of going, or the state of a process in which the variables are attributable to a constant system of chance causes. (For detailed discussion on the multiple definitions, see ANSI/ASQC Standard A3-1987, "Definitions, Symbols, Formulas, and Tables for Control Charts.") One definition of quality assurance is: All the planned and systematic activities are implemented within the quality system that can be demonstrated to provide confidence that a product or service will fulfill requirements for quality. One definition for quality control is: the operational techniques and activities used to fulfill requirements for quality. Often, however, *quality assurance* and *quality control* are used interchangeably, referring to the actions performed to ensure the quality of a product, service, or process.

Quality audit A systematic, independent examination and review to determine whether quality activities and related results comply with planned arrangements and whether these arrangements are implemented effectively and are suitable to achieve the objectives.

Quality control *See* Quality assurance/quality control.

Quality costs *See* Cost of poor quality.

Quality function deployment A structured method in which customer requirements are translated into appropriate technical requirements for each stage of product development and production. The QFD process is often referred to as listening to the voice of the customer.

Quality management (QM) The application of a quality management system in managing a process to achieve maximum customer satisfaction at the lowest overall cost to the organization while continuing to improve the process.

*Quality manual The chief document for standard operating procedures, processes, and specifications quality management system. The manual serves as a permanent reference guide for the implementation and maintenance of the quality management system described by the manual.

*Quality plan The plan that integrates quality philosophies into an organization's environment. It includes specific continuous improvement strategies and actions. Plans are developed at the departmental, group, plant, division, and company levels. Lower-level plans should support the company's strategic objectives. A quality plan emphasizes defect prevention through continuous improvement rather than defect detection.

*Quality records Written verification that a company's methods, systems, and processes were performed according to the quality system documentation, such as inspection or test results, internal audit results, and calibration data.

Quality trilogy Juran's three-pronged approach to managing for quality. The three legs are quality planning (developing the products and processes required to meet customer needs), quality control (meeting product and process goals), and quality improvement (achieving unprecedented levels of performance).

*Quick changeover The rapid switch of tooling and fixtures on a piece of equipment.

Random sampling Sample units are selected in such a manner that all combinations of n units under consideration have an equal chance of being selected as the sample.

Range chart A control chart in which the subject group, R, is used to evaluate the stability of the variability within a process.

R chart See Range chart.

Red bead experiment An experiment developed by W. Edwards Deming to illustrate that it is impossible to put employees in rank order of performance for the coming year based on their performance during the past year because performance differences must be attributed to the system, not to employees. Four thousand red and white beads (20% red) in a jar and six people are needed for the experiment. The participants' goal is to produce white beads, because the customer will not accept red beads. One person begins by stirring the beads and then, blindfolded, selects a sample of 50 beads. That person hands the jar to the next person, who repeats the process, and so on. When everyone has his or her sample, the number of red beads for each is counted. The limits of variation between employees that can be attributed to the system are calculated. Everyone will fall within the calculated limits or variation that could arise from the system. The calculations will show that there is no evidence one person will be a better performer than another in the future. The experiment shows that it would be a waste of management's time to try to find out why, say, John produced four red beads and Jane produced 15; instead, management should improve the system, making it possible for everyone to produce more white beads.

Registration to standards A process in which an accredited, independent third-party organization conducts an on-site audit of a company's operations against the requirements of the standard to which the company wants to be registered. Upon successful completion of the audit, the company receives a certificate indicating that it has met the standard requirements.

Reliability The probability of a product performing its intended function under stated conditions without failure for a given period of time.

*Repair Corrective action to a damaged product so that the product will fulfill the original specifications.

*Repeatability When a piece of equipment, process, person, or system is able to perform, with little or no variation, the same activity over and over again.

*Reproducibility When similar pieces of equipment, processes, people, or systems are able to repeat their performance with little or no variation between them.

*Rework Action taken on nonconforming products or services to allow them to meet the original specifications.

*Root cause The ultimate reason behind a nonconformance.

Sample standard deviation chart A control chart in which the subgroup standard deviation, s, is used to evaluate the stability of the variability within a process.

s Chart See Sample standard deviation chart.

Seven tools of quality Tools that help organizations understand their processes in order to improve them. The tools are the cause-and-effect diagram, check sheet, control chart, flowchart, histogram, Pareto chart, and scatter diagram. (See individual entries.)

Shewhart cycle *See* Plan-do-study-act cycle.

Shewhart, Walter A. (deceased) Referred to as the father of statistical quality control, Dr. Shewhart brought together the disciplines of statistics, engineering, and economics. He described the basic principles of this new discipline in his book *Economic Control of Quality of Manufactured Product*. Shewhart, ASQC's first Honorary Member, was best known for creating the control chart. Shewhart worked for Western Electric and AT&T/Bell Telephone Laboratories in addition to lecturing and consulting on quality control.

*****Single Minute Exchange of Dies (SMED)** A technique pioneered by Shigeo Shingo that stressed modifying changeovers in order to reduce changeover times to 10 minutes or less. SMED facilitates agile, flexible production.

*****Single piece flow** *See* One piece flow.

Six Sigma A methodology that provides businesses with the tools to improve the capability of their business processes. The increase in performance and the decrease in process variation lead to defect reduction and improvement in profits, employee morale, and quality of product.

SPC *See* Statistical process control.

Special causes Causes of variation that arise because of special circumstances. They are not an inherent part of a process. Special causes are also referred to as *assignable causes*. (*See also* Common causes.)

Specification A document that states the requirements to which a given product or service must conform.

SQC *See* Statistical quality control.

Standard deviation A computed measure of variability indicating the spread of the data set around the mean.

Statistical process control The application of statistical techniques to control a process. Often the term *statistical quality control* is used interchangeably with *statistical process control*.

Statistical quality control The application of statistical techniques to control quality. Often the term *statistical process control* is used interchangeably with *statistical quality control*, although statistical quality control includes acceptance sampling as well as statistical process control.

*****Suppliers** Persons who provide materials, parts, or services directly to manufacturers.

*****Supply chain** A supply chain is the network of organizations involved in the movement of materials, information, and money as raw materials flow from their source through production until they are delivered as a finished product or service to the final customer.

*****System** The activities, people, and processes that work together to accomplish a specific goal or objective.

*****Takt time** How often a single part should be produced based on customer demand. Takt times are the starting point for line balancing.

Taguchi, Genichi The executive director of the American Supplier Institute, the director of the Japan Industrial Technology Institute, and an honorary professor at Nanjing Institute of Technology in China. Taguchi is well known for developing a methodology to improve quality and reduce costs, which in the United States is referred to as the *Taguchi Methods*. He also developed the quality loss function.

Taguchi Methods™ The American Supplier Institute's trademarked term for the quality engineering methodology developed by Genichi Taguchi. In this engineering approach to quality control,

Taguchi calls for off-line quality control, online quality control, and a system of experimental design to improve quality and reduce costs.

Tampering Action taken to compensate for variation within the control limits of a stable system. Tampering increases rather than decreases variation, as evidenced in the funnel experiment.

Top-management commitment Participation of the highest-level officials in their organization's quality improvement efforts. Their participation includes establishing and serving on a quality committee, establishing quality policies and goals, deploying those goals to lower levels of the organization, providing the resources and training that the lower levels need to achieve the goals, participating in quality improvement teams, reviewing progress organization-wide, recognizing those who have performed well, and revising the current reward system to reflect the importance of achieving the quality goals.

Total quality management A term initially coined in 1985 by the Naval Air Systems Command to describe its Japanese-style management approach to quality improvement. Since then, total quality management (TQM) has taken on many meanings. Simply put, TQM is a management approach to long-term success through customer satisfaction. TQM is based on the participation of all members of an organization in improving processes, products, services, and the culture they work in. TQM benefits all organization members and society. The methods for implementing this approach are found in the teachings of such quality control leaders as Philip Crosby, W. Edwards Deming, Armand V. Feigenbaum, Kaoru Ishikawa, and J. M. Juran.

TQM *See* Total quality management.

Type I error An incorrect decision to reject something (such as a statistical hypothesis or a lot of products) when it is acceptable.

Type II error An incorrect decision to accept something when it is unacceptable.

u Chart *See* Nonconformities-per-unit chart.

UCL *See* Upper control limit.

Upper control limit Control limit for points above the central line in a control chart.

*Value-added activities The activities in a process or system that transform raw materials, parts, or components, etc., into a usable product or service for the customer.

*Value analysis The process of analyzing the value stream in order to determine which activities add value and which do not. Opportunities for improvements are also sought.

*Value stream The activities that take place in order to provide a product or service to a customer.

Variables data Measurement information. Control charts based on variables data include average ($\overline{X}$) chart, range (R) chart, and sample standard deviation chart.

Variation A change in data, a characteristic, or a function that is caused by one of four factors: special causes, common causes, tampering, or structural variation.

Vision statement A statement summarizing an organization's values, mission, and future direction for its employees and customers.

*Waste Any activity or action that fails to add value to the product or service being provided to the customer.

*Work-in-process The unfinished parts or products waiting in a manufacturing line for processing and completion.

$\overline{X}$ chart Average chart.

Zero defects A performance standard developed by Philip Crosby to address a dual attitude in the workplace: people are willing to accept imperfection in some areas, while, in other areas, they expect the number of defects to be zero. This dual attitude had developed because of the conditioning that people are human and humans make mistakes. However, the zero defects methodology states that if people commit themselves to watching details and avoiding errors, they can move closer to the goal of zero defects. The performance standard that must be set is "zero defects," not "close enough."

Answers to Selected Problems

Chapter 10

14. Try $i = 0.0009$; $h = \dfrac{0.006}{0.0009} + 1 = 8$.

   Use $i = 0.0009$, $h = 8$.

   1st cell boundary $= 0.6535 - \dfrac{i}{2}$

   $= 0.6535 - \dfrac{0.0009}{2} = 0.65305$

Cell Boundaries	Cell Midpoints	Frequency
0.65305–0.65395	0.6535	/
0.65395–0.65485	0.6544	///// ///// ///// ///
0.65485–0.65575	0.6559	/
0.65575–0.65665	0.6562	
0.65665–0.65755	0.6571	///
0.65755–0.65845	0.658	///// ///
0.65845–0.65935	0.6589	///// ///
0.65935–0.66025	0.6598	///

16. $R = X_h - X_i = 150 - 125 = 25$

   $h = \dfrac{R}{i} + 1 = \dfrac{25}{i} + 1$

   When $i = 3, h = 9$

   $\quad\quad i = 5, h = 6 \leftarrow$ Choose

   $\quad\quad i = 7, h = 5$

   Lowest midpoint $= 125$

   Boundaries of first cell: 125 ± 2.5

17. $\overline{\overline{X}} = 16$, $UCL_x = 20$, $LCL_x = 12$
    $\overline{R} = 7$, $UCL_R = 15$, $LCL_R = 0$

18. $\overline{\overline{X}} = 50.2$, $UCL_x = 50.7$, $LCL_x = 49.7$
    $\overline{R} = 0.7$, $UCL_R = 1.6$, $LCL_R = 0$

19. $n = 8$, $\overline{\overline{X}} = 3.02$, $\overline{R} = 0.10$
    $UCL_X = 3.02 + (0.373)(0.10) = 3.06$,
    $UCL_r = 1.864(0.10) = 0.19$
    $LCL_X = 3.02 - (0.373)(0.10) = 2.98$,
    $LCL_r = 0.136(0.10) = 0.01$

24. $\sigma = 3$, $6\sigma = 18$, $C_p = 0.44$, $C_{pk} = 0.33$

25. $\sigma = 0.3$, $6\sigma = 1.8$, $C_p = 0.6$, $C_{pk} = 0.3$

26. $\sigma = 0.04$, $6\sigma = 0.24$, $C_p = 0.42$, $C_{pk} = 0.25$

Appendix 5

1. $\overline{p} = 0.004$, $UCL = 0.0174$, $LCL = 0$
2. $\overline{p} = 0.038$, $UCL = 0.085$, $LCL = 0$
5. $\overline{p} = 0.0405$, $\overline{np} = 12$, $UCL = 22$, $LCL = 2$
7. $\overline{c} = 6$, $UCL = 13$, $LCL = 0$
8. $\overline{u} = 2.345$, $UCL = 3$, $LCL = 1$

Bibliography

Abarca, D. "Making the Most of Internal Audits." *Quality Digest*, February 1999, pp. 26–28.

Adam, P., and R. VandeWater. "Benchmarking and the Botton Line: Translating BPR into Bottom-Line Results." *Industrial Engineering*, February 1995, pp. 24–26.

Adcock, S. "FAA Orders Fix on Older 737s." *Newsday*, May 8, 1998.

Aeppel, Timothy. "On the Factory Floors, Top Workers Hide Secrets to Success." *The Wall Street Journal*, July 3, 2003.

Aft, L. *Quality Improvement Using Statistical Process Control*. New York: Harcourt Brace Jovanovich, 1988.

Alsup, F., and R. Watson. *Practical Statistical Process Control*. New York: Van Nostrand Reinhold, 1993.

Alukal, G. "Keeping Lean Alive." *Quality Progress*, October 2006, pp. 67–69.

AMA Management Briefing. *World Class Quality*. New York: AMA Publications Division, 1990.

American Society for Quality, P.O. Box 3005, Milwaukee, WI 53201-3005.

Bacus, H. "Liability: Trying Times." *Nation's Business*, February 1986, pp. 22–28.

Balestracci, D. "Resolutions: Weight and . . . Budget? (Pt. 2)." *Quality Digest*, February 2007, p. 20.

Balestracci, D. "Why Deming Was Such a Curmudgeon." *Quality Digest*, October 2006, p. 19.

Bamford, J. "Order in the Court." *Forbes*, January 27, 1986, pp. 46–47.

Bergamini, D. *Mathematics*. New York: Time, 1963.

Bernowski, K., and B. Stratton. "How Do People Use the Baldrige Award Criteria?" *Quality Progress*, May 1995, pp. 43–47.

Berry, Thomas. *Managing the Total Quality Transformation*. Milwaukee, WI: ASQC Quality Press, 1991.

Besterfield, D. *Quality Control*, 4th ed. Englewood Cliffs, NJ: Prentice Hall, 1994.

Biesada, A. "Strategic Benchmarking." *Financial World*, September 29, 1992, pp. 30–36.

Bishara, R., and M. Wyrick. "A Systematic Approach to Quality Assurance Auditing." *Quality Progress*, December 1994, pp. 67–69.

Borsai, T., B. Ludovico, and G. Dzialas. "ISO 13485: A Path to the Global Market." *Quality Digest*, July 2007, pp. 35–39.

Bossert, J. "Lean Manufacturing and Six Sigma—Synergy Made in Heaven." *Quality Progress*, July 2003, pp. 31–32.

Bothe, D. "SPC for Short Production Runs." *Quality*, December 1988, pp. 58–59.

Bovet, S. F. "Use TQM, Benchmarking to Improve Productivity." *Public Relations Journal*, January 1994, p. 7.

Brachulis, J. "Implementing Strategic Goal Deployment." *IIE Solutions*, August 1998, pp. 25–29.

Breyfogle, F., and B. Meadows. "Bottom-Line Success with Six Sigma." *Quality Progress*, May 2001, pp. 101–104.

Brocka, B., and M. Brocka. *Quality Management: Implementing the Best Ideas of the Masters*. Homewood, IL: Business One Irwin, 1992.

Brown, R. "Zero Defects the Easy Way with Target Area Control." *Modern Machine Shop*, July 1966, p. 19.

Bruder, K. "Public Benchmarking: A Practical Approach." *Public Press*, September 1994, pp. 9–14.

Brumm, E. "Managing Records for ISO 9000 Compliance." *Quality Progress*, January 1995, pp. 73–77.

Bunch, R. "AQP Awards Promote Business Results." *Quality Progress*, June 2003, pp. 31–36.

Burr, Irving W. *Statistical Quality Control Methods*. New York: Marcel Dekker, 1976.

Butz, H. "Strategic Planning: The Missing Link in TQM." *Quality Progress*, May 1995, pp. 105–108.

Byrnes, Daniel. "Exploring the World of ISO 9000." *Quality*, October 1992, pp. 19–31.

Byron, E. "Beauty, Prestige and Worry Lines." *The Wall Street Journal*, August 20, 2007.

Callison, J. "Change Is Good: MMS Rolls Up Sleeves, Helps Streamline Production." *Cincinnati Enquirer*, May 19, 2002.

Campanella, J., ed. *Principles of Quality Costs*. Milwaukee, WI: ASQC Quality Press, 1990.

Camperi, J. A. "Vendor Approval and Audits in Total Quality Management." *Food Technology*, September 1994, pp. 160–62.

Carson, P. P. "Deming versus Traditional Management Theorists on Goal Setting: Can Both Be Right?" *Business Horizons*, September 1993, pp. 79–84.

Chowdhury, S. *The Ice Cream Maker*. New York: Currency Doubleday, 2005.

Cochran, C. "Breaking Down the Walls." *Quality Digest*, May 2003, pp. 43–47.

Cochran, C. "High-Impact Auditing." *Quality Digest*, August 2003, pp. 39–46.

Cochran, C. "Six Problem-Solving Fundamentals." *Quality Digest*, September 2002, pp. 29–34.

Cochran, C. "The Ten Biggest Quality Mistakes." *Quality Digest*, February 2007, pp. 28–32.

Collier, D., S. Goldstein, and D. Wilson. "A Thing of the Past? The 1992 Model May Have Been the Best Baldrige Model for Organizational Performance." *Quality Progress*, October 2002, pp. 97–103.

Conner, G. *Lean Manufacturing for the Small Shop*. Dearborn, MI: Society of Manufacturing Engineers, 2001.

Cook, B. M. "Quality: The Pioneers Survey the Landscape." *Industry Week*, October 21, 1991, pp. 68–73.

Cooper, N., and P. Noonan. "Do Teams and Six Sigma Go Together?" *Quality Progress*, June 2003, pp. 25–31.

Crago, M. "Meeting Patient Expectations." *Quality Progress*, September 2002, pp. 41–43.

Crawford-Mason, C. "Deming and Me." *Quality Progress*, September 2002, pp. 45–48.

Crosby, P. B. *Cutting the Cost of Quality: The Defect Prevention Workbook for Managers*. Boston: Industrial Education Institute, 1967.

Crosby, P. B. *The Eternally Successful Organization: The Art of Corporate Wellness*. New York: New American Library, 1988.

Crosby, P. B. *Quality Is Free*. New York: Penguin Books, 1979.

Crosby, P. B. *Quality Is Free: The Art of Making Quality Certain*. New York: McGraw-Hill, 1979.

Crosby, P. B. *Quality without Tears: The Art of Hassle-Free Management*. New York: McGraw-Hill, 1979.

Cross, C. "Everything but the Kitchen Sink." *Industrial Engineer*, April 2007, pp. 32–37.

Cross, C. "Fighting for Air." *Industrial Engineer*, February 2007, pp. 30–35.

Crownover, D. "Baldrige: It's Easy, Free, and It Works." *Quality Progress*, July 2003, pp. 37–41.

Crownover, D. *Take It to the Next Level*. Dallas: NextLevel Press, 1999.

Cullen, C. "Short Run SPC Re-emerges." *Quality*, April 1995, p. 44.

Daniels, S. "Check Out This Baldrige Winner." *Quality Progress*, August 2002, pp. 41–47.

Daniels, S. "From One-Man Show to Baldrige Recipient." *Quality Progress*, July 2007, Volume 40, No. 7, p. 50–55.

Daniels, S. "The Little Hot Dog Stand That Could." *Quality Progress*, September 2002, pp. 66–71.

Darden, W., W. Babin, M. Griffin, and R. Coulter. "Investigation of Products Liability Attitudes and Opinions: A Consumer Perspective." *Journal of Consumer Affairs*, June 22, 1994.

Davis, P. M. "New Emphasis on Product Warnings." *Design News*, August 6, 1990, p. 150.

Davis, P. M. "The Right Prescription for Product Tampering." *Design News*, January 23, 1989, p. 224.

Day, C. R. "Benchmarking's First Law: Know Thyself." *Industry Week*, February 17, 1992, p. 70.

Day, R. G. *Quality Function Deployment*. Milwaukee, Wisconsin: ASQ Quality Press, 1993.

Dearing, J. "ISO 9001: Could It Be Better." *Quality Progress*, February 2007, pp. 23–27.

DeFoe, J. A. "The Tip of the Iceberg." *Quality Progress*, May 2001, pp. 29–37.

Deming, W. Edwards. *The New Economics*. Cambridge, MA: MIT CAES, 1993.

Deming, W. Edwards. *Out of the Crisis*. Cambridge, MA: MIT Press, 1986.

Dentzer, S. "The Product Liability Debate." *Newsweek*, September 10, 1984, pp. 54–57.

DeToro, I. "The Ten Pitfalls of Benchmarking." *Quality Progress*, January 1995, pp. 61–63.

DeVor, R., T. Chang, and J. Sutherland. *Statistical Quality Design and Control*. New York: Macmillan, 1992.

Disney, L. "Wave of Relief." *Industrial Engineer*, February 2007, pp. 24–29.

Dobyns, L., and C. Crawford-Mason. *Quality or Else: The Revolution in World Business*. Boston: Houghton Mifflin, 1991.

Duncan, A. *Quality Control and Industrial Statistics*. Homewood, IL: Irwin, 1974.

Dusharme, M. "RFID Tunes into Supply Chain Management." *Quality Digest*, October 2006, pp. 37–41.

Eaton, B. "Cessna's Approach to Internal Quality Audits." *IIE Solutions, Industrial Engineering*, June 1995, pp. 12–16.

Ellison, S. "Inside Campbell's Big Bet: Heating Up Condensed Soup." *The Wall Street Journal*, July 31, 2003, p. A1.

Engle, P. "When Inventory Goes Missing." *Industrial Engineer*, November 2006, p. 22.

Ericson, J. "Lean Inspection through Supplier Partnership." *Quality Progress*, November 2006, pp. 36–42.

Eureka, W. E., and N. E. Ryan. *The Customer Driven Company*. Dearborn, MI: ASI Press, 1988.

Farahmand, K., R. Becerra, and J. Greene. "ISO 9000 Certification: Johnson Controls' Inside Story." *Industrial Engineering*, September 1994, pp. 22–23.

Feigenbaum, Armand V. "Changing Concepts and Management of Quality Worldwide." *Quality Progress*, December 1997, pp. 43–47.

Feigenbaum, A. V. *Total Quality Control*. New York: McGraw-Hill, 1983.

Feigenbaum, A. V. "The Future of Quality Management." *Quality Digest*, May 1998, pp. 24–30.

Feigenbaum, A. V. "How to Manage for Quality in Today's Economy." *Quality Progress*, May 2001, pp. 26–27.

Feigenbaum, A. V. "The International Growth of Quality." *Quality Progress*, February 2007, pp. 36–40.

Feiler, B. "The Therapist at the Table." *Gourmet*, October, 2002, pp. 234–38.

Fenn, D. *Alpha Dogs: How Your Small Business Can Become a Leader of the Pack*. New York: Collins, 2005.

"The First among Equals." *Quality Digest*, June, 1999.

Foster, S. T. *Managing Quality: Integrating the Supply Chain*. Upper Saddle River, Prentice Hall, 2007.

Franco, V. R. "Adopting Six Sigma." *Quality Digest*, June 2001, pp. 28–32.

Frank, M. *How to Run a Successful Meeting in Half the Time*. Upper Saddle River, NJ: Prentice Hall, 1989.

Frum, D. "Crash!" *Forbes*, November 8, 1993, p. 62.

Galpin, D., R. Dooley, J. Parker, and R. Bell. "Assess Remaining Component Life with Three Level Approach." *Power*, August 1990, pp. 69–72.

Galpin, T. *The Human Side of Change*. San Francisco Jossey-Bass, 1996.

Gardner, R. A. "Resolving the Process Paradox." *Quality Progress*, March 2001, pp. 51–59.

George, M. *Lean Six Sigma*. New York: McGraw-Hill, 2002.

Gerling, A. "How Jury Decided How Much the Coffee Spill Was Worth." *The Wall Street Journal*, September 4, 1994.

Gest, T. "Product Paranoia." *U.S. News & World Report*, February 24, 1992, pp. 67–69.

Geyelin, M. "Product Liability Suits Fare Worse Now." *The Wall Street Journal*, July 12, 1994.

Ghattas, R. G., and S. L. McKee. *Practical Project Management*. Upper Saddle River, NJ: Prentice Hall, 2001.

Gitlow, H. S. *Planning for Quality, Productivity and Competitive Position*. Homewood, IL: Business One Irwin, 1990.

Gitlow, H. S., and S. J. Gitlow. *The Deming Guide to Quality and Competitive Position*. Englewood Cliffs, NJ: Prentice Hall, 1987.

Goetsch, D. L. *Effective Supervision*. Upper Saddle River, NJ: Prentice Hall, 2002.

Goetsch, D. L., and S. B. Davis. *ISO 14000 Environmental Management*. Upper Saddle River, NJ: Prentice Hall, 2001.

Goetsch, D. L., and S. B. Davis. *Understanding and Implementing ISO 9000 and ISO Standards*. Upper Saddle River, NJ: Prentice Hall, 1998.

Goodden, R. "Product Reliability Considerations Empower Quality Professionals." *Quality*, April 1995, p. 108.

Goodden, R. "Reduce the Impact of Product Liability on Your Organization." *Quality Progress*, January 1995, pp. 85–88.

Gooden, R. L. "How a Good Quality Management System Can Limit Lawsuits." *Quality Progress*, June 2001, pp. 55–59.

Gooden, R. L. *Product Liability Prevention: A Strategic Guide*. Milwaukee: ASQ Quality Press, 2000.

Gordon, D. "AS9000: One Approval Accepted Everywhere." *Quality Digest*, June 2003, pp. 37–40.

Grahn, D. "The Five Drivers of Total Quality." *Quality Progress*, January 1995, pp. 65–70.

Grant, E., and T. Lang. "Why Product-Liability and Medical Malpractice Lawsuits Are So Numerous in the United States." *Quality Progress*, December 1994, pp. 63–65.

Grant, E., and R. Leavenworth. *Statistical Quality Control*. New York: McGraw-Hill, 1988.

Grant, P. "A Great Step Backwards." *Quality*, May 1985, p. 58.

Greasley, A. *Operations Management*. John Wiley & Sons, 2006.

Greene, R. "2001 Baldrige Award Winner Profile." *Quality Digest*, September 2002, pp. 51–53.

Greene, R. "The Tort Reform Quagmire." *Forbes*, August 11, 1986, pp. 76–79.

Greenfield, M. "Process Mapping's Next Step." *Quality Progress*, September 2002, pp. 50–55.

Hare, L. "SPC: From Chaos to Wiping the Floor." *Quality Progress*, July 2003, pp. 58–63.

Hare, L., R. Hoerl, J. Hromi, and R. Snee. "The Role of Statistical Thinking in Management." *Quality Progress*, February 1995, pp. 53–59.

Harrington, J. "Creating New Middle Managers." *Quality Digest*, August 2002, p. 14.

Harrington, J. "Managing Resistance to Change." *Quality Digest*, September 2002, p. 16.

Harrington, H. "Hit the Nail, Not the Thumb." *Quality Digest*, October 2006, p. 16.

Harrington, H. "Harrington's Wheel of Fortune." *Quality Digest*, February 2007, p. 16.

Harry, M., and R. Schroeder. *Six Sigma: The Breakthrough Management Strategy Revolutionizing the World's Top Corporations*. New York: Doubleday, 2000.

Haupt, H. "Getting Credit for Service." *Quality Progress*, November 2006, pp. 51–55.

Hayward, S. *Churchill on Leadership*. New York: Forum, 1997.

Heldt, J. J. "Quality Pays." *Quality*, November 1988, pp. 26–27.

Heldt, J. J., and D. J. Costa. *Quality Pays*. Wheaton, IL: Hitchcock Publishing, 1988.

Himelstein, L. "Monkey See, Monkey Sue." *BusinessWeek*, February 7, 1994, pp. 112–13.

Hockman, K., R. Grenville, and S. Jackson. "Road Map to ISO 9000 Registration." *Quality Progress*, May 1994, pp. 39–42.

Houghton, P. "Improving Pharmacy Service." *Quality Digest*, October 2006, pp. 49–54.

Hoyer, R. W., and B. B. Hoyer. "What Is Quality?" *Quality Progress*, July 2001, pp. 52–62.

Hutchins, G. "The State of Quality Auditing." *Quality Progress*, March 2001, pp. 25–29.

Hutchinson, E. E. "The Road to TL 9000: From the Bell Breakup to Today." *Quality Progress*, January 2001, pp. 33–37.

Hutton, D. W. *From Baldrige to the Bottom Line*. Milwaukee: ASQ Quality Press, 2000.

Iannello, P. "Unplugged . . . Untangled . . . and Informed." *Quality Digest*, February 2007, pp. 54–59.

"ICU Checklist System Cuts Patients' Stay in Half." *Wall Street Journal*, August 6, 2003, p. D9.

Ireson, W., and C. Coombs. *Handbook of Reliability Engineering and Management*. New York: McGraw-Hill, 1988.

Ishikawa, K. *Guide to Quality Control*, rev. ed. White Plains, NY: Kraus International Publications, 1982.

Ishikawa, K. *What Is Total Quality Control? The Japanese Way*. Englewood Cliffs, NJ: Prentice Hall, 1985.

Iyer, S. "Don't Forget the People." *Quality Progress*, October 2006, pp. 60–66.

Japan Human Relations Associations, eds. *Kaizen Teian 1*, Portland, OR: Productivity Press, 1992.

Johnson, K. "Print Perfect." *Quality Progress*, July 2003, pp. 48–56.

Johnson, W. "AS9100: On Course and Gaining." *Quality Digest*, February 2007, pp. 43–48.

Jones, D. M. "Merging Quality Cultures in Contract Manufacturing." *Quality Progress*, February 2007, pp. 41–46.

Juran, J. M. *Juran on Leadership for Quality: An Executive Handbook*. New York: Free Press, 1989.

Juran, J. M. *Juran on Planning for Quality*. New York: Free Press, 1988.

Juran, J. M. *Juran on Quality by Design: The New Steps for Planning Quality into Goods and Services*. New York: Free Press, 1992.

Juran, J. M. "The Quality Trilogy." *Quality Progress*, August 1986, pp. 19–24.

Juran, J. M., and F. M. Gryna. *Quality Planning and Analysis: From Product Development through Usage*. New York: McGraw-Hill, 1970.

Kackar, Raghu. "Taguchi's Quality Philosophy: Analysis and Commentary." *Quality Progress*, December 1986, pp. 21–29.

Kanholm, J. "New and Improved ISO 9000:2000." *Quality Digest*, October 1999, pp. 28–32.

Kaplan, R., and D. Norton. *The Balanced Scorecard*. Boston: Harvard Business School Press, 1996.

Kececioglu, D. *Reliability and Life Testing Handbook*. Englewood Cliffs, NJ: Prentice Hall, 1993.

"Keep It Trendy," *Forbes*, July 18, 1994.

Keller, C. "QOS—A Simple Method for Big or Small." *Quality Progress*, July 2003, pp. 28–31.

Ketola, J., and K. Roberts. *ISO 9001:2000 In a Nutshell*, 2d ed. Chico, CA: Paton Press, 2001.

Ketola, J., and K. Roberts. "Transition Planning for ISO 9001:2000." *Quality Digest*, March 2001, pp. 24–28.

King, R. "So What's the Law, Already?" *Forbes*, May 19, 1986, pp. 70–72.

Kirscht, R. "Quality and Outsourcing." *Quality Digest*, July 2007, pp. 40–43.

Kolarik, W. J. *Creating Quality: Concepts, Systems, Strategies and Tools*. New York: McGraw-Hill, 1995.

Lancaster, J. *Making Time: Lillian Moller Gilbreth*. Boston: Northeastern University Press, 2004.

Landro, L. "Hospitals Seek Cure for Mishaps." *The Wall Street Journal*, July 31, 2003, p. D3.

Landro, L. "Sick of Hospital Treatment? New Forums Let You Rate Care." *The Wall Street Journal*, August 1, 2003, p. D3.

Lareau, W. *Office Kaizen*. Milwaukee, WI: ASQ Quality Press, 2003.

Lawton, R. "Balance Your Balanced Scorecard." *Quality Progress*, March 2002, pp. 66–71.

Lazalier, M. "Coax, Don't Squeeze." *Industrial Engineer*, April 2007, pp. 26–31.

Leibfried, K. H. *Benchmarking: A Tool for Continuous Improvement*. New York: Harper Business, 1992.

Levinson, W. A. "ISO 9000 at the Front Line." *Quality Progress*, March 2001, pp. 33–36.

Lindborg, H. "Get Rid of Clutter." *Quality Progress*, February 2007, p. 50.

London, C. "Strategic Planning for Business Excellence." *Quality Progress*, August 2002, pp. 26–33.

Mader, D. "How to Identify and Select Lean Six Sigma Projects." *Quality Progress*, July 2007, pp. 58–60.

"Making Things: The Essence and Evolution of the Toyota Production System." Toyota, Inc. Special Report.

Malcolm Baldrige National Quality Award, U.S. Department of Commerce, Technology Administration, National Institute of Standards and Technology, Gaithersburg, MD.

Manos, A. "The Benefits of Kaizen and Kaizen Events." *Quality Progress*, February 2007, p. 47.

Marcus, A. "Limits on Personal-Injury Suits Urged." *The Wall Street Journal*, April 23, 1991.

Marsh, S. "Executive Level Planning." *Quality Digest*, August 2003, p. 22.

Martin, S. "Courting Disaster: A Huge Jury Award Sends the Wrong Message." *Bicycling*, March 1994, p. 136.

Mascitelli, R. "Design's Due Process." *Industrial Engineer*, April 2007, pp. 38–43.

Mathews, J. "The Cost of Quality." *Newsweek*, September 7, 1992, pp. 48–49.

Matthews, C. "Linking the Supply Chain to TQM." *Quality Progress*, November 2006, pp. 29–34.

Mauro, T. "Frivolous or Not, Lawsuits Get Attention." *USA Today*, February 23, 1995.

McElroy, A., and I. Fruchtman. "Use Statistical Analysis to Predict Equipment Reliability." *Power*, October 1992, pp. 39–46.

McGuire, P. "The Impact of Product Liability," Report 908. The Conference Board, 1988.

McManus, K. "No Time for Projects." *Industrial Engineer*, November 2006, p. 20.

McManus, K. "The Last Great Fad." *Industrial Engineer*, December 2006, p. 18.

McManus, K. "The Pull of Lean." *Industrial Engineer*, April 2007, p. 20.

Mehta, M. and K. Rampura. "Squeezing Out Extra Value." *Industrial Engineer*, December 2006, pp. 29–35.

Meier, B. "Court Rejects Coupon Settlement in Suit Over G.M. Pickup Trucks." *The New York Times*, April 18, 1995.

Milas, G. "How to Develop a Meaningful Employee Recognition Program." *Quality Progress*, May 1995, pp. 139–42.

Miller, I., and J. Freund. *Probability and Statistics for Engineers*. Englewood Cliffs, NJ: Prentice Hall, 1977.

Miller, J. R., and J. S. Morris. "Is Quality Free or Profitable?" *Quality Progress*, January 2000, pp. 50–53.

Miscikowski, D., and E. Stein. "Empowering Employees to Pull the Quality Trigger." *Quality Progress*, October 2006, pp. 43–48.

Moen, R., T. Nolan, and L. Provost. *Improving Quality through Planned Experimentation*. New York: McGraw-Hill, 1991.

Montgomery, D. *Introduction to Statistical Quality Control*. New York: John Wiley & Sons, 2001.

Moran, J. W., and P. C. La Londe. "ASQ Certification Program Gains Wider Recognition." *Quality Progress*, April 2000, pp. 29–41.

Munoz, J., and C. Nielsen. "SPC: What Data Should I Collect? What Charts Should I Use?" *Quality Progress*, January 1991, pp. 50–52.

Munro, R. A. "Linking Six Sigma with QS-9000." *Quality Progress*, May 2000, pp. 47–53.

Munro, S. "The Last Bastion of Inefficiency." *Industrial Engineer*, November 2006, pp. 34–39.

Nakhai, B., and J. Neves. "The Deming, Baldrige, and European Quality Awards." *Quality Progress*, April 1994, pp. 33–37.

Nash, M., S. Poling, and S. Ward. "Six Sigma Speed." *Industrial Engineer*, November 2006, pp. 40–44.

Neave, H. *The Deming Dimension*. Knoxville, TN: SPC Press, 1990.

"Needed: A Backup for Ma Bell." *U.S. News & World Report*, September 30, 1991, p. 22.

Nesbitt, T. "Flowcharting Business Processes." *Quality*, March 1993, pp. 34–38.

Neuscheler-Fritsch, D., and R. Norris. "Capturing Financial Benefits from Six Sigma." *Quality Progress*, May 2001, pp. 39–44.

Nutter, J. "Designing with Product Liability in Mind." *Machine Design*, May 24, 1984, pp. 57–60.

Okes, D. "Complexity Theory Simplifies Choices." *Quality Progress*, July 2003, pp. 35–37.

Okes, D. "Organize Your Quality Tool Belt." *Quality Progress*, July 2002, pp. 25–29.

Orsini, J. "What's Up Down Under?" *Quality Progress*, January 1995, pp. 57–59.

Palady, P. "The Status Quo's Failure in Problem-Solving." *Quality Progress*, August 2002, pp. 34–39.

Palmes, P. "How to Fail the ISO 9001 Driver's Test." *Quality Progress*, October 2006, pp. 33–36.

Parker, D., and A. M. Parker. "Turning the Frown Upside Down." *Quality Progress*, October 2006, p. 88.

Parsons, C. "The Big Spill: Hot Java and Life in the Fast Lane." *Gannett News Services*, October 25, 1994.

Paton, S. "Juran: A Lifetime of Quality." *Quality Digest*, August 2002, pp. 19–23.

Pearson, T. A. "Measure for Six Sigma Success." *Quality Progress*, February 2001, pp. 35–40.

Perry, M. "The Fish(bone) Tale." *Quality Progress*, November 2006, p. 88.

Petroski, H. "The Merits of Colossal Failure." *Discover*, May 1994, pp. 74–82.

Phillips-Donaldson, D. "On Leadership." *Quality Progress*, August 2002, pp. 24–25.

Pittle, D. "Product Safety: There's No Substitute for Safer Design." *Trial*, October 1991, pp. 110–14.

Pond, R. *Fundamentals of Statistical Quality Control*. New York: Merrill, 1994.

PQ Systems. *Applying Design of Experiments Using DOEpack*. Dayton, OH: PQ Systems, 2001.

Press, A., G. Carroll, and S. Waldman. "Are Lawyers Burning America?" *Newsweek*, March 20, 1995, pp. 30–35.

Prevette, S. "Systems Thinking—An Uncommon Answer." *Quality Progress*, July 2003, pp. 32–35.

Pritts, B. A. "Industry-wide Shakeout." *Quality Progress*, January 2001, pp. 61–64.

"Product Liability." *Business Insurance*, October 4, 1993, p. 36.

"Product Liability: The Consumer's Stake." *Consumer Reports*, June 1984, pp. 336–39.

"Product Liability Advice: Surpass Design Expectations." *Design News*, May 17, 1993, pp. 31–35.

Pyzdek, T. *What Every Engineer Should Know about Quality Control*. New York: Marcel Dekker, 1989.

Pyzdek, T. "Managing Metric Madness." *Quality Digest*, February 2007, p. 22.

"Quality Glossary." *Quality Progress*, July 2002, pp. 43–61.

"Quality Initiative Deemed a Success." *IIE Solutions*, February 2002, p. 18.

Ramberg, J. S. "Six Sigma: Fad or Fundamental?" *Quality Digest*, May 2000, pp. 28–32.

Redmond, M. "60 Minutes to a Solution." *Quality Progress*, February 2007, p. 80.

Reid, R. D. "From Deming to ISO 9001:2000." *Quality Progress*, June 2001, pp. 66–70.

Reid, R. D. "Why QS 9000 Was Developed and What's in Its Future." *Quality Progress*, April 2000, pp. 115–17.

Reid, R. "Developing the Voluntary Healthcare Standard." *Quality Progress*, November 2006, pp. 68–71.

Reidenbach, R., and R. Goeke. "Six Sigma, Value and Competitive Strategy." *Quality Progress*, July 2007, pp. 45–49.

RFID Study Group. "Challenges in RFID Enabled Supply Chain Management." *Quality Progress*, November 2006, pp. 23–28.

Rice, C. M. "How to Conduct an Internal Audit and Still Have Friends." *Quality Progress*, June 1994, pp. 39–40.

Rienzo, T. F. "Planning Deming Management for Service Organizations." *Business Horizons*, May 1993, pp. 19–29.

Robinson, C., ed. *How to Plan an Audit*. Milwaukee, WI: ASQC Press, 1987.

Robitaille, D. "Inhale, Exhale, Transition." *Quality Digest*, May 2003, pp. 29–32.

Robitaille, D. "The Basics of Internal Auditing." *Quality Digest*, June, 2007, pp. 42–46.

Rooney, J., L. Heuvel, and D. Lorenzo. "Reduce Human Error." *Quality Progress*, September 2002, pp. 27–36.

Rowland, F. "Liability of Product Packaging." *Design News*, February 1, 1993, p. 120.

Rowland, F. "Product Warnings Updated." *Design News*, March 9, 1992, p. 168.

Rowland, J. R. "Should Congress Ease the Product Liability Law?" *American Legion*, April 1993, p. 10.

Roy, R. *A Primer on the Taguchi Method*. New York: Van Nostrand Reinhold, 1990.

Roy, R. K. "Sixteen Steps to Improvement." *Quality Digest*, June 2001, pp. 24–27.

Russell, J. P. "Auditing ISO 9001:2000." *Quality Progress*, July 2001, pp. 147–48.

Russell, J. *Quality Management Benchmark Assessment*. Milwaukee, WI: 1991. ASQC Press, 1991.

Russell, J. "Quality Management Benchmark Assessment." *Quality Progress*, May 1995, pp. 57–61.

Russell, J. P. "Know and Follow ISO 19011's Auditing Principles." *Quality Progress*, February 2007, pp. 29–34.

Salot, S. "Controlling Heavy Metals (and Other Nasty Stuff)." *Quality Digest*, October 2006, pp. 32–36.

Savell, L. "Who's Liable When the Product Is Information?" *Editor and Publisher*, August 28, 1993, pp. 35–36.

Saxena, R. "The Changing Chain." *Industrial Engineer*, April 2007, p. 24.

Schwinn, D. R. "Six Sigma and More: On Not Losing Sight of the Big Picture." *Quality E-line*, May 2, 2001, www.pqsystems.com.

Scovronek, J. "Reliability Sample Testing, A Case History." *Quality Progress*, February 2001, pp. 43–45.

Sedlock, R. "Under One Roof." *Quality Progress*, July 2007, p. 80.

Sharrock, R. A. "Well-Appointed Establishment." *Industrial Engineer*, April 2007, pp. 44–48.

Shewhart, Walter. *Economic Control of Quality of Manufactured Product*. New York: Van Nostrand Reinhold, 1931, p. 6. Reissued in 1980 by the American Society for Quality.

Shipley, D. "ISO 9000 Makes Integrated Systems User Friendly." *Quality Progress*, July 2003, pp. 26–28.

Shiu, M., and M. Tu. "QFD's Evolution in Japan and the West." *Quality Progress*, July 2007, pp. 30–38.

"Short-Run SPC Re-emerges." QEI Speaker Interview. *Quality*, April 1995, p. 44.

Smith, G. *Statistical Process Control and Quality Improvement*. New York: Merrill, 1991.

Smith, K. "Quality Conversation with Sister Mary Jean Ryan." *Quality Digest*, July 2003, pp. 45–46.

Smith, K. "Six Sigma at Ford Revisited." *Quality Digest*, June 2003, pp. 28–32.

Smith, L. "Would You Like Standardization with That?" *Quality Digest*, February 2007, pp. 10–11.

Smith, R. "The Benchmarking Boom." *Human Resources Focus*, April 1994, pp. 1–6.

Snee, R. D. "Dealing with the Achilles' Heel of Six Sigma Initiatives." *Quality Progress*, March 2001, pp. 66–72.

Spanyi, A., and M. Wurtzel. "Six Sigma for the Rest of Us." *Quality Digest*, July 2003, pp. 22–26.

Spearman, M. "Realities of Risk: Supply Chains Must Scrutinize the Unknown." *Industrial Engineer*, February 2007, pp. 36–41.

Spendolini, M. *The Benchmarking Book*. New York: Amcom, 1992.

Spigener, J. B., and P. A. Angelo. "What Would Deming Say?" *Quality Progress*, March 2001, pp. 61–64.

Srikanth, M., and S. Robertson. *Measurements for Effective Decision Making*. Wallingford, CT: Spectrum Publishing Co., 1995.

Stamatis, D. H. "Who Needs Six Sigma, Anyway?" *Quality Digest*, May 2000, pp. 33–38.

Stein, P. "By Their Measures Shall Ye Know Them." *Quality Progress*, May 2001, pp. 72–74.

Stevens, T. "Dr. Deming: Management Today Does Not Know What Its Job Is." *Industry Week*, January 17, 1994, pp. 20–28.

Stevenson, W. *Operations Management*, 7th ed. New York: McGraw-Hill Irwin, 2002.

Summers, D. *Quality*, 3d ed. Upper Saddle River, NJ: Prentice Hall, 2003.

Surak, J. G. "Quality in Commercial Food Processing." *Quality Progress*, February 1999, pp. 25–29.

Sussland, W. "Connecting the Planners and Doers." *Quality Progress*, June 2002, pp. 55–61.

Szuchman, P. "The Pampered Move." *The Wall Street Journal*, August 6, 2003, p. W1.

Taguchi, Genichi. *Introduction to Quality Engineering*. Dearborn, MI: American Supplier Institute, 1986.

Thorpe, J., and W. Middendorf. *What Every Engineer Should Know about Product Liability*. New York: Marcel Dekker, 1979.

Tobias, R. *Applied Reliability*. New York: Van Nostrand Reinhold, 1986.

"The Toyota Production System: Leaner Manufacturing for a Greener Planet." Toyota Motor Corporation, 1998.

Travalini, M. "The Evolution of a Quality Culture." *Quality Progress*, May 2001, pp. 105–8.

Traver, R. "Nine-Step Process Solves Product Variability Problems." *Quality*, April 1995, p. 94.

Traver, R. "Pre-Control: A Good Alternative to X-bar and R Charts." *Quality Progress*, September 1985, pp. 11–13.

Tsuda, Y., and M. Tribus. "Planning the Quality Visit." *Quality Progress*, April 1991, pp. 30–34.

"U.S. FAA: FAA Orders Immediate Inspection for High-Time Boeing 737s, Extends Inspection Order." *M2 Press Wire*, May 11, 1998.

Van Patten, J. "A Second Look at 5S." *Quality Progress*, October 2006, pp. 55–59.

Vardeman, S. *Statistics for Engineering Problem Solving*. Boston: PWS Publishing, 1994.

Vardeman, S., and J. Jobe. *Statistical Quality Assurance Methods for Engineers*. New York: John Wiley & Sons, Inc., 1999.

Verseput, R. "Digging into DOE." *Quality Digest*, June 2001, pp. 33–36.

Wade, J. *Utility versus Risk: On the Nature of Strict Tort Liability for Products*. 44 Miss. L. J. 825.

Walker, H. "The Innovation Process and Quality Tools." *Quality Progress*, July 2007, pp. 18–22.

Walpole, R., and R. Myers. *Probability and Statistics for Engineers and Scientists*. New York: Macmillan, 1989.

Walters, J. "The Benchmarking Craze." *Governing*, April 1994, pp. 33–37.

Walton, M. *Deming Management at Work*. New York: Pedigree Books, 1991.

Walton, M. *The Deming Management Method*. New York: Putnam, 1986.

Watson, G. "Digital Hammers and Electronic Nails—Tools of the Next Generation." *Quality Progress*, July 1998, pp. 21–26.

Watson, R. "Modified Pre-Control." *Quality*, October 1992, p. 61.

Weimer, G. A. "Benchmarking Maps the Route to Quality." *Industry Week*, July 20, 1992, pp. 54–55.

Welborn, C. "Using FMEA to Assess Outsourcing Risk." *Quality Progress*, August 2007, pp. 17–21.

Wendt, L., and H. Kogan. *Give the Lady What She Wants*. South Bend, IN: AND Books, 1997.

West, J. E. "Implementing ISO 9001:2000, Early Feedback Indicates Six Areas of Challenge." *Quality Progress*, May 2001, pp. 65–70.

West, J., G. Haworth, D. Arter, K. Harvey, P. Naish, and J. Green. "Should You Transition to ISO 9001:2000?" *Quality Progress*, September 2002, pp. 58–65.

Westcott, R. "Customers: A Love/Hate Relationship?" *Quality Progress*, July 2002, pp. 35–41.

Westcott, R. "Overlooked and Underutilized: ISO 9004." *Quality Digest*, July 2003, pp. 49–52.

Westcott, R. "Maximize the Use of Your Abilities." *Quality Progress*, October 2006, pp. 70–71.

Westcott, R. "Continual Innovation and Reinvention." *Quality Progress*, July 2007, pp. 56–57.

"What Is Supply Chain Management?" IQC, Inc., August 2002 Newsletter.

Wheeler, D. *Advanced Topics in Statistical Process Control*. Knoxville, TN: SPC Press, 1995.

Wheeler, D. *Short Run SPC*. Knoxville, TN: SPC Press, 1991.

Wheeler, D. *Understanding Industrial Experimentation*. Knoxville, TN: SPC Press, 1990.

Wheeler, D. *Understanding Variation: The Key to Managing Chaos*. Knoxville, TN: SPC Press, 1993.

Wheeler, D., and D. Chambers. *Understanding Statistical Process Control*. Knoxville, TN: SPC Press, 1992.

Wiesendanger, B. "Benchmarking for Beginners." *Sales and Marketing Management*, November 1992, pp. 59–64.

Williams, B. "Calibration Management in the ISO/IEC 17025 Accredited Facility." *Quality Digest*, June, 2007, pp. 37–41.

Wilson, L. *Eight-Step Process to Successful ISO 9000 Implementation.* Milwaukee, WI: ASQC Press, 1996.

Winslow, R. "Hospitals' Weak Systems Hurt Patients, Study Says." *The Wall Street Journal*, July 5, 1995.

Winter, D. "Valuing Technology." *Ward's Auto World*, March 2002, p. 5

Wysocki, B. "Hospitals Cut ER Waits." *The Wall Street Journal*, July 3, 2002.

Yasuda, Yuzo. *40 Years, 20 Million Ideas: The Toyota Suggestion System.*

Zaciewski, R. "Attribute Charts Are Alive and Kicking." *Quality*, March 1992, pp. 8–10.

Zollers, F. E. "Product Liability Reform: What Happened to the Crisis?" *Business Horizons*, September 1990, pp. 47–52.

Index